Sensor Network Methodologies for Smart Applications

Salahddine Krit
Ibn Zohr University, Morocco

Valentina Emilia Bălaş
Aurel Vlaicu University of Arad, Romania

Mohamed Elhoseny
Mansoura University, Egypt

Rachid Benlamri
Lakehead University, Canada

Marius M. Bălaş
Aurel Vlaicu University of Arad, Romania

A volume in the Advances in
Information Security, Privacy, and
Ethics (AISPE) Book Series

IGI Global
PUBLISHER of TIMELY KNOWLEDGE

Published in the United States of America by
    IGI Global
    Information Science Reference (an imprint of IGI Global)
    701 E. Chocolate Avenue
    Hershey PA, USA 17033
    Tel: 717-533-8845
    Fax: 717-533-8661
    E-mail: cust@igi-global.com
    Web site: http://www.igi-global.com

Library of Congress Cataloging-in-Publication Data

Names: Krit, Salahddine, 1976- editor. | Balas, Valentina Emilia, editor. |
    Elhoseny, Mohamed, editor. | Benlamri, Rachid, editor. | Bălaş, Marius
    M., 1956- editor.
Title: Sensor network methodologies for smart applications / Salahddine
    Krit, Valentina Emilia Bălaş, Mohamed Elhoseny, Rachid Benlamri, and
    Marius M. Bălaş, editors.
Description: Hershey, PA : Information Science Reference, an imprint of IGI
    Global, [2020] | Includes bibliographical references and index. |
    Summary: "This book contains research on the methods of intelligent
    systems and technologies and their various applications within
    sustainable development practices"-- Provided by publisher.
Identifiers: LCCN 2020002712 (print) | LCCN 2020002713 (ebook) | ISBN
    9781799843818 (hardcover) | ISBN 9781799853312 (paperback) | ISBN
    9781799843825 (ebook)
Subjects: LCSH: Sensor networks--Industrial applications. | Sustainable
    development.
Classification: LCC TK7872.D48 .S4345 2020 (print) | LCC TK7872.D48
    (ebook) | DDC 006.2/5--dc23
LC record available at https://lccn.loc.gov/2020002712
LC ebook record available at https://lccn.loc.gov/2020002713

This book is published in the IGI Global book series Advances in Information Security, Privacy, and Ethics
(AISPE) (ISSN: 1948-9730; eISSN: 1948-9749)

British Cataloguing in Publication Data
A Cataloguing in Publication record for this book is available from the British Library.

All work contributed to this book is new, previously-unpublished material.
The views expressed in this book are those of the authors, but not necessarily of the publisher.

For electronic access to this publication, please contact: eresources@igi-global.com.

Advances in Information Security, Privacy, and Ethics (AISPE) Book Series

ISSN:1948-9730
EISSN:1948-9749

Editor-in-Chief: *Manish Gupta* State University of New York, USA

MISSION

As digital technologies become more pervasive in everyday life and the Internet is utilized in ever increasing ways by both private and public entities, concern over digital threats becomes more prevalent.

The **Advances in Information Security, Privacy, & Ethics (AISPE) Book Series** provides cutting-edge research on the protection and misuse of information and technology across various industries and settings. Comprised of scholarly research on topics such as identity management, cryptography, system security, authentication, and data protection, this book series is ideal for reference by IT professionals, academicians, and upper-level students.

COVERAGE

- Computer ethics
- Cookies
- CIA Triad of Information Security
- Privacy Issues of Social Networking
- Tracking Cookies
- Electronic Mail Security
- Internet Governance
- Security Classifications
- Information Security Standards
- Access Control

IGI Global is currently accepting manuscripts for publication within this series. To submit a proposal for a volume in this series, please contact our Acquisition Editors at Acquisitions@igi-global.com or visit: http://www.igi-global.com/publish/.

Titles in this Series

For a list of additional titles in this series, please visit:
http://www.igi-global.com/book-series/advances-information-security-privacy-ethics/37157

Privacy Concerns Surrounding Personal Information Sharing on Health and Fitness Mobile Apps
Devjani Sen (Independent Researcher, Canada) and Rukhsana Ahmed (University at Albany, SUNY, USA)
Information Science Reference • © 2020 • 300pp • H/C (ISBN: 9781799834878) • US $215.00

Safety and Security Issues in Technical Infrastructures
David Rehak (VSB – Technical University of Ostrava, Czech Republic) Ales Bernatik (VSB – Technical University of Ostrava, Czech Republic) Zdenek Dvorak (University of Zilina, Slovakia) and Martin Hromada (Tomas Bata University in Zlin, Czech Republic)
Information Science Reference • © 2020 • 499pp • H/C (ISBN: 9781799830597) • US $195.00

Cybersecurity Incident Planning and Preparation for Organizations
Akashdeep Bhardwaj (University of Petroleum and Energy Studies, Dehradun, India) and Varun Sapra (University of Petroleum and Energy Studies, India)
Information Science Reference • © 2020 • 300pp • H/C (ISBN: 9781799834915) • US $215.00

Blockchain Applications in IoT Security
Harshita Patel (KLEF, Vaddeswaram, Guntur, Andhra Pradesh, India) and Ghanshyam Singh Thakur (MANIT, Bhopal, Madhya Pradesh, India)
Information Science Reference • © 2020 • 300pp • H/C (ISBN: 9781799824145) • US $215.00

Modern Theories and Practices for Cyber Ethics and Security Compliance
Winfred Yaokumah (University of Ghana, Ghana) Muttukrishnan Rajarajan (City University of London, UK) Jamal-Deen Abdulai (University of Ghana, Ghana) Isaac Wiafe (University of Ghana, Ghana) and Ferdinand Apietu Katsriku (University of Ghana, Ghana)
Information Science Reference • © 2020 • 302pp • H/C (ISBN: 9781799831495) • US $200.00

701 East Chocolate Avenue, Hershey, PA 17033, USA
Tel: 717-533-8845 x100 • Fax: 717-533-8661
E-Mail: cust@igi-global.com • www.igi-global.com

Table of Contents

Preface xiv

Acknowledgment xix

Chapter 1
A Review on FPGA-Based Digital Filters for De-Noising ECG Signal 1
  Seema Nayak, Department of Electronics and Communication
    Engineering, IIMT College of Engineering, Greater Noida, India
  Manoj Nayak, Department of Mechanical Engineering, FET, Manav
    Rachnan International Institute of Research and Studies, Faridabad,
    India
  Pankaj Pathak, G. L. Bajaj Institute of Technology and Management,
    India

Chapter 2
Artificial Intelligence Based on Biological Neurons: Constructing Neural
Circuits for IoT 25
  Rinat Galiautdinov, Independent Researcher, Italy

Chapter 3
Hybrid Genetic Approach for Solving Fuzzy Graph Coloring Problem 54
  Mohamed Amine Basmassi, ISO Laboratory, Faculty of Sciences, Ibn
    Tofail University, Morocco
  Sidina Boudaakat, SSDIA Laboratory, ENSET Mohammedia, Hassan II
    University of Casablanca, Morocco
  Lamia Benameur, LIROSA Laboratory, Faculty of Sciences, Abdelmalik
    Essaadi University, Morocco
  Omar Bouattane, SSDIA Laboratory, ENSET Mohammedia, Hassan II
    University of Casablanca, Morocco
  Ahmed Rebbani, SSDIA Laboratory, ENSET Mohammedia, Hassan II
    University of Casablanca, Morocco
  Jihane Alami Chentoufi, ISO Laboratory, Faculty of Sciences, Ibn Tofail
    University, Morocco

Chapter 4
IOT Technology, Applications, and Challenges: A Contemporary Survey and
Classification...65
    Asha Gowda Karegowda, Siddaganga Institute of Technology,
        Tumakuru, India
    Devika G., Government Engineering College, KRPET, Mandya, India
    Ramya Shree T. P., Siddaganga Institute of Technology, Tumakuru, India

Chapter 5
Nano-Biosensors Tech and IPM in Plant Protection to Respond to Climate
Change Challenges in Morocco 114
    Wafaa Mokhtari, Institut Agronomique et Vétérinaire Hassan II,
        Morocco & Complexe Horticole d'Agadir, Morocco
    Mohamed Achouri, Institut Agronomique et Vétérinaire Hassan II,
        Morocco & Complexe horticole d'Agadir, Morocco
    Abdellah Remah, Institut Agronomique et Vétérinaire Hassan II,
        Morocco & Complexe Horticole d'Agadir, Morocco
    Noureddine Chtaina, Institut Agronomique et Vétérinaire Hassan II,
        Morocco
    Hassan Boubaker, Ibn Zohr University, Morocco

Chapter 6
Reinstate Authentication of Nodes in Sensor Network....................................130
    Ambika N., Department of Computer Applications, SSMRV College,
        Bangalore, India

Chapter 7
Comprehensive Ontological Model for Senior Wellness Activity Recognition
in Smart Homes 148
    Hajar Khallouki, Department of Software Engineering, Lakehead
        University, Canada
    Rachid Benlamri, Department of Software Engineering, Lakehead
        University, Canada
    Abdulsalalm Yassine, Department of Software Engineering, Lakehead
        University, Canada

Chapter 8
Sliding Mode Control for PV Grid-Connected System With Energy
Storage 168
    Saloua Marhraoui, Mohammed V University in Rabat, Morocco
    Ahmed Abbou, Mohammed V University in Rabat, Morocco
    Zineb Cabrane, Mohammed V University in Rabat, Morocco
    Salahddine Krit, Ibn Zohr University, Morocco

Chapter 9
Comparison Analysis of MAC Protocols for Wireless Sensor Networks: A
Comprehensive Survey 200
> *Tapaswini Samant, Kalinga Institute of Industrial Technology, India*
> *Yelithoti Sravana Kumar, Kalinga Institute of Industrial Technology,*
>   *India*
> *Swati Swayamsiddha, Kalinga Institute of Industrial Technology, India*

Compilation of References 219

Related References.. 245

About the Contributors 272

Index.. 278

Detailed Table of Contents

Preface xiv

Acknowledgment xix

Chapter 1
A Review on FPGA-Based Digital Filters for De-Noising ECG Signal 1

    Seema Nayak, Department of Electronics and Communication
        Engineering, IIMT College of Engineering, Greater Noida, India
    Manoj Nayak, Department of Mechanical Engineering, FET, Manav
        Rachnan International Institute of Research and Studies, Faridabad,
        India
    Pankaj Pathak, G. L. Bajaj Institute of Technology and Management,
        India

This chapter gives an overview of synthesis and analysis of digital filters on FPGA for denoising ECG signal, which provides clinical information related to heart diseases. Various types of IIR and FIR filtration techniques used for noise removal are also discussed. Many developments in the medical system technology gave birth to monitoring systems based on programmable logic devices (PLDs). Although not new to the realm of programmable devices, field programmable gate arrays (FPGAs) are becoming increasingly popular for rapid prototyping of designs with the aid of software simulation and synthesis. They are reprogrammable silicon chips, configured to implement customized hardware and are highly desirable for implementation of digital filters. The extensive literature review of various types of noise in ECG signals, filtering techniques for noise removal, and FPGA implementation are presented in this chapter.

Chapter 2
Artificial Intelligence Based on Biological Neurons: Constructing Neural
Circuits for IoT 25
Rinat Galiautdinov, Independent Researcher, Italy

The chapter describes the new approach in artificial intelligence based on simulated biological neurons and creation of the neural circuits for the sphere of IoT which represent the next generation of artificial intelligence and IoT. Unlike existing technical devices for implementing a neuron based on classical nodes oriented to binary processing, the proposed path is based on simulation of biological neurons, creation of biologically close neural circuits where every device will implement the function of either a sensor or a "muscle" in the frame of the home-based live AI and IoT. The research demonstrates the developed nervous circuit constructor and its usage in building of the AI (neural circuit) for IoT.

Chapter 3
Hybrid Genetic Approach for Solving Fuzzy Graph Coloring Problem 54
Mohamed Amine Basmassi, ISO Laboratory, Faculty of Sciences, Ibn
Tofail University, Morocco
Sidina Boudaakat, SSDIA Laboratory, ENSET Mohammedia, Hassan II
University of Casablanca, Morocco
Lamia Benameur, LIROSA Laboratory, Faculty of Sciences, Abdelmalik
Essaadi University, Morocco
Omar Bouattane, SSDIA Laboratory, ENSET Mohammedia, Hassan II
University of Casablanca, Morocco
Ahmed Rebbani, SSDIA Laboratory, ENSET Mohammedia, Hassan II
University of Casablanca, Morocco
Jihane Alami Chentoufi, ISO Laboratory, Faculty of Sciences, Ibn Tofail
University, Morocco

A hybrid genetic approach (HGA) is proposed to solve the fuzzy graph coloring problem. The proposed approach integrates a number of new features, such as an adapted greedy sequential algorithm, which is integrated in genetic algorithm to increase the quality of chromosomes and improve the rate of convergence toward the chromatic number. Moreover, an upper bound is used to generate the initial population in order to reduce the search space. Experiments on a set of five well-known DIMACS benchmark instances show that the proposed approach achieves competitive results and succeeds in finding the global optimal solution rapidly for complex fuzzy graph.

Chapter 4
IOT Technology, Applications, and Challenges: A Contemporary Survey and
Classification 65

Asha Gowda Karegowda, Siddaganga Institute of Technology,
Tumakuru, India
Devika G., Government Engineering College, KRPET, Mandya, India
Ramya Shree T. P., Siddaganga Institute of Technology, Tumakuru, India

The world we in is virtually becoming smaller since living and nonliving things are connected to the internet. Internet of things, or IoT, is a system of interconnected things, each with unique identifiers (UIDs) and the ability to exchange data without the need of human intervention. The rapid growth of IoT is considered the next wave for enhancing services in almost all sectors of life, at low cost and time. This chapter presents IoT in a broader context, in terms of its growth, IoT operating systems, architecture, and future trends of IoT. The major contribution is detailed information of umpteen IoT applications. The various benefits of IoT, matter of concerns with respect to IoT, scope of research work are also discussed. The integration of various technologies is the main enabling factor of IoT, yielding more benefits to society as a whole. Also, supports in understanding implementation technologies and the major applications of their domain where IoT plays a vital role and future problems for next 20 years are also explicated.

Chapter 5
Nano-Biosensors Tech and IPM in Plant Protection to Respond to Climate
Change Challenges in Morocco 114

Wafaa Mokhtari, Institut Agronomique et Vétérinaire Hassan II,
Morocco & Complexe Horticole d'Agadir, Morocco
Mohamed Achouri, Institut Agronomique et Vétérinaire Hassan II,
Morocco & Complexe horticole d'Agadir, Morocco
Abdellah Remah, Institut Agronomique et Vétérinaire Hassan II,
Morocco & Complexe Horticole d'Agadir, Morocco
Noureddine Chtaina, Institut Agronomique et Vétérinaire Hassan II,
Morocco
Hassan Boubaker, Ibn Zohr University, Morocco

In this chapter, the authors introduce two research axes: Part A, nano-biosensors as ad-hoc technologies designed to meet plant diagnostic sensitivity and specificity needs at point of care, and Part B, the study of the interaction of drought and infection stresses in crops investigating bio-control potential antagonists in developing integrated approach (IPM) for disease control measures in crops system. The first part will be revising most used nano-biosensors in plant pathogens detection using different platforms in greenhouses, on-field, and during postharvest. A special focus will be on optical and voltametric immuno/DNA sensors application in plant protection. The last part will present case studies of using nanoparticles functionalized with antibody/ DNA for detecting pathogenic Pseudomonas sp, mosaic viruses, Botrytis cinereal, and Fusarium mycotoxins (DON). The second part will be interpreting experimental results of a case study on evaluating bio-control efficacy of local Trichoderma spp. using root dips treatment in Fusarium solani-green beans pathosystem as a model.

Chapter 6

Reinstate Authentication of Nodes in Sensor Network....................................130

Ambika N., Department of Computer Applications, SSMRV College, Bangalore, India

Transmitting data using the wireless medium provides a greater opportunity for the adversary to introduce different kinds of attacks into the network. Securing these monitoring devices deployed in unattended environments is a challenging task for the investigator. One of the preliminary practices adopted to ensure security is authentication. The chapter uses a new authentication procedure to evaluate the nodes. It uses sensors of a different caliber. Static and mobile nodes are used to bring the thought into play. The mobile nodes in the network demand vigorous defense mechanisms. The chapter defines a new authentication algorithm engaging itself to validate the communicating parties. The work is devised to substitute new nodes into the cluster to foster prolonged working of the network. By adopting the proposed protocol, backward and forward secrecy is preserved in the network. The work assures the security of the nodes by minimizing forge and replay attacks.

Chapter 7

Comprehensive Ontological Model for Senior Wellness Activity Recognition in Smart Homes 148

Hajar Khallouki, Department of Software Engineering, Lakehead University, Canada
Rachid Benlamri, Department of Software Engineering, Lakehead University, Canada
Abdulsalalm Yassine, Department of Software Engineering, Lakehead University, Canada

There are several works in the field of smart homes for healthcare, with different types of sensors used to monitor medical, behavioral and environmental parameters for patients. In the context of smart home for the elderly, the use of sensors needs to be adapted to respect the privacy of elders and to work passively without the need for caregiver assistance. Most research in this area focused on activity recognition (e.g. eating, sleeping, watching TV, etc.) which may be defined as the identification of a sequence of actions (e.g. using microwave, lying down, etc.). In this chapter, we propose a comprehensive ontological model for well-being activity recognition in smart home. Our approach takes into account different aspects of the well-being context such as patient profile, object being used to perform the activity, the time of running the activity, its location, etc. In order to validate the proposed ontology and reason on it, we perform a set of queries and inference rules.

Chapter 8
Sliding Mode Control for PV Grid-Connected System With Energy Storage 168
Saloua Marhraoui, Mohammed V University in Rabat, Morocco
Ahmed Abbou, Mohammed V University in Rabat, Morocco
Zineb Cabrane, Mohammed V University in Rabat, Morocco
Salahddine Krit, Ibn Zohr University, Morocco

We need to solve the problem due to the nonlinearity and power fluctuation in the photovoltaic (PV) connected storage system and grid; for that, the authors develop an algorithm to obtain the maximum power point tracking (MPPT) via control of the duty cycle of DC/DC boost converter. Consequently, they design an MPPT based on the second-order sliding mode control. Next, generating the law control founded on the Lyapunov theory can augment the robustness and stability of the PV connected grid. Then, they add a battery energy storage system (BESS) with a control management algorithm in the DC/DC side to eliminate any fluctuation of the output power of the PV system because of the temperature and irradiation variation. On the grid side, they control the DC/AC inverter side by the three-phase voltage source inverter control (VSIC) as a charge controller for the grid parameters.

Chapter 9
Comparison Analysis of MAC Protocols for Wireless Sensor Networks: A Comprehensive Survey 200
Tapaswini Samant, Kalinga Institute of Industrial Technology, India
Yelithoti Sravana Kumar, Kalinga Institute of Industrial Technology, India
Swati Swayamsiddha, Kalinga Institute of Industrial Technology, India

Wireless sensor networks (WSN) are rapidly emerging as an interesting and challenging area of research in the field of communication engineering. This review work is different from other state-of-the-art literature as the MAC protocols discussed here are applicable both for homogeneous and heterogeneous networks. Performances like energy efficiency, cost optimization, throughput, bandwidth utilization, and scalability of the sensor network depend on MAC protocols, which are application-based. In the study, the authors have surveyed different MAC protocols with different merits and demerits. Based on the study, it is very hard to recommend any particular protocol as a standard for implementation as these are exclusively application dependent. The work can be further extended in terms of hybrid protocols, which may carry the advantages of the respective protocols along with energy-efficient criteria for practical implementation. Further cooperative WSN communication can be used for internet of things (IoT)-based systems, where the node placements and multi-operations concepts are of main concern.

Compilation of References 219

Related References 245

About the Contributors 272

Index 278

Preface

The book chapters cover a selection of innovative research on theory, methods and techniques in the areas of Wireless Sensor Networks (WSN), smart homes, Internet of Things (IoT), smart grid technology, smart sensors in healthcare, Nano-biosensors technology, and other sensor-based intelligent systems. The book addresses many research challenges related to sensor networks and smart technology modeling, development, adoption and deployment in various science and engineering applications. It also addresses a number of societal and ethical aspects for deploying and using sensor networks in our daily life. Furthermore, the book covers both fundamental and technological research that is needed to achieve effective integration of computing methods, artificial intelligent, sensor data, and networking infrastructure to build sustainable systems that have impact on the quality of life.

The list of topics covered by the chapters of this book are:

Wireless Sensor Networks
Network Security
Internet of Things (IoT)
Artificial Intelligence Applications
Big Data and Cloud Computing
Smart Technologies
Sensors & Ad-hoc Networks
Data Security
Machine Learning
Optimization Algorithms
Smart Cities Applications
Sensor-Based Intelligent Systems
Sensor Networks & IoT for Healthcare

The book enables both junior researchers and expert to review the state of the art in the field of Sensor Network Methodology and Smart Applications. It provides them with valuable theoretical and technical knowledge to develop new research ideas and projects in order to advance the research in this emerging field.

Below, we describe the research contributions of each chapter in this book.

CHAPTER 1: A REVIEW ON FPGA-BASED DIGITAL FILTERS FOR DE-NOISING ECG SIGNAL

This chapter gives an overview of synthesis and analysis of digital filters on FPGA for denoising ECG signal, which provides clinical information related to heart diseases. Various types of IIR and FIR filtration techniques are used for noise removal. Many developments in the medical system technology gave birth to monitoring systems based on Programmable Logic Devices (PLDs).

CHAPTER 2: ARTIFICIAL INTELLIGENCE BASED ON BIOLOGICAL NEURONS – CONSTRUCTING NEURAL CIRCUITS FOR IOT

This chapter describes the new approach in Artificial Intelligence based on simulated biological neurons and creation of neural circuits for the sphere of IoT. This is an important topic in the creation of next generation products emerging from integrating Artificial Intelligence and IoT.

CHAPTER 3: HYBRID GENETIC APPROACH FOR SOLVING FUZZY GRAPH COLORING PROBLEM

In this chapter, a hybrid genetic approach (HGA) is proposed to solve the fuzzy graph coloring problem. The proposed approach integrates a number of new features, such as, adapted greedy sequential algorithm, which is integrated in a genetic algorithm to increase the quality of chromosomes and improve the rate of convergence toward the chromatic number.

CHAPTER 4: IOT TECHNOLOGY, APPLICATIONS, AND CHALLENGES – A CONTEMPORARY SURVEY AND CLASSIFICATION

This chapter presents a comprehensive survey of state of the art in IoT technology and its applications. It also addresses the main research challenges related to IoT development, deployment and its usage. The focus of the chapter is on the rapid growth of IoT as next wave for enhancing services in almost every sector of our life, showing its advantages on timely access to data at lower cost.

CHAPTER 5: NANO-BIOSENSORS TECH AND IPM IN PLANT PROTECTION TO RESPOND CLIMATE CHANGE CHALLENGES IN MOROCCO

In this chapter, the authors investigated an interesting topic related to the use of nano-biosensors as ad-hoc technologies designed to meet plant diagnostic sensitivity and specificity needs at point of care. They also studied the interaction of drought and infection stresses in crops, investigating bio-control potential antagonists in developing integrated approach (IPM) for disease control measures in crops system.

CHAPTER 6: REINSTATE AUTHENTICATION OF NODES IN SENSOR NETWORK

This chapter demonstrates that transmitting data using the wireless medium provides a greater opportunity for the adversary to introduce different kinds of attacks into the network. The authors argue that securing these monitoring devices, deployed in unattended environments, is a challenging task for the investigator. One of the preliminary practices adopted to ensure security is authentication. The proposed work uses a new authentication procedure to evaluate the nodes.

CHAPTER 7: COMPREHENSIVE ONTOLOGICAL MODEL FOR SENIOR WELLNESS ACTIVITY RECOGNITION IN SMART HOMES

In this chapter, the authors describe state of the art in the field of smart homes for healthcare, deploying different types of sensors to monitor medical, behavioral and environmental parameters for patients. In the context of smart home for the elderly, the use of sensors needs to be adapted to respect the privacy of elders and to work passively without the need for caregiver assistance. Most research in this area focused on activity recognition (e.g. eating, sleeping, watching TV, etc.) which may be defined as the identification of a sequence of actions (e.g. using microwave, lying down, etc.). In this chapter, the authors propose a comprehensive ontological model for well-being activity recognition in smart home.

CHAPTER 8: SLIDING MODE CONTROL FOR PV GRID-CONNECTED SYSTEM WITH ENERGY STORAGE

In this chapter the authors discuss the need for solving the problem caused by the nonlinearity and power fluctuation in the Photovoltaic (PV) connected storage system and grid. To address this issue, the authors developed an algorithm to obtain the Maximum Power Point Tracking (MPPT) via control of the duty cycle of DC/DC boost converter.

CHAPTER 9: COMPARISON ANALYSIS OF MAC PROTOCOLS FOR WIRELESS SENSOR NETWORKS – A COMPREHENSIVE SURVEY

Wireless Sensor Network (WSN) is rapidly emerging as an interesting and challenging area of research in the field of communication engineering. In this chapter, the authors present a literature, that is different from other state-of-the-art. They argue that the MAC protocols discussed in this chapter are applicable for both homogeneous and heterogeneous networks. Performances like energy efficiency, cost optimization, throughput, bandwidth utilization and scalability of the sensor network depend on MAC protocols which are application-based. In this research, the authors surveyed different MAC protocols with different merits and demerits. Their main finding is that it is very hard to recommend any particular protocol as a standard for implementation, as these are exclusively application dependent.

The work described in each of the above-mentioned chapters is a modest contribution in a fast expanding research area. The aim of the research community is to develop the technology needed for tomorrow's smart cities driven by highly connected cooperative intelligent systems fed by real-time sensor data. However, to achieve the concept of a smart home and smart city, more research work is needed to address many open, yet unresolved research challenges. These are mainly related to technology standards adoption, security, privacy, interoperability, and system scalability, just to name a few. Nevertheless, it should be noted that the research in this field has provided new opportunities for economic growth, health and wellness, efficiency energy management and saving, green and safe environment, and autonomous transportation systems. The main goal is to promote sustainable development of cities.

Acknowledgment

First I would like to thank my colleagues, co-editors of the book, Professors Valentina Bălaş, Mohamed Elhoseny, Rachid Benlamri and Marius Bălaş for not giving up on their dreams of writing this book. Also, I would like to thank the many people who have helped me learn and practice both the art and science of networking throughout the years.

My special thanks to the reviewers for taking the time and efforts reviewing and improving the quality of the research presented in the book chapters.

Finally, I would like to thank the publisher, IGI Global Group, and all editorial team who significantly helped us put this project together. Also, special thanks to the Ibn Zohr University Agadir Morocco, Aurel Vlaicu, University of Arad Romania, Faculty of Computers and Information, Mansoura University Egypt, and Lakehead University Canada for their support.

Chapter 1

A Review on FPGA–Based Digital Filters for De–Noising ECG Signal

Seema Nayak
Department of Electronics and Communication Engineering, IIMT College of Engineering, Greater Noida, India

Manoj Nayak
Department of Mechanical Engineering, FET, Manav Rachnan International Institute of Research and Studies, Faridabad, India

Pankaj Pathak
ⓘ https://orcid.org/0000-0003-2642-1500
G. L. Bajaj Institute of Technology and Management, India

ABSTRACT

This chapter gives an overview of synthesis and analysis of digital filters on FPGA for denoising ECG signal, which provides clinical information related to heart diseases. Various types of IIR and FIR filtration techniques used for noise removal are also discussed. Many developments in the medical system technology gave birth to monitoring systems based on programmable logic devices (PLDs). Although not new to the realm of programmable devices, field programmable gate arrays (FPGAs) are becoming increasingly popular for rapid prototyping of designs with the aid of software simulation and synthesis. They are reprogrammable silicon chips, configured to implement customized hardware and are highly desirable for implementation of digital filters. The extensive literature review of various types of noise in ECG signals, filtering techniques for noise removal, and FPGA implementation are presented in this chapter.

DOI: 10.4018/978-1-7998-4381-8.ch001

OVERVIEW OF ECG

Electrocardiogram (ECG) signal is the most important electrical signal in the field of medical science which has a great need to be processed before further analysis for diagnostic and research purpose. An electrocardiogram is a display of the electrical activity of the heart (cardiac) muscle as obtained from the surface of the skin. Specific feature extraction from an ECG with precise computation and interpretation is always a challenging task. For the correct diagnosis of the heart, the ECG signal should be free from the noise/artifacts. It is necessary to remove these noises/artifacts from the signal for further processing and analysis.

Normally, the frequency range of an ECG signal is of 0.05–100 Hz and its amplitude ranges from 1–10 mV. The ECG signal is characterized by five peaks and valleys labelled by the letters P, Q, R, S, T as shown in Figure 1 and in some cases (especially in infants) another peak called U, may also be seen. The performance of ECG analyzing system depends mainly on the accurate and reliable detection of the QRS complex, as well as T and P waves. The P-wave represents the activation of the upper chambers of the heart, the atria, while the QRS complex and T-wave represent the excitation of the ventricles or the lower chamber of the heart. The detection of the QRS complex is the most important task in automatic ECG signal analysis. Once the QRS complex has been identified a more detailed examination of ECG signal including the heart rate, the ST segment etc. can be performed.

The electrical activity of the heart is generally sensed by monitoring electrodes placed on the skin surface. Unfortunately, other artifactual signals of similar frequency and often larger amplitude reach the skin surface and mix with the ECG signals. The challenge lies in the removal of these frequencies from the ECG signal for its usefulness.

ECG signals are often contaminated by noise from diverse resources which degrade ECG signals significantly. The ECG signal need to be noise free or low noisy (Zaman et al. 2012). As a clean ECG signal gives full detailed information about the electrophysiology of the heart diseases and ischemic changes that may occur. ECG signal contains high and low frequency noise components that must be filtered before further analysis (Tawfik and Kamal 2010).

Typical examples of noise interference in an ECG signal are:

- Power line interference
- Electrode contact noise.
- Motion artifacts.
- Muscle contraction.
- Base line drift.
- Instrumentation noise generated by electronic devices.
- Electrosurgical noise.

Figure 1. Typical ECG signal

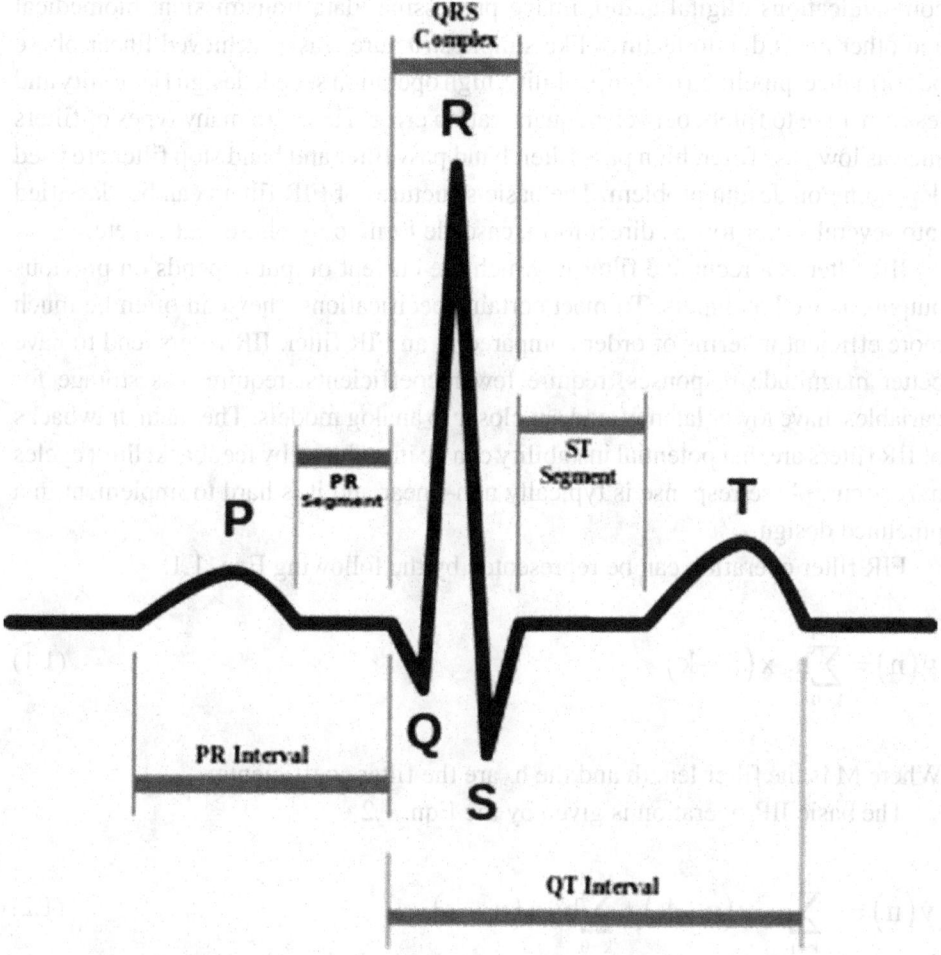

DIGITAL FILTERS

In signal processing, digital filters play an important role to remove the unwanted signals such as noise or extract the important part of the signals that lie within certain frequency range. There are various methods to remove noise in the ECG signal which may involve the Infinite Impulse response (IIR) or Finite Impulse Response (FIR) filter having their own advantages and disadvantages. Digital filter is the pre-eminent solution that caters the noise reduction up to satisfactory level and is best suited for ECG analysis by improving the quality of ECG signal. Digital filters are programmable hence help to reduce the design cycle. Moreover, they perform noiseless mathematical operations at each intermediate step in transforming. FIR filters are widely used in many digital signal processing application areas such as communications, digital audio, image processing, data transmission, biomedical and other areas due to features like simple structure, easily achieved linear-phase performance, pipelined design, stability, high operation speed, design flexibility and less sensitive to filter coefficient quantization error. There are many types of filters such as low pass filter, high pass filter, band pass filter and band stop filter are used depending on design problem. The basic structures of FIR filters can be classified into several major forms: direct form, cascade form, poly phase, lattice, etc.

IIR filter is a recursive filter in which the current output depends on previous outputs as well as inputs. To meet certain specifications, they can often be much more efficient in terms of order compared to an FIR filter. IIR filters tend to have better magnitude responses, require fewer coefficients, require less storage for variables, have lower latency, and are closer to analog models. The main drawbacks of IIR filters are that potential instability can be introduced by feedback, limit cycles may occur, phase response is typically non-linear and it is hard to implement in a pipelined design.

FIR filter operation can be represented by the following Eqn. 1.1.

$$y(n) = \sum_{k=0}^{M-1} h_k \, x(n-k) \tag{1.1}$$

Where M is the filter length and the h_k are the filter coefficients.

The basic IIR operation is given by the Eqn. 1.2

$$y(n) = -\sum_{k=1}^{N} a_k \, y(n-k) + \sum_{k=0}^{M-1} b_k \, x(n-k) \tag{1.2}$$

Where M is the maximum input delay, b_k are the numerator coefficients; N is the maximum output delay, and a_k are the denominator coefficients.

From Eqn. 1.1 and 1.2, computation of y[n] involves calculation of y[n-1], y[n-2],.....y[n-N] and x[n], x[n-1], x[n-2],.......x[n-M] hence there is a requirement of basic elements like delay or storage, multipliers and adders (subtraction is considered as addition). Computations of y[n] can be arranged in different ways to give the same difference equation, which leads to different structures for realization of discrete-time LTI systems.

An implementation can be represented using either a block diagram or a signal flow graph. Operations in block diagram for realization of digital filters are represented in Figure 2. An adder can generally deal with more than two sequences, a multiplier

Figure 2. Basic operations in block diagram

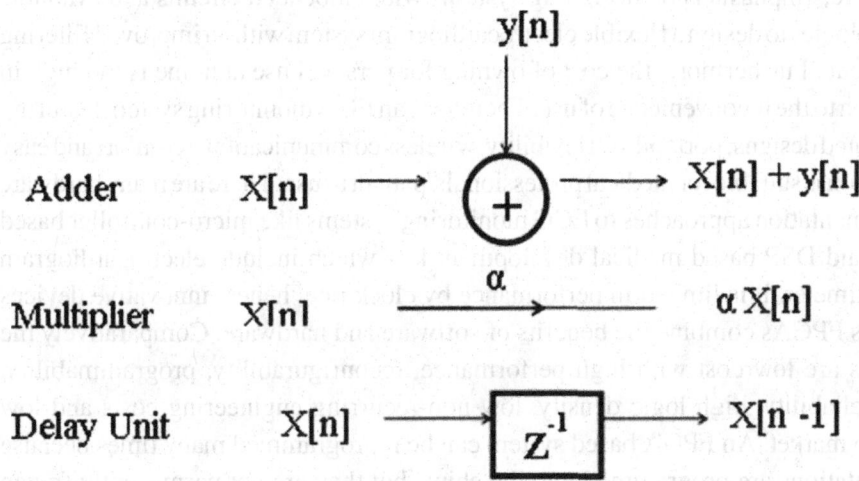

usually has the computational cost and thus it is desired to reduce the number of multipliers in different systems. Delay Unit can be implemented by providing a storage register for each unit delay in digital implementation. Structures used for FIR filters are Direct and Cascade form where the structures used for IIR Filter are Direct Form-I, II, Cascade form and Parallel form. Different digital filter structures are available to eliminate the diverse forms of noise sources.

FPGAs and ASICs based filters are implemented with a parallel-pipelined architecture which enhances the overall performance. Digital filter implementation in FPGAs and other Very Large Scale Integration (VLSI) implementations allow for higher sampling rates and lower cost than that available from traditional Digital Signal Processing (DSP) chips. Digital filters are used extensively in all areas of electronics industry because digital filter have the potential to attain much better signal to noise ratio than analog filter and at each intermediate stage the analog filters add more noise to signal. Digital filters have emerged as a strong option for removing noise, shaping spectrum, and minimizing inter symbol interference in communication architecture. These filters have become popular because their precise reproducibility allows designing engineers to achieve performance level that are difficult to obtain in analog filter. FIR filter is mostly used in DSP application due to its advantages over IIR filter.

The traditional ECG diagnostic products from different companies are bulky and impractical for patients to use in long term health monitoring applications. Power is a significant design constraint for implementing efficient portable biomedical applications. Electrocardiogram lacks flexibility and neither it is compatible with the

Personal Computer (PC) and communications standards nor can it be easily upgraded, therefore, emphasis is on designing systems with embedded circuits and available technologies to design a flexible electrocardiogram system with an improved filtering technique. Furthermore, the cost of owning for personal use at home is too high in addition to the inconvenience of use. Therefore, an ECG monitoring system featuring integrated designs, portability, flexibility, wireless communications, comfort and easy operation is suitable for medical professionals/patient to use. There are many hardware implementation approaches to ECG monitoring systems like micro-controller based ECG and DSP based medical development kits which include electrocardiogram and oximeter. It is limited in performance by clock rate, hence, innovative devices such as FPGAs combine the benefits of software and hardware. Comparatively the FPGAs are low cost with high performance, reconfigurability, programmability, high reliability, high logic density, low non-recurring engineering costs and low time to market. An FPGA based system can be reprogrammed many times because computations are programmed into the chips, but they are not permanently frozen at a time of manufacturing process.

FIELD PROGRAMMABLE GATE ARRAYS (FPGAs)

FPGAs are pre-fabricated silicon devices that can be electrically programmed in the field to become almost any kind of digital circuit or system (Farooq et al. 2012). FPGAs provide cheaper solution and faster time to market as compared to Application Specific Integrated Circuits (ASIC) which normally require a lot of resources in terms of time and money to obtain first device. For varying requirements, a portion of FPGA can be partially reconfigured while the rest of an FPGA is still running. Any future updates in the final product can be easily upgraded by simply downloading a new application bit stream. FPGAs comprise of programmable logic blocks which implement logic functions, programmable routing that connects these logic functions, I/O blocks that are connected to logic blocks through routing interconnect and that make off-chip connections. FPGA becomes more flexible due to programmable logic and routing interconnections, which are its main advantages and also the major cause of its draw back (Brown et al. 2005). Flexible nature of FPGAs makes them significantly larger, slower, and more power consuming than their ASIC counterparts. But despite these disadvantages, FPGAs present a compelling alternative for digital system implementation due to their less time to market and low volume cost.

The major advantages of FPGAs are versatility, flexibility, huge performance gain for some applications and re-useable hardware designs. Some recent FPGAs include DSP features, such as ALTERA and Xilinx which makes FPGAs more attractive for DSP algorithm implementations. FPGAs are well suited for the implementation of fixed-point digital signal processing algorithms. However, the advancement in process technology has enabled and necessitated a number of developments in the basic FPGA architecture. These developments are aimed at further improvement in the overall efficiency of FPGAs so that the gap between FPGAs and ASICs might be reduced.

FPGAs have become increasingly important to the electronics industry. They have the potential for higher performance and lower power consumption than microprocessors and compared with application specific integrated circuits (ASICs), offer lower non-recurrent engineering (NRE) costs, reduced development time, easier debugging and reduced risk. Since modern FPGAs can meet many of the performance requirements of ASICs, they are being increasingly used in their place. The architecture of FPGA (Figure 3) consists of many logic modules which are placed in array structure and these modules are configurable at site and are therefore called as Configurable Logic Blocks (CLBs). The channels between the CLBs are used for routing. The arrays of the CLBs are surrounded by programmable I/O modules and connected via programmable interconnects. The architecture of various FPGAs differs from vendor to vendor and is characterized by its structure, content of logic block and routing resources. FPGAs now deliver ASIC-like density and performance, while their flexibility and operational characteristics offer distinct advantages over their ASIC counterparts. The designers are inclined more towards FPGAs as modern architectures with embedded processors, memory blocks and Digital Signal Processors (DSPs) are emerging in FPGA.

FPGAs are the fastest growing segment of the semiconductor industry and are becoming increasingly popular for rapid prototyping of designs with the aid of software simulation and synthesis. Software synthesis tools translate high-level language descriptions of the implementation into formats that may be loaded directly into the FPGAs. An increasing number of design changes through software synthesis become more cost effective than similar changes done for hardware prototypes. In addition, the implementation may be constructed on existing hardware to help further reduce the cost. It is well known that FPGAs are widely used in the implementation of fast digital systems for retrieval, processing, storage, and transmission of data. Verilog and Very High Speed Integrated Circuit HDL (VHDL) are hardware description languages to design digital logic using FPGAs and CPLDs. FPGAs provide optimal device utilization through conservation of board space and system power.

Figure 3. Overview of FPGA architecture (Farooq et al. 2012)

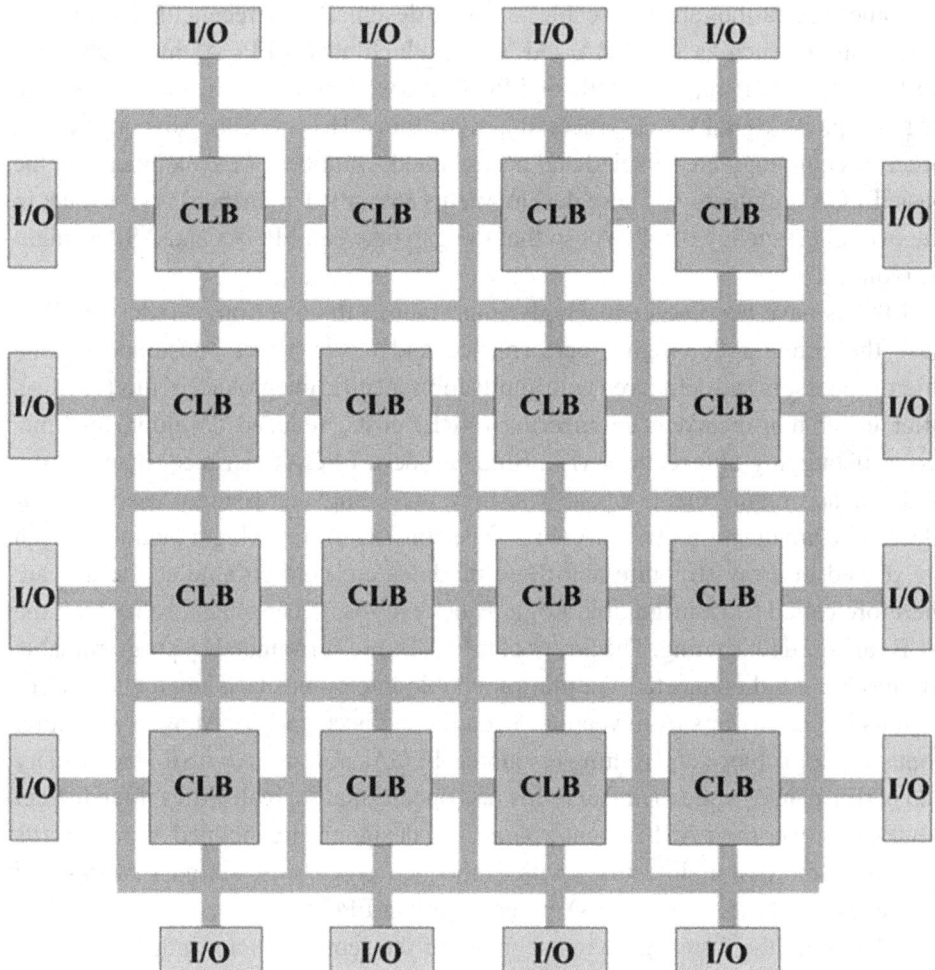

Advantages of Using FPGAs Based Design

The advantages of using FPGA based designs are summarized with following concern (Kolawole et al. 2015)

- **Performance**: Due to the advantage of hardware parallelism, task is accomplished in a single clock cycle. Control of inputs and outputs at hardware level in FPGA, provides faster response.

- **Time to market**: FPGA technology offers flexibility, rapid prototyping capabilities and without going through the long fabrication process of custom ASIC design, an idea or concept can be tested and verified.
- **Cost**: The non-recurring engineering expense of FPGA-based hardware solutions is much less as compared to custom ASIC design.
- **Reliability**: The software tools provide the programming environment and the FPGA circuitry is truly a "hard" implementation of program execution. As FPGAs do not use operating systems, reliability concerns get minimized with its true parallel execution and deterministic hardware dedicated to every task.
- **Long-term maintenance**- The FPGA chips are field-upgradable and do not require the time and expense for its redesign. These chips can keep up with future modifications that might be necessary with the changes over time and do not cause maintenance and forward-compatibility challenges.

FPGAs suffer in terms of area, performance and power consumption relative to the many more customized alternatives such as full custom design, ASICs, Mask-Programmable Gate Arrays (MPGAs). Measurements of the FPGA to ASIC gap are useful for both the FPGA designers and architects who aim to narrow this gap and the system designers who select the implementation platform for their design (Kuon and Rose 2007). Figure 4 presents a rough comparison of different solutions used to reduce the drawbacks of FPGAs and ASICs.

Figure 4. Comparison of different solutions used to reduce ASIC and FPGA drawbacks (Farooq et al. 2012)

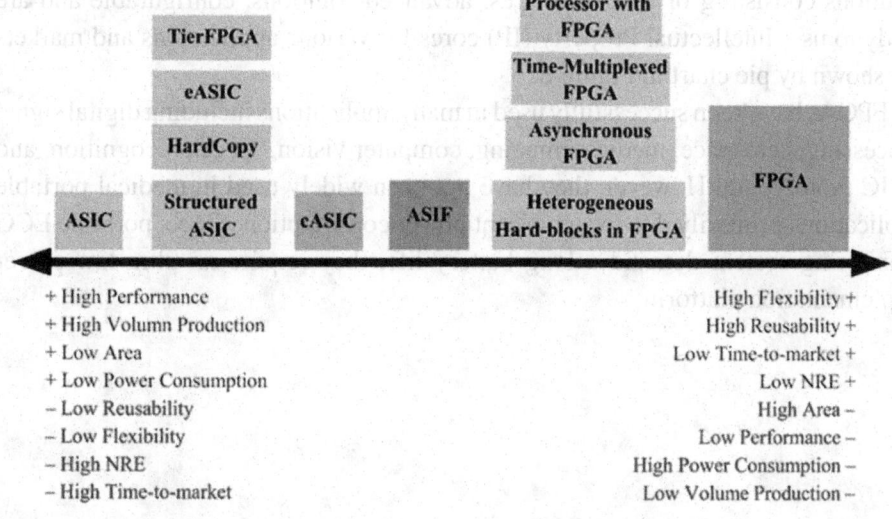

The disadvantages of FPGA are primarily related to the limited number of logic operation that can be implemented on a particular device. The increase in density of FPGAs in the future will certainly expand the design space available to the designer and make several constraints less severe in case of MACs.

FPGAs Market Trends

FPGA technology is growing continuously to gain momentum, and the worldwide FPGA market which was valued at USD 5.08 billion in 2012 is expected to grow to USD 8.95 billion by 2019, growing at a Compound Annual Growth Rate (CAGR) of 8.5% from 2013 to 2019 as per the new market report published in Transparency Market Research (2013). Since their invention by Xilinx in 1984, FPGAs have gone from being simple glue logic chips to actually replacing custom ASICs and processors for signal processing and control applications. Profiling the top FPGA companies for 2013, Altera and Xilinx continue to dominate the market for general purpose programmable logic as per the report of Source Tech 411 (2013) about top FPGA companies for 2013 dated 28th April 2013 (Grover and Soni 2014).

The other FPGA companies like Actel, Lattice, Lucent, Quick logic, Cypress etc. had their market share in 1998 and 2012 as shown in Figure 5. These two companies comprise approximately 90% market share (Xilinx 47%, Altera 41%) in 2013 with combined revenues in excess of $4.5B and a market cap over $20B. The FPGA market is expected to increase to $7.9 billion by 2020. Expansion in new applications and lower NRE cost per ASIC design are the key drivers to grow lucratively (Research and markets, 2015).

FPGAs are an ideal fit for many different markets due to their field-programmability. Many industry leaders in the field of FPGAs especially Xilinx provide comprehensive solutions consisting of FPGA devices, advanced solutions, configurable and are ready to use. Intellectual Property (IP) cores for various applications and markets are shown by pie chart in Figure 6.

FPGAs have been successfully used in many applications including digital signal processing, aerospace, medical imaging, computer vision, speech recognition, and ASIC prototyping. However, they have not been widely used in medical portable applications primarily due to significant power consumption. Since, portable ECG monitoring system demand a long battery life, they require an ultra-low power implementation platform.

Figure 5. FPGAs market share

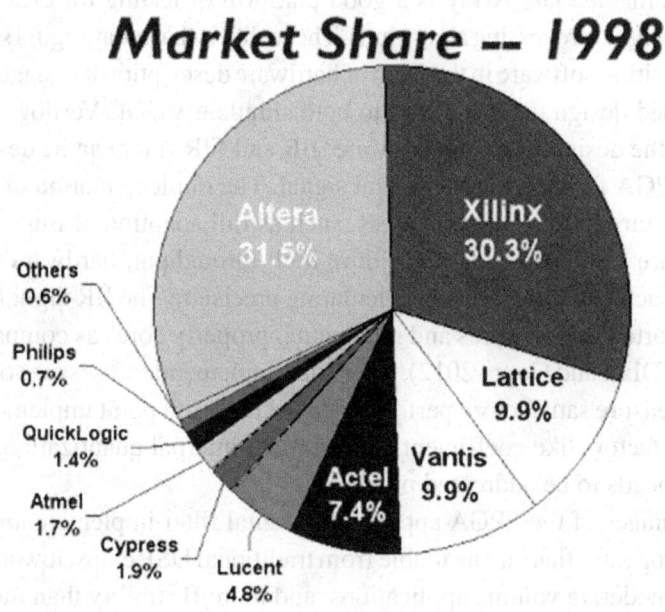

Figure 6. Application of FPGA

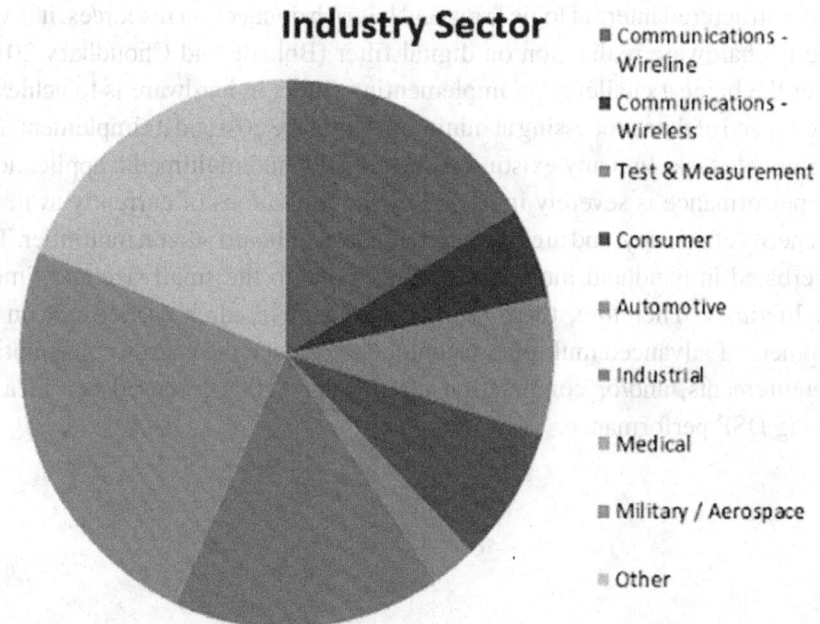

Filter Implementation Techniques in FPGAs

Field Programmable Gate Array is a good platform of testing for evaluating and implementing signal processing algorithms. The reality is that today digital systems are designed by writing software in the form of hardware description languages (HDLs). Computer-aided design tools are used to both simulate VHDL/Verilog design and to synthesize the design to actual hardware. IIR and FIR filter can be designed and realized on FPGA for filtering the digital signal. The implementation of IIR filters in FPGA structure has set of advantages, such as full adaption of implemented in FPGA structure to the filtering algorithm, high throughput, hardware utilization effectiveness, achieving high rate of calculating precision. The IIR digital filters are not well supported by softwares and intellectual property cores as compared to the FIR filtering (Dixit and Gupta 2012). There are implementations issues of IIR filter in FPGA. To ensure satisfactory performances of the fixed point implementation of IIR filter, the factors like coefficient quantization, internal quantization, overflow and stability needs to be addressed properly.

The advantages of the FPGA approach to digital filter implementation include higher sampling rates than are available from traditional DSP chips, lower costs than an ASIC for moderate volume applications, and more flexibility than the alternate approaches. Since many current FPGA architectures are in-system programmable, the configuration of the device may be changed to implement different functionality if required. In designing digital filters, it requires a lot of multipliers, adders and registers. The multipliers and adders can be formed using the logical array. Since FPGA is a structured internal logic array and has rich connection resources, it is very suitable for hardware realization on digital filter (Bokade and Choudhary 2015). However the biggest challenge in implementing filters in hardware is to achieve a specified speed of data processing at minimum hardware cost and its implementation is a formidable task. In many existing real-time DSP and multimedia applications, system performance is severely impacted by the limitations of currently available speed, energy efficiency, and area requirement of an onboard silicon multiplier. This is exacerbated in handheld multimedia devices due to the small size and limited battery lifetimes. Therefore, there has been a lot of research carried out on the development of advanced multiplier techniques to reduce the energy consumption, area requirements, and/or computation time. Wang (2007) focused key idea for improving DSP performance.

Challenges in FPGAs Based Digital Systems

FPGA has become an attractive implementation solution in the modern digital systems due to its reconfigurable architecture, ease of design and flexibility, better performance and low NRE cost. The fundamental challenges and issues in FPGAs involve programmable routing circuit design and architecture. The logic block architecture of an FPGA is also extremely important because it has a dramatic effect on how much programmable routing is required. A crucial goal in the evolution of FPGAs is the reduction of the area, performance, and power penalty of using the devices versus ASICs. One possible approach to reaching this goal is to integrate more hard blocks into FPGAs. As fabrication technology progresses, it is clear that power consumption, both dynamic and static, has become a serious issue. It is important for architects to continue to think of high-level architectural methods to reduce power consumption. Before FPGAs can be effectively used in portable electronics, issues related to the static power consumption of routing must be addressed. However, significant factors and issues of concern in FPGA system are area, delay, power consumption, radiation effects on FPGAs and process variation. Kuon and Rose, et al. (2007) has performed a thorough comparison between FPGAs or similar devices and ASICs in the past.

LITERATURE SURVEY

The literature study was helpful in finding the knowledge gaps and the research needs. The literature review of various types of noise in ECG signals, filtering techniques for noise removal and FPGA implementation are presented in modular form in the following sub-sections.

Review on Acquisition of ECG Signal

An extensive study of different approaches for using ECG signal for denosing before processing is done. Many researchers have developed or simulated circuits in different manner for acquisition of ECG signal.

Bansal et al. (2009) has developed a prototype of a single channel ECG monitor system which includes an analog system and a FM transceiver pair interfaced through sound port of the computer. The developed hardware consists of a cascaded amplifier, an active band pass filter and the right leg drive circuits. Further the amplified and filtered output is fed to a cost effective FM Transmitter. The receiver unit receives the signal in wireless mode and is interfaced in a simple manner to the laptop through the sound port. The real time acquired data is viewed and filtered using

MATLAB software. The noise present in the output of the ECG system is removed using online software digital filters. The developed ECG system is also interfaced to the computer using the DAQ card (National Instruments (NI)). The ECG wave is recorded and analyzed using LabVIEW software. The system is cost effective and has the capability to accurately measure cardiac response and transfer it in real time to enable remote consultation between doctors and patients.

Spinelli et al. (2001) presented a biopotential amplifier for single supply operation which uses a Driven Right Leg Circuit (DRL) to drive the patient's body to a DC common mode voltage, centring biopotential signals with respect to the amplifier's input voltage range. The described circuit is suited for low consumption, battery-powered applications, requiring a single battery and avoiding switching voltage inverters to achieve dual supplies. A biopotential amplifier with a gain of 60 dB and a DC input range of ±200 mV implemented using low power operational amplifiers. The circuit tested successfully with real biomedical signals. An ECG record picked up by the proposed amplifier and was digitized by a general-purpose microcontroller and transmitted to the computer using an optic link.

Morales et al. (2011) presented an electrocardiogram acquisition system based on a Field Programmable Analog Array (FPAA) device, which has been adapted to the requirements of the signal shape and/or medical specifications. They acquired real ECG signals using a three-lead configuration and detailed study of its technical features has been carried out, showing good suitability for ECG acquisition. Further digital ECG signal processing is performed on a Field Programmable Gate Array (FPGA) device, which has allowed different digital configurations to be tested, including FIR and wavelet filtering and identification of wave features such as the QRS complex. The pairing of FPAA and FPGA devices confirmed a compact and versatile bio-signal acquisition platform. The platform has shown very good performance with different types of electrodes, and has also demonstrated the dynamic reconfiguration capabilities of the devices, which enable, for the tuning of gain and bandwidth as required by different input conditions and ECG application requirements. The analysed performance parameters provide values such as 102 dB CMRR (Common-Mode Rejection Ratio), 14-bit ADC resolution, and 75 dB SNR, among other parameters, thus satisfying the minimum requirements for clinical use.

Sharma et al. (2012) designed and implemented a variable gain amplifier for biomedical signal. Variable gain provides the facility to increase or decrease the gain depending upon the acquiring signal and the same amplifier hardware can be used for acquiring various biomedical signals. The circuit was simulated on Multisim and the prototype version was built on PCB.

Wang et al. (2012) introduced the design of real time ECG signal acquisition circuit with the DSP chip TMS320VC5509A in detail. Such system has advantages of small volume, little power consumption, low cost and real-time processing. The

ECG signal is filtered using analog filter, amplified and then converted from analog to digital field. Finally, it is displayed in the LCD (Liquid Crystal Display) after digital low-pass filtering in DSP.

Bansal (2013) designed various 50 Hz active Notch filter topologies in P-Spice for computer based compact and enhanced signal acquisition system. Based on their amplitude and phase response add on the Notch filter hardware was developed and used with real time computer based ECG system having amplifier system which includes biosensors, cascade amplifier, right leg drive circuit, analogue active filter and 50 Hz notch filter.

Review on Denoising ECG Signal Using Digital Filters

In order to provide a clear ECG signal for the doctor and improve the precision of analysis and diagnosis, several methods to denoise ECG signals have been discussed. For processing of ECG signal, researchers have used ECG signal collected from MIT-BIH database which is widely accepted in the world and from acquired real time ECG signal of human subject using acquisition circuit which has been already discussed. Acquisition of ECG signal using actual circuit is a challenging task and complex method. ECG signals stored in database are noise free and easy to use for the research. The problem with the research work, which uses the databases, are inconvenient for testing the ECG signal as a biometric feature because there is no sufficient data samples for learning and testing (Tawfik and Kamal 2010). Therefore, researcher have used various filtering techniques to remove low, high frequency and 50 Hz noise component contaminated in ECG signal during the course of research.

Tragardh and Schlegel (2006) came out with a review report on high frequency ECG. The standard ECG is by convention limited to 0.05-100 Hz, but higher frequencies are also present in the ECG signal is also an established fact. With high-resolution technology, it is possible to record and analyze these higher frequencies. The highest amplitudes of the high-frequency components are found within the QRS complex. The term "high frequency" has been used by several investigators to refer to the process of recording ECGs with an extended bandwidth of up to 1000 Hz. Several investigators had tried to analyze HF-QRS (high frequency) with the hope that additional features seen in the QRS complex would provide information enhancing the diagnostic value of the ECG.

Sadabadi et al. (2007) conducted a mathematical method based on two sets of information, which are dominant scale of QRS complexes and their domain. The task is proposed by using a varying-length window that is moving over the whole signals. Both the high frequency (EMG noise) and low frequency (base-line wandering) removal tasks are evaluated for manually corrupted ECG signals and are validated for actual recorded ECG signals. Their proposed method was suitable due to its simplicity, fast implementation, and preservation of characteristics of ECG waves. At the same time, there were some difficulties due to pre-stage detection of QRS complexes and specification of algorithm's parameters for varying morphology cases.

Sayadi and Shamsollahi (2008) presented efficient denoising and lossy compression schemes for ECG signals based on a modified extended Kalman filter structure (EKF). They extended their previous work on two-dimensional EKF structure and modified its governing equations to a 17-dimensional case. The performances of the proposed method were evaluated using standard denoising with SNR and compression efficiency with compression ratio, percentage area difference, and the weighted diagnostic distortion measures. Several Massachusetts Institute of Technology–Beth Israel Deaconess Medical Center (MIT–BIH) ECG databases were used for performance evaluation. The denoising results showed, an average SNR improvement of 10.16 dB which, is 1.8 dB more than the next benchmark methods such as EKF2; for compression efficiency, the algorithm extended to include more than five Gaussian kernels showed typical average CR of 11.37:1 with WDD<1.73% consequently. The proposed framework is suitable for a hybrid system that integrates the proposed algorithmic approaches for clean ECG data storage or transmission scenarios with high output SNRs, high CRs, and low distortions.

Kaur et al. (20 11) implemented different methods for performance evaluation to remove the noise based on parameter i.e. Power Spectral density (PSD), Average Power and Signal to Noise Ratio (SNR). They found the IIR Zero phase filtering is an efficient method for the removal of Baseline wander from ECG signal based on the reduction of PSD value as the Baseline drift reduced effectively. They found the Average Power and SNR are also greater as compared to others. The order of filter required was very low, therefore, the complexity and computational load required is very less as compared to others. They proved that IIR Zero phase filtering is the best method among the various proposed techniques.

Zaman et al. (2012) analyzed theoretically and empirically the different de-noising methods to denoise the ECG signal. They used MATLAB to simulate and evaluate ideal ECG signal and noisy ECG signal offline. The analysis of existing methods has been compared based on different parameters. They used three methods for de-noising like Notch filter, Sgolay filter and Wavelet. After comparing their performances, they considered that Sgolay filter is the best for recovering ECG signal, because it gives better output and small error value of 2.50%. On the other hand the Notch filter and wavelet has the error of 30.33% and 30.56% respectively, which are higher than the Sgolay filter. An accuracy of 97.50%, 69.67% and 69.44% are found for Sgolay, Notch and Wavelet respectively. Therefore, Sgolay filter with highest accuracy considered to be the best method for de-noising the ECG signal.

Chandrakar et al. (2013) studied Finite Impulse Response (FIR) filter based on various Windows and Infinite Impulse Response (IIR) filters for noise removal of ECG signal. The results showed that Kaiser Window based FIR filter is best to remove artifacts from ECG signal whereas the Rectangular Window showed the distorted signal.

Joshi et al. (2013) implemented several filtration techniques such as low pass, high pass, band pass and Notch filter to remove the noise from signal. The results of the implementation showed that the receptive filter removed noise specifically meant for it. The Notch filter showed great results in suppression of power line interferences, which were clear from Fast Fourier Transform (FFT) and PSD of the filtered signal. They also used moving averaging filter which has shown very good efficiency in smoothing out the waveform and suppressing 50 Hz power line noise but could not clear the baseline wandering. Thus the combination of band pass filter and moving averaging filter were highly efficient in clearing considerable noise along with baseline wander.

Barik and Bagha (2013) has proposed second order infinite impulse response Notch filter, adaptive notch filtering technique with Least Mean Square (LMS) algorithm and Discrete Wavelet Transform (DWT) method for the removal of power line interference from ECG signal. They used different ECG signals from MIT-BIH arrhythmia database with added power-line interference noise which is common in ECG signal. Further, the results were analyzed in MATLAB, based on two synthesis parameters MSE and SNR. The IIR Notch filter, LMS algorithm applied directly to the non-stationary signal like ECG has shown more ringing effect. Db4 wavelet transform showed the output SNR value 97.60% with respect to other IIR, LMS, Haar, Db2, Db3. The performance of Daubechies wavelet transform showed good results compared to other methods, to estimate the better quality de-noising of ECG signal.

Bortolon and Christov (2014) has proposed and tested the high-frequency noise filter, based on the method of Savitzky-Golay with a dynamic adjustment of the filtering window inside the QRS interval. It suppressed sufficiently the EMG noise, with minimal distortion of high frequency content of the QRS complex. Electromyographic (EMG) noise is constantly present in stress test electrocardiographic recordings, due to physical exercise. They studied 106 patients: age 63±10 years, 45 males and acquired digital 12-lead electrocardiograms during stress ECG test using veloergometer. Median recording duration of 7.08 minutes and a mean number of 669 RR intervals. The considered EMG filter was tested, and the noise suppression out of QRS is at least 2.5 times higher than in QRS. A comparison of the results produced significant decrease of standard deviation of PCA index in QRS interval.

Review on FPGAs Based Digital Filter Design

Due to the advancement in Very Large Scale Integration (VLSI) Technology, realization of digital filters is done in Application Specific Integrated Circuits (ASIC) and Field Programmable Gate Arrays (FPGAs) platforms. Researchers actively worked towards an approach to design and implementation of digital filter algorithms based on FPGAs due to its inherent advantages in its design technology. The digital filters require a lot of multipliers, adders and registers. The multipliers and adders can be formed using logical arrays. Since FPGA is a structured internal logic array and with rich connection resources, it is more suitable for hardware realization of digital filter. Several works pertaining to the implementation of digital filter in a FPGA platform are as follow.

Chou et al. (1993) described an approach to the implementation of digital filter algorithms based on Field Programmable Gate Arrays (FPGAs). The digital filtering algorithms are most commonly implemented using general purpose digital signal processing chips for audio applications, or special purpose digital filtering chips and Application Specific Integrated Circuits (ASICs) for higher sampling rates. The advantages of the FPGA approach to digital filter implementation include higher sampling rates than are available from traditional DSP chips, lower costs than an ASIC for moderate volume applications, and more flexibility than the alternate approaches.

Kuon et al. (2007) reviewed and surveyed on the historical development of programmable logic devices, the fundamental programming technologies that the programmability is built on, and then described the basic understandings gleaned from research on architectures. FPGA architecture has a dramatic effect on the quality of the final device's speed performance, area efficiency, and power consumption. The key elements of modern commercial FPGA architecture were included with prospective future trends in the field. Survey has explored many issues in the complex and rapidly evolving world of pre-fabricated FPGA architectures and assessed a significant difference of the use of various technologies. They found FPGA possess certain characteristics that distinguish from the rest. The comparison of FPGA vs. ASIC are summarized in Table 1.1 with each row indicating the particular metric being compared: area consumed critical path delay, and dynamic power.

Ravi Kumar (2012) carried out digital filtering with low pass FIR architecture to filter the 50 Hz coupled noise and other high frequency noises. The filtered signal is subjected to Short Time Fourier transform by which lot of inferences can be drawn by medical experts. A recorded ECG signal used as test input to test the modules implemented on FPGA. The Modelsim Xilinx Edition and Xilinx Integrated Software Environment used simulation and synthesis respectively. The Xilinx Chipscope tool used to test the results and the logic run on FPGA with a Xilinx Spartan-3 Family

Table 1. The FPGA: ASIC gap

Metric	Soft logic only	Soft logic & DSP	Soft logic & Memory	Soft logic & DSP &Memory
Area	35	25	33	18
Delay	3.4	3.5	3.5	3.0
Dynamic power	14	12	14	7.1

FPGA development board. ECG monitor system used to collect, store, playback, wireless transmission integrated into a FPGA chip, so that it greatly reduced the development of analog circuits, reduced development costs, research and design cycle. The filters have a good application value, according to their study.

Islam et al. (2012) proposed the architecture of a programmable digital IIR filter based Xilinx FPGA board. In this architecture gate level design has been used to analyze the impulse response of the IIR filter. FPGAs are increasingly being promoted in signal processing applications with their tractability, parallelism, high speed, and fast time-to-market. Digital filter is one of the important contents of digital signal processing. The characteristic of frequency selection in lower order in comparison with FIR, IIR digital filters are widely applied in modern signal processing systems. Hardware description languages such as Verilog differ from software programming languages because of the propagation of time and signal dependencies (sensitivity). The IIR filters are widely used in digital signal processing with the help of programmable digital processors. But during the realization of large order filters the speed, cost, and flexibility is affected because of complex computations. Therefore, the implementation of 2nd order IIR filters on FPGAs is quite appropriate and relevant as it gives enhanced speed, which they implemented in XC2S150 processor FPGA. This is due to the fact that the hardware implementation of lot of multipliers can be done on FPGA which is a limitation in case of programmable digital processors. The above approach gives a better performance than the common filter structures in terms of speed of operation, cost, and power consumption and also worked in real time processing of any digital signal.

Dixit and Gupta (2012) proposed the implementation and simulation of IIR filter using Xilinx System Generator software and the simulink environment in MATLAB on an FPGA. They found that the overflow and quantisation effects considerably for stability while designing filter from the given specifications. The speed of computation was greatly increased by implementing a filter on an FPGA, rather than a conventional DSP processor. They suggested the parallel processing capability of the FPGA greatly increases the speed of operation in the implementation of the digital filter and cautioned filters to be carefully designed and guarded against quantisation effects and overflow errors.

Chauhan et al. (2013) analysed FPGA implementation, cost of FIR filter based on various windowing techniques. The filter has designed and simulated using MATLAB 7.6. In implementation, the amount of multiplier and adder circuits plays an important role in chip area, speed and cost. All window techniques have been compared in terms of number of adders and multipliers. Their results showed that performances of Blackman & Hanning window techniques are better as compared to Rectangular, Hamming and Bartlett. The performance has been further improved using Kaiser and Equiripple window methods. Equiripple window based FIR filter has shown 28.97% reduction in multiplier and 29.08% in adder as compared to Kaiser to provide cost effective solution for DSP applications.

Kasetwar and Gulhane (2013) presented a study on implementation of adaptive power line interference canceller for ECG signals and recommended Least mean-square (LMS) algorithm for implementation of adaptive power line interference canceller using FPGA. They proposed Notch filters and adaptive interference cancellers for power line interference. They found Notch filters were ineffective, whenever the power line frequency was not stable or not precisely known. A mismatch between the suppression band and the power line frequency might lead to inadequate reduction of the power line interference. Therefore, adaptive interference cancellers are beneficial.

LMS algorithm is commonly used on adaptive filtering since it enables the design of modular systolic architectures. Use of systolic architectures in the implementation of LMS algorithm resulted in reduced area and high speed of operation.

Adaptive interference cancellers implemented using general-purpose microcontrollers or digital signal processors. They found that these solutions are not optimum since they have low processing speed and sequential execution due to the fact that adaptive cancellers are implemented with a software program. They offered an alternative solution i.e. the implementation of the adaptive controller with FPGA. The fundamental advantage that FPGAs offers is the parallel implementation of ECG signal processing algorithms without losing processing speed. It also provides additional control and management features in monitoring the ECG process.

Pawar and Bhaskar (2013) used high pass filtering of ECG signal to remove the baseline wander which distorts the low frequency segment of ECG signal is S-T segment carrying the information related to heart attack. The results of different filter design techniques like equiripple, least square and various windowing methods are compared in MATLAB using SNR and correlation coefficient were analyzed to find the degree of mismatch between noisy ECG and filtered ECG. The designed FIR filter with Kaiser Window worked excellent in removing baseline wandering and then implemented on FPGA platform. They found that FPGA implementation of FIR high pass filter with Kaiser Window technique requires less hardware resources, less time (7.247 nsec) and less power consumption (83mW) than any other design methods.

Yadav and Mehra (2014) examined the optimal implementation cost performance of various IIR Filters, which are relevant for real time application. These low pass IIR filters designed and analyzed by FDATool and the implementation cost has analyzed on the basis of filter order, multiplier, adder, and input samples. Analysis results showed that Elliptical IIR filter offer important advantages over their substantially lower computational or hardware complexity and system latency.

Bokde and Choudhary (2015) used low pass FIR filter to filter the noise in ECG signal. The research aimed for implementation option that satisfies the requirement on flexibility and portability such as speed enhancement and hardware cost. FIR serial architecture technique was used for the hardware design in order to minimize the hardware resource. The output of the filter is compared with the ECG signal before filtering by plotting the signal in time domain and frequency domain using MATLAB for analysis. As discussed, the literature survey has showed that the high end FPGA has a huge throughput advantage but it comes with a hardware costs.

This chapter explains the Electrocardiogram (ECG) signal and its significance in medical science. Various types of noises contaminated in the signal and the methods of noise removal are discussed. History of Field Programmable Gate Arrays is also presented. Further, a comprehensive review of the published literature related to research topic has been carried out.

CONCLUSION

General purpose DSP implementations often lacks the performance necessary for moderate sampling rates, and ASIC approaches are limited in flexibility and may not be cost effective. A Verilog implementation of FPGA based digital filters produces appreciable results because of various benefits like low power consumption, higher efficiency, faster etc. Various approaches which are surveyed are most effective for implementations with the constraints of low cost and low power. The modern architecture of FPGA reduces the number of multipliers and suggesting the small chip area. Also it illustrated that the FPGA approach is both flexible and provides performance comparable or superior to traditional approaches because of the programmability of this technology.

REFERENCES

Bansal, D. (2013). Design Of 50 Hz Notch Filter Circuits For Better Detection of Online ECG. *International Journal of Biomedical Engineering and Technology, 13*(1), 30–48. doi:10.1504/IJBET.2013.057712

Bansal, D., Khan, M., & Salhan, A. K. (2009). A Computer Based Wireless System For Online Acquisition, Monitoring and Digital Processing of ECG Waveforms. *Computers in Biology and Medicine, 39*(4), 361-367.

Barick, S., & Bagha, S. (2013). Removal of 50Hz Power Line Interference For Quality Diagnosis of ECG Signal. *International Journal of Engineering Science and Technology, 5*(05), 1149–1155.

Bokde, P. R., & Choudhari, N. K. (2015). Implementation of Digital Filter on FPGA For ECG Signal Processing. *International Journal of Emerging Technology and Innovative Engineering, 1*(2), 175–181.

Bortolan, G., & Christov, I. (2014). Dynamic Filtration of High-Frequency Noise in ECG Signal. *Computing in Cardiology, 41*, 1089-1092.

Chandrakar, B., Yadav, O. P., & Chandra, V. K. (2013). A Survey of Noise Removal Techniques For ECG Signals. *International Journal of Advanced Research in Computer and Communication Engineering, 2*(3), 1354–1357.

Chauhan, S., Sharma, L., & Mehra, R. (2013). Cost Analysis of Digital FIR Filter Using Different Window Techniques. *International Journal of Electrical, Electronics and Data Communication, 1*(6), 25-29.

Dixit, H. V., & Gupta, V. (2012). IIR Filters Using System Generator For FPGA Implementation. *International Journal of Engineering Research and Applications, 2*(5), 303–306.

Farooq, U., Marrakchi, Z., & Mehrez, Z. (2012). *Tree Based Heterogeneous FPGA Architecture* (Vol. 16). Springer. doi:10.1007/978-1-4614-3594-5

Grover, N., & Soni, M.K. (2014). Design of FPGA Based 32-bit Floating Point Arithmetic Unit and Verification of Its VHDL Code Using MATLAB. *International Journal of Information Engineering and Electronic Business, 1*, 1-14. DOI: doi:10.5815/ijieeb.2014.01.01

Islam, R., Sarker, R., Saha, S., & Nokib Uddin, A. F. M. (2012). *Design of A Programmable Digital IIR Filter Based on FPGA*. IEEE. doi:10.1109/ICIEV.2012.6317409

Joshi, P. J., Patkar, V. P., Pawar, A. B., Patil, P. B., Bagal, U. R., & Mokal, B. D. (2013). ECG Denoising Using MATLAB. *International Journal of Scientific and Engineering Research, 4*(5), 1401–1405.

Kasetwar, A. R., & Gulhane, S. M. (2013). Adaptive Power Line Interference Canceller: A Survey. *International Journal of Advances in Engineering and Technology, 5*(2), 319–326.

Kaur, M., & Singh, B. (2011). Comparisons of Different Approaches For Removal of Baseline Wander From ECG Signal. *International Journal of Computer Applications*, 30-36.

Kolawole, E. S., Ali, W. H., Cofie, P., Fuller, J., Tolliver, C., & Obiomon, P. (2015). Design and Implementation of Low- Pass, High- Pass And Band- Pass Finite Impulse Response (FIR) Filters Using FPGA. *Circuit and System, 6*(02), 30–48. doi:10.4236/cs.2015.62004

Kuon, I., & Rose, J. (2007). Measuring The Gap Between FPGAs and ASICs. *IEEE Transactions on Computer-Aided Design of Integrated Circuits and Systems, 26*(2), 203–215. doi:10.1109/TCAD.2006.884574

Kuon, I., Tessier, R., & Rose, J. (2007). FPGA Architecture: Survey and Challenges. *Electronic Design Automation, 2*(2), 135–253. doi:10.1561/1000000005

Morales, D. P., Garcia, A., Castillo, E., Carvajal, M. A., Banqueri, J., & Palma, A. J. (2011). Flexible ECG Acquisition System Based on Analog and Digital Reconfigurable Devices. *Sensors and Actuators. A, Physical, 165*(2), 261–270. doi:10.1016/j.sna.2010.10.008

Pawar, D.J., & Bhaskar, P.C. (2013). FPGA Based FIR Filter Design for Enhancement of ECG Signal by Minimizing Base-line Drift Interference. *International Journal of Current Engineering and Technology*, 1775-1778.

Ravikumar, M. (2012). Electrocardiogram Signal Processing on FPGA For Emerging Healthcare Applications. *International Journal of Electronics Signals and Systems, 1*(3), 91–96.

Sadabadia, H., Ghasemia, M., & Ghaffaria, A. (2007). *A Mathematical Algorithm For ECG Signal denoising Using Window Analysis*. Biomed Pap Med Fac Univ Palacky Olomouc Czech Repub.

Sharma, S., Kumar, G., Miishra, D. K., & Mohapatra, D. (2012). Design and Implementation of a Variable Gain Amplifier for Biomedical Signal Acquisition. *International Journal of Advanced Research in Computer Science and Software Engineering*, 2(2), 193–198.

Spinelli, E.M., Martinez, N. H., & Mayosky, M. A. (2001). A Single Supply Biopotential Amplifier. *Medical Engineering & Physics, 23*, 235-238.

Tawfik, M. M., Selim, H., & Kamal, T. (2010). Human Identification Using QT Signal and QRS Complex of the ECG. *Journal of Electronic and Electrical Engineering, 3*(1), 383–387.

Tragardh, E., & Schlegel, T. T. (2006). *High-Frequency ECG*. Academic Press.

Wang, P., & Zhigang, L. V. (2012). Design of A Simple 3-Lead ECG Acquisition System Based on MSP430F149. *International Conference on Computer and Automation Engineering*, 44, 86–91.

Wang, Y. (2007). Multiplierless CSD Techniques for High Performance FPGA Implementation of Digital Filters (PhD Thesis). University of Oklahoma, Graduate College.

Yadav, S.K., & Mehra, R. (2014). Analysis of Different IIR Filter based on Implementation Cost Performance. *International Journal of Engineering and Advanced Technology, 3*(4), 267-270.

Zaman, M. T. U., Hossain, D., Arefin, M. T., Rahman, M. A., Islam, S. N., & Haque, A. K. M. F. (2012). Comparative Analysis of De-Noising on ECG Signal. *International Journal of Emerging Technology and Advanced Engineering*, 2(11), 479–486.

Chapter 2
Artificial Intelligence Based on Biological Neurons:
Constructing Neural Circuits for IoT

Rinat Galiautdinov

https://orcid.org/0000-0001-9557-5250

Independent Researcher, Italy

ABSTRACT

The chapter describes the new approach in artificial intelligence based on simulated biological neurons and creation of the neural circuits for the sphere of IoT which represent the next generation of artificial intelligence and IoT. Unlike existing technical devices for implementing a neuron based on classical nodes oriented to binary processing, the proposed path is based on simulation of biological neurons, creation of biologically close neural circuits where every device will implement the function of either a sensor or a "muscle" in the frame of the home-based live AI and IoT. The research demonstrates the developed nervous circuit constructor and its usage in building of the AI (neural circuit) for IoT.

INTRODUCTION

Although the concept of the "Internet of Things" (IoT) has been around for a long time, it, especially in the light of our rapid technological development, is constantly evolving. We can say that IoT is the embodiment of the gradual merger of the physical and digital worlds, as data is collected from an ever-growing number of devices and then combined into so-called "big data." The number of such devices of the "Internet of things", according to experts and analysts, will reach 50 billion by 2020.

DOI: 10.4018/978-1-7998-4381-8.ch002

However, when you try to transfer the data collected by IoT devices to a centralized storage, such as a cloud, there is a problem with the delay in their transmission. In many respects, even though the connection speed is constantly increasing, the characteristics of this process do not correspond to the available data growth. If you transfer the "raw" data, that is, unprocessed, all in a row, the delay will increase and, therefore, the overall system performance will suffer.

Data processing is one of those areas in which AI can make a significant contribution. In addition, it opens the way to the introduction of technological innovations in various fields, from optimizing the movement of urban transport to improving public safety and improving the provision of financial services.

Implementing AIoT requires components that can cope with complex and diverse conditions at the edge of the network. The periphery, as you know, can be literally anything - from airborne vehicles and aircraft to factories or oil installations in the desert. All this requires a flexible and adaptable approach to the production of components to solve this problem. An important point is that AI promises to eliminate the influence of the human factor on decision-making as much as possible. This puts more pressure on system integrators: they need to provide special control over the quality of the functioning of the system, since an accident in systems with artificial intelligence does not always have an obvious culprit or a visible reason.

Another difficulty we face with is related to the fact that we always have to tune something in the settings and in some cases we might want the devices to work in one mode in another in the other one, so as a result we'd constantly have to spend our time on tuning, changing, updated and doing lots of work. Besides it would take lots of time to read the instructions, to learn how to use some software, etc.

One more problem is enclosed in the fact that what we call "Artificial Intelligence" is not really the AI, it provides extremely narrow functionality which only allows to select the proper "answer" based on multi-criteria condition. Such the AI can not really think, evolve with time. Such the AI will require the constant installation of the new and new modules for processing new tasks and the abilities of such the AI can't be compared with the abilities of even the most primitive creature.

A good example of the scope of limited AI is the recognition of text, images and speech, which we can implement using neural networks and machine learning. During training, such artificial intelligence remembers thousands, if not millions, of various iterations of data and is able to correctly determine the image or an object located in the zone of its action. No matter how complex the predictions of such an AI become, it is still limited by a narrow function. If something goes beyond the given parameters, the AI becomes almost useless. For example, artificial intelligence, trained to recognize written numbers, can master this task and easily push people out of this sphere of activity, because it will work more efficiently, without fatigue and interruptions, but it will be completely useless if it is given to it without retraining such tasks as e.g. letter identification.

As for the concept of border (peripheral) computing, the initial idea of IoT was that the data for processing and subsequent analysis was sent to some central device or to the cloud. However, as the number of devices increases exponentially, many applications have already reached the limit of their capabilities, and all this large amount of data transferred back and forth leads to problems with unacceptable delays in decision making and response.

Border computing solves this problem by processing "big data" directly on the edge of the network. Thus, the device can independently determine what needs to be sent to the cloud and what can be filtered out like digital garbage. In fact, this concept offers the movement of computing power to the "edge" of the network - to where the Internet connects to various devices.

And here we come to another problem which is enclosed in neural computation.

A model approach to research allows us to overcome the limitations and difficulties that arise when setting up a laboratory experiment, due to the possibility of conducting so-called numerical experiments, and to study the response of the system under study to changes in its parameters and initial conditions.

In this regard, computer simulation is widely used in all natural sciences. Neuroscience or the science of the brain, whose task is to study the functioning of the brain and nervous system. The brain is a complex object consisting of a large number of different types of cells, including the main signal cells - neurons (cells that generate and transmit electro-chemical impulses that can form networks through contacts called synapses), glial cells that regulate metabolism, blood vessel cells, etc. Modeling such systems, complex in internal connections and large in the number of elements, using modern personal computers is extremely difficult, due to the large th computing capacity derived models.

However, the use of supercomputer technologies allows the use of more diverse modeling methods. One of such the methods is called large-scale modeling. Large-scale modeling is one of the directions in supercomputer modeling. This method is intended for the development and conduct of numerical experiments with global computer models of multidimensional systems in which macro and micro models that simulate the interconnected functioning of multilevel systems are integrated. This direction arose relatively recently due to significant progress in the technology of manufacturing microcircuits, parallel computing, and the increased processing power of supercomputer systems, which became available with the advent of specialized software. Large-scale modeling is based on the principle of hierarchical reduction, which assumes that any complex system consists of hierarchically subordinate subsystems (levels of organization). A high-level organization system consists of lower-level systems, and a combination of low-level organization systems forms a higher-level system. Application of this principle to modeling in neuroscience allows us to represent the brain in the form of several 4 interacting independently described subsystems. The hierarchy of the model allows you to achieve the level of detail required by research, by increasing or decreasing the number of organization levels considered. However, with an increase in the number of levels of organization, the number of parameters describing the system increases, which greatly complicates the task of creating a realistic model that reproduces the phenomena observed in a laboratory experiment. An increase in the number of model parameters leads to an increase in the amount of input data required to determine them, data that are difficult to measure and often do not have a sufficient degree of accuracy. In this regard, abstraction is used when creating models - an approach that allows you to discard parameters that are unimportant for research in the framework of the task, with the aim of solving which the model was developed. Thus, the task of abstracting is to preserve only what is important for the construction and analysis of models at different levels of the organization without losing the convenience of manipulation. The considered modeling method provided researchers with a set of neural network simulators that can greatly simplify research in the field of neuroscience.

Additionally to that it's very important to have the ability of virtual construction of neural circuit both for researching goals and the applied ones in the sphere of Artificial Intelligence and IoT. As a result the author describes the major features of such the neural constructor and represents it, showing how it was applied in simulation of the neural circuit of Aplysia(the mollusk) and Planarian(Tricladida) and how it could be used in the next generation of the IoT where all the smart devices are represented as the sensors and the constructed neural circuit processes the signals the way the nervous system processes the signals coming from the sensors of any live creature.

PROBLEMS

The Limits of The IoT Technology

IoT devices in their pure form collect data with only small or specifically specified amounts of computation. For further analysis, data is sent to the cloud. However, in such premises not all data has the same value. Take, for example, video materials for a security system: the system needs frames in which people or objects move, while images of an unchanging background are clearly not of particular interest. Sending all the data obtained during shooting to the cloud for analysis will lead to the occupation of the transmission channel bandwidth, which could be used to greater advantage.

Computing Power and Harsh Work

The transfer or implementation of AI to the periphery may require a lot of computing resources. Standard storage and memory devices will help provide the required performance, but the problem is that commercially available components of this type are generally poorly adapted to work in the harsh environments typically found in borderline applications. For example, when monitoring traffic at the location of IoT devices, cyclical changes in temperature are possible during the transition from day to night and from summer to winter. In addition, automotive systems must withstand shock and vibration, while industrial systems must withstand increased levels of pollution, etc.

Disadvantage of the Artificial Neuron

Currently in the AI sphere we use so called Artificial Neuron, which is basically represents extremely primitive and limited ability of the biological neuron.

Let's consider some of the disadvantages of usage of the artificial neuron vs biological neuron (here and after under the term "artificial neuron" we should understand the classical representation of the artificial neuron and under the term "biological neuron" we should understand the simulated biological neuron, for example the one which is represented in the Neural constructor of Rinat Galiautdinov):

- A weak signal can't initiate a reaction of an artificial neuron. However if we consider biological neuron, in certain conditions the weak signal can initiate the reaction, as an example it could be seen in the processes called "Summation", "Long-Term Potentiation". Here's the simple example which illustrates this: a dripping tap. A single drop of water initiate extremely weak signal which might repeat not so frequently. So if we have only one sensor which is responsible only for listening of the sounds then it will not activate the system based on the artificial neurons used at the moment simply because such the signal want pass to another neuron. However if we use the simulated biological neurons (represented in the Neural constructor of Rinat Galiautdinov) then we will see that a single drop of water will not initiate a signal, however the series of weak and repeatable signals will initiate the signal and eventually it will be passed to another neuron. Such the process is called "Summation" and we can simulate such the process only if we use biological neurons.
- The artificial neuron can't distinguish the power of signals. So the artificial neuron there is no difference between the power of signals for such the processes as "dripping tap" and our memories of some positive or negative

29

feelings we had during the day. However all that changes when we consider a biological neuron: in some cases it can't or can initiate a signal and in the other cases even a week signal can immediately initiate a reaction if there was a strong enough signal during a day. In our life we constantly face with such the examples: let's say you found 5 dollars, it creates strong enough positive signal in your brain and then during the day you will memorize this even from time to time. Or you spilled coffee in the morning: it created a strong enough negative signal and from time to time you will memorize this event during the day. Most probably you will forget about these events next day or the day after that. Such the process is called "Long-Term Potentiation" (LTP) and it can be simulated only if we use biological neurons.

- The artificial neurons can't automatically change the architecture of the neural network, they can only change the weights. The biological neurons can change the architecture of the neural network. So the artificial network built on the basis of the biological neurons can evolve with time and adjust itself to the new environment.

- The neural network based on the artificial neurons can't really include and use the new devices/sensors which this network does not know: simply speaking the neural network's developers never taught the network to use some new unknown device/sensor. In the same situation the neural network built on the basis of the neural circuits based on the biological neurons can include the new device/sensor into the system and learn how to use it.

BACKGROUND

Artificial Neuron

The artificial or programming neuron used in Computer Science partially simulates the biological neuron. Such the artificial neuron receives the number of the signals as the input data and each of these signals is in fact the output of another neuron. Each input gets multiplied by the appropriate weight(simulating the synaptic strength) then we can sum all the values and define the level of neuron activation. The final result of this operation would be either 0 or 1 (Michie D., Spiegelhalter D., Taylor C., & Campbell J., 1994).

There are different kind of the neural networks but all of them are based on the above described configuration. There are multiple input signals for the artificial neuron: $x_1, x_2,...,x_n$. These input signals correspond to the input signal in the synapses of biological neurons. Each signal gets multiplied by the appropriate weight $w_1, w_2,..,w_n$, and then all they gets redirected to the summation block marked with a symbol $\sum$. Each weight corresponds to the power of a single biological synapse. The summation block which corresponds to the body of the biological element, arithmetically sums the inputs and creates the output R.

Figure 1. Illustrates the Artificial neuron

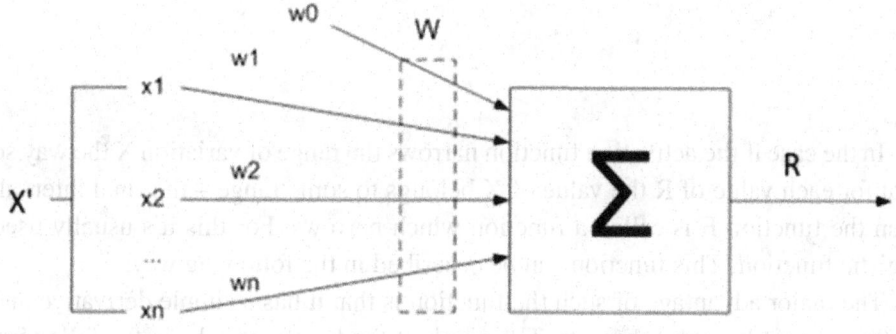

Such the description can be defined with the following formula:

$$R = \sum_{i=1}^{n} W_i X_i + W_0 .$$

Where:

W_0 – is a bias

W_i – is the weight of the ith neuron

x_i - is the exit of the ith neuron

n – is the number of the neurons, which serve as the input for the processing neuron. And this was already described in the article of Abbott L.F. and Kepler T.B. (1990).

The signal W_0 which has a name "bias" represents the shift limit function. This signal allows you to shift the origin of the activation function, which subsequently leads to an increase in the learning speed. This signal is added to each neuron, it learns like all other scales, and its feature is that it connects to the +1 signal, and not to the output of the previous neuron. The received signal R gets processed by the activation function and returns the output signal X (Figure 2).

Figure 2. Illustrates the artificial neuron with the activation function

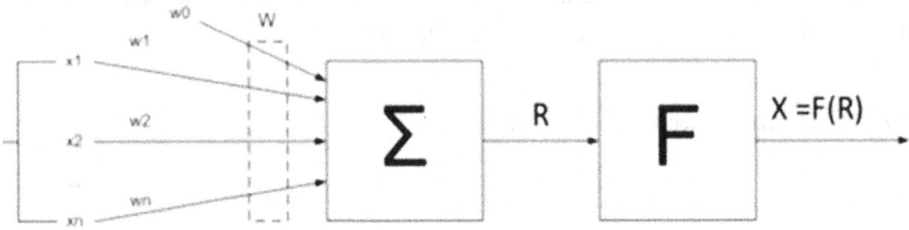

In the case if the activation function narrows the range of variation X the way so that for each value of R the value of X belongs to some range – the final interval, then the function F is called a function which narrows. For this it's usually used logistic function. This function can be described in the following way:

The major advantage of such the function is that it has a simple derivative and differentiates along the abscissa. The graph of the function looks in the following way: (Figure 3).

Figure 3. Type of logistic/sigmoidal activation function

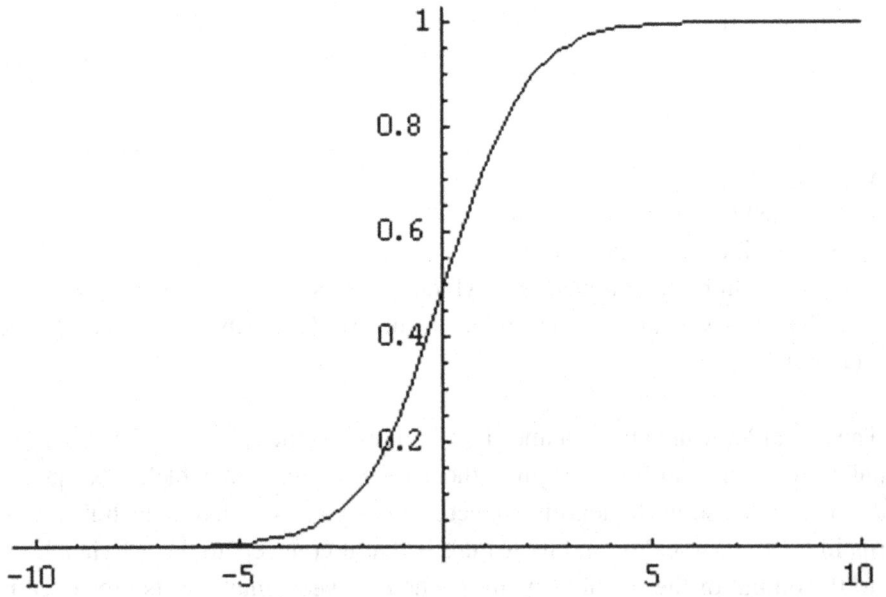

The function increases the weak signals and reduces "too strong" signals.

Another function that is also often used is hyperbolic tangent. It resembles a sigmoid in shape and is often used by biologists as a mathematical model of nerve cell activation. It looks in the following way:

Like the logistic function, the hyperbolic tangent is S-shaped, but it is symmetrical with respect to the origin, and at the point of $R = 0$ the value of the output signal $X = 0$

The graph shows that this function unlike logistic one accepts the values of the different signs, what could be a beneficial for a certain type of neural networks.

Figure 4.

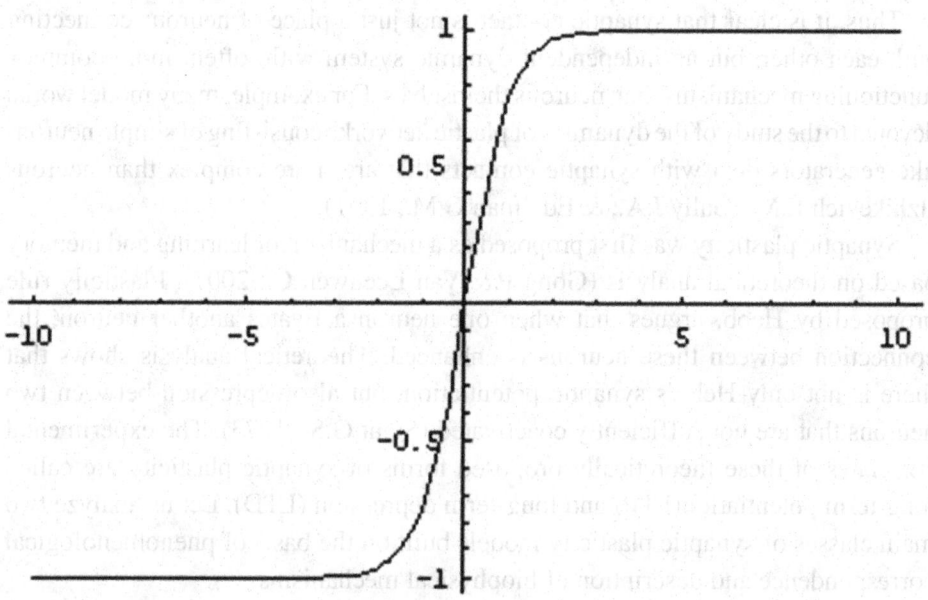

The considered model of an artificial neuron ignores many properties of a biological neuron. For example, it does not take into account time delays that affect the dynamics of the system. Input signals immediately generate the source. But despite this, artificial neural networks composed of the considered neurons reveal the properties that are inherent in the biological system (Abdelbar A. M. & S. M. Hedetniemi, 1998).

Models of Synaptic Plasticity

Unlike an isolated cell, network-connected neural oscillators capable of forming various kinds of connections between themselves, the nature of which will determine the properties of the network as a whole. One of the characteristic features of a neural network is the ability to learn, which is manifested in the modification of the parameters of elements or connections in response to external stimulation. This modification can lead to a change in the dynamic behavior of the network as different input stimuli are presented. On the other hand, the spread of impulse activity in such networks affects the current state of interelement connections.

Most of the cells of the nervous system possess plastic connections (Nicholls J.G. et al, 2001).

Thus, it is clear that synaptic contact is not just a place of neurons connecting with each other, but an independent dynamic system with, often, more complex functioning mechanisms than neurons themselves. For example, many model works devoted to the study of the dynamics of plastic networks consisting of simple neuron-like generators deal with synaptic contacts that are more complex than neurons (Izhikevich E.M., Gally J.A., & Edelman G.M., 1991).

Synaptic plasticity was first proposed as a mechanism for learning and memory based on theoretical analysis (Gong P.& Van Leeuwen C., 2007). Plasticity rule proposed by Hebb, argues that when one neuron activates another neuron, the connection between these neurons is enhanced. Theoretical analysis shows that there is not only Hebb's synaptic potentiation, but also depression between two neurons that are not sufficiently coactivated (Stent G.S., 1973). The experimental correlates of these theoretically proposed forms of synaptic plasticity are called long-term potentiation(LTP) and long-term depression (LTD). Let us analyze two main classes of synaptic plasticity models built on the basis of phenomenological correspondence and description of biophysical mechanisms.

Phenomenological models are characterized by a description of the process that regulates synaptic plasticity as a "black box". The Black Box accepts a set of variables as input, and produces output changes in synaptic efficiency. There are two different classes of phenomenological models (Morrison A. et al., 2008) that change the efficiency of signal transmission through the synapse depending on either the frequency of the pulses or the ratio of the times of the appearance of pulses, and differ in the type of input variables.

Many of the phenomenological models of synaptic plasticity that have been proposed in recent years, based on the dependence of the properties of synaptic transmission on the pulse frequency (Dayan P. & Abbott L.F., 2001). In these models, it is assumed that the frequency of pre-synaptic and postsynaptic pulses measured over a period of time determines the sign and magnitude of synaptic plasticity. This rule can be formulated as follows:

$$\frac{dW_i}{dt} = f\left(x_i, y, W_i, \ldots\right).$$

Where:
W_i - synapse efficiency
i, x_i - pulse frequency of the pre-synaptic neuron,
y is the frequency of the postsynaptic neuron.
A simple example of a frequency-based model is as follows:

$$\frac{dW_i}{dt} = \eta\left(x_i - x_0\right)\left(y - y_0\right).$$

Where:
η is the learning rate, which is assumed to be relatively low,
x_0, y_0 are some constants.

The discovery of the dependence of synaptic efficiency on the ratios of pulse arrival times at the synapse (Spike-timing dependent plasticity (STDP)) aroused interest in creating a new class of models. Most of these models depend only on the relative time between pairs of pulses; however, models have recently appeared that depend on the arrival times of three pulses (Pfister J.P. & Gerstner W., 2006).

Under certain assumptions about the statistics of the appearance of presynaptic and postsynaptic pulses, and the duration of pulsed overlap, these models can be averaged and reduced to frequency models (Kempter R. et al., 1991).

In the simplest case, the STDP effect is described by a simple curve, as shown in Figure 5.

The direct sequence of impulses "presynaptic-postsynaptic"

Figure 5. Illustrates STDP curve

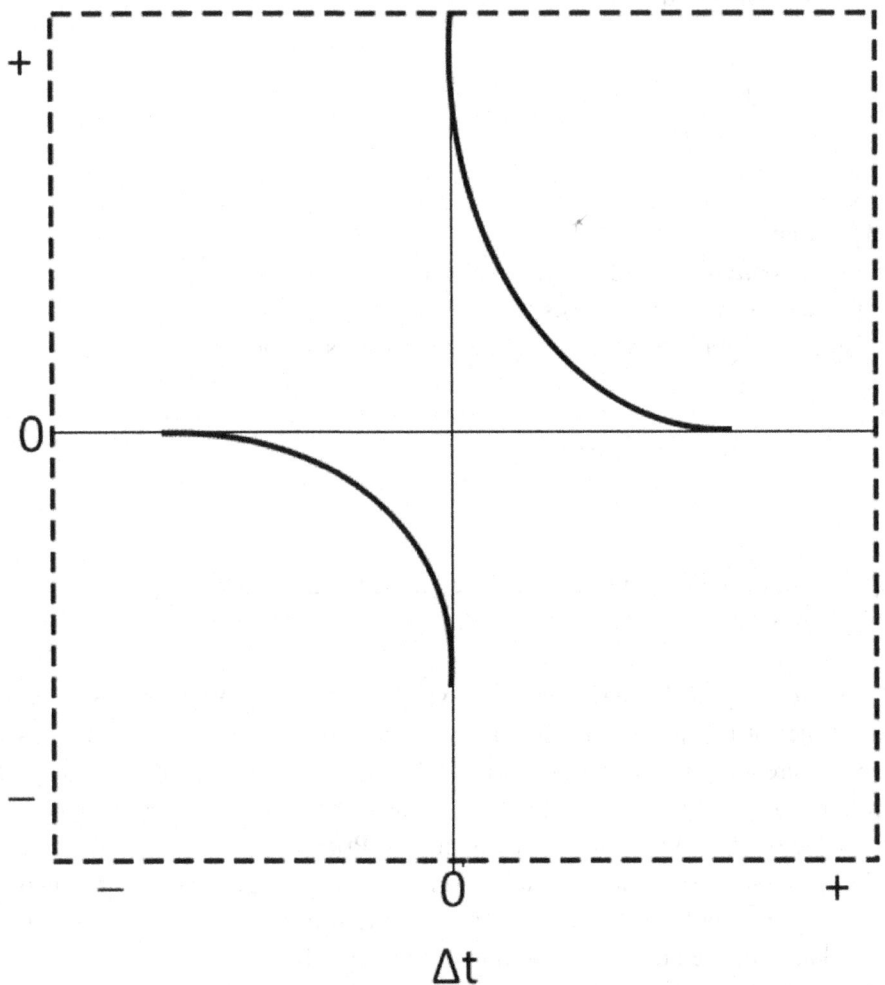

leads to an increase in bond weight (LTP potentiation effect). If the order of appearance of the pulses at the synapse is reversed, then the effect of depression (LTD) occurs. STDP also depends on many other factors, such as the frequency of activation of neurons at which pairs of impulses arise (Markram H. et al., 1997), the level of local postsynaptic depolarization, or the initial state of synapses. Biophysical models, in contrast to phenomenological models, are based on modeling biochemical and physiological processes that lead to synaptic plasticity. A simple dynamic system that implements the hypothesis of controlling bond strength by changing the concentration of calcium is:

$$\frac{dW_i}{dt} = \eta(Ca)(O(Ca) - lW_i).$$

Another important type of synaptic plasticity is short-term frequency-dependent plasticity, at which the signal transmission efficiency depends on the frequency of the pulses followed. A model of such plasticity was proposed in the article of Tsodyks M. et al. (2000). It describes the kinetics of the release of a neurotransmitter (synaptic resources) using a 4th-order dynamic system. Short-term ductility depends on incoming pulses at time scales of tens of milliseconds. Using this model, in particular, it was possible to describe the mechanism of occurrence of population burst discharges in neural networks (Tsodyks M. et al., 2000). In various aspects of neurodynamics, other models of short-term plasticity have been proposed that have varying degrees of biophysical detailing (Abbott L.F. et al., 1997). Relying, as a rule, on the description of specific experimental phenomena, plasticity models remain essentially poorly studied from the point of view of nonlinear dynamics of transmission and conversion of pulsed excitations between neurons via plastic synaptic connections.

Formally, the short-term plasticity model discussed above is written as follows:

$$\frac{dx}{dt} = \frac{z}{\tau_{rec}} - \sum_{i=1}^{N} ux\delta(t - t_i) \quad \cdot \quad \frac{dy}{dt} = \frac{-y}{\tau_1} + \sum_{i=1}^{N} ux\delta(t - t_i) \quad \cdot \quad \frac{dz}{dt} = \frac{y}{\tau_1} - \frac{z}{\tau_{rec}} \quad \cdot$$

$$\frac{du}{dt} = -\frac{u}{\tau_{fac}} + \sum_{i=1}^{N} U(1 - u)\delta(t - t_i)$$

Where:

ti – the time of occurrence of the i-th presynaptic impulse,

N – total number of impulses,

x, y, z – variables describing the shares of synaptic resources in restored, active and inactive states, respectively.

u – a variable responsible for synaptic depression or exacerbation.

$\tau 1, \tau rec, \tau fac$ - characteristic times of synapse dynamics.

The type of synapse is determined by a set of parameters, for example, for the exciting type of communication between two exciting neurons, it is proposed to take the synapse parameters:

A=1.8 mV, U=0.5, τ_{rec}=800 ms, τ_{fac}=0 ms, τ_1=3 ms.

The postsynaptic current in this model is described using the weighted synaptic variable y.

Each presynaptic impulse causes a jump in the variable y, after which y drops to zero in a relatively short time. The effect of synaptic depression (a decrease in the effective binding strength) is manifested in the fact that with each subsequent input pulse, the value of the jump in the variable y decreases if the interval between presynaptic pulses is not large enough for the synapse to return to its original state. Therefore, the higher the frequency of presynaptic pulses, the stronger and faster the strength of the bond decreases. Moreover, for the case of a periodic input action, the stationary value of the jumps is determined by the nonlinear function of the frequency of the input action.

In addition to synaptic plasticity, the dynamic organization of inter-element interactions in networks of neural oscillators is based on the so-called structural plasticity, which consists in the formation and destruction of connections between network elements. So in the process of operation of a neural system, the architecture of inter-element communications can change depending on the dynamics of the entire network. When recording new information, new relationships may reflect other propagation paths of pulsed signals. In conclusion, it should be noted that the basis of many computational properties, amazing efficiency and noise immunity in the processing of information and the high adaptive performance of neural networks is the dynamic organization of inter-element interactions, which is provided, inter alia, by various mechanisms of synaptic plasticity.

SOLUTIONS

Artificial Intelligence Platform

Speaking of a symbiosis called AIoT, we usually mean an AI platform located on the periphery of the network. Typically, this decision takes the form of a small industrial computer (IPC) with an integrated industrial-class processor. However, for real-time data analysis, such a processor needs adequate support in the form of flash memory and a disk drive.

Memory and Data Storage

To solve the problems of implementing AI in borderline applications, as mentioned above, industrial-grade data and memory storage devices are needed. First of all, it is necessary to study and identify the risks present in each specific place of data collection. This will allow the components to be implemented in accordance with the clear requirements of a particular application.

Overview of Modern Neural Network Simulators

Due to the increasing availability of computing resources, studies using computer modeling are becoming increasingly popular.

The use of a model approach really looks promising, since in modern neuroscience there are a number of issues that can only be solved using modeling. However, there is a problem that the results of such studies are difficult to verify and reproduce. To this end, carefully tested, documented simulators of neural networks are being developed for a wide range of users. These software tools allow you to standardize the code, which simplifies the interaction of research groups, and also contain built-in parallel programming tools, ensuring the availability of modern information technology.

These simulators are widely used in the construction of large-scale models of neural networks. There are several types of simulators:

- Simulators with a simple neuron model: PCSIM, NEST, Brain and NCS.
- Simulators with a neuron model consisting of several compartments: NEURON, GENESIS, SPLIT and MOOSE
- Event driven simulators: MVASPIKE
- Dynamic systems analysis systems: XPP
- Neural constructor simulator of Rinat Galiautdinov, which uses the simulated biological neurons and allows to build neural circuits and explore their behavior.

Along with the software for modeling cellular networks, simulators with hardware-implemented neural networks are widely used.

An example of such simulators is FACETS, a platform simulating the operation of approximately 106 neurons located on several connected boards, each of which houses analog network cores (ANCs), which are the main element of the FACETS architecture, which consists of neurons and synaptic connections (on average, each neuron this system has with 1 thousand contacts with others).

NEST

NEST is a software simulator used to model networks, biologically realistic elements and relationships. The program is optimized to simulate large neural networks and is currently capable of processing a model consisting of 100,000 elements (neurons) and approximately one billion connections (synapses) between them. In the simulation environment, a descending (top-down) approach to the description of the neural network is implemented, which is a kind of large-scale modeling. In accordance with

Figure 6. Illustrates example of a structured model of a neural network(a) and its representation (b)

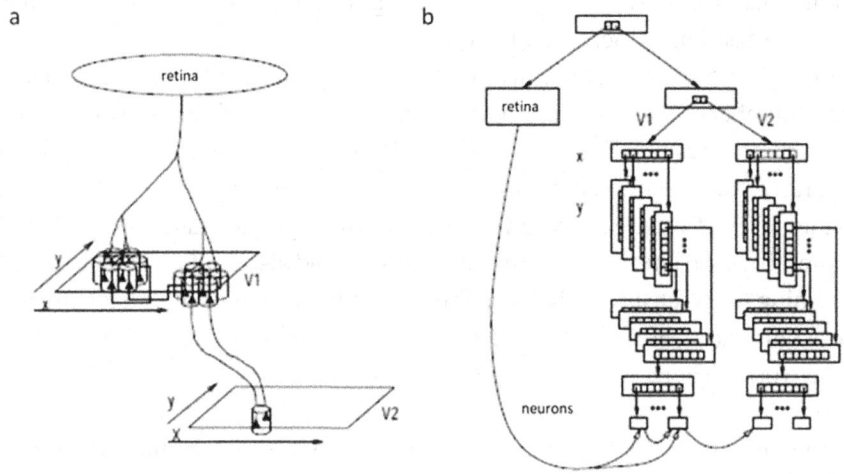

the principle of hierarchy, neural networks are considered as multi-level structures that can be represented in the form of trees.

Obviously, the depicted neural network contains a large number of components with numerous common connections. The entire network is subdivided into structures that describe various levels of system organization: the retina and two model brain regions, V1 and V2. The model diagram in NEST is shown in Figure 6 (b). A network model is constructed using nodes and links. The nodes can be neurons, devices, and subnets that can exchange (receive and send) events of various types, for example, spikes or currents. The NEST software package contains built-in models of neurons, devices, synapses, and the use of the modular principle of application organization allows the user to create their own models.

Most of the implemented neuron models consist of a small number of compartments, which reduces the degree of detail of the described biophysical processes and morphological features of a single cell. Neurons form a network through the so-called synaptic contacts, for each of which its own dynamics can be determined.

The simplest model of communication between nodes is characterized by weight (determines the strength of interaction between nodes) and delay time (time required for transition signal from one node to another). The simulator has built-in synapse models with implemented mechanisms of plasticity, synaptic depression and recovery, and also provides functions for creating various topological communication schemes and structuring large networks.

The main components of the system: the core and the interpreter of the simulation system. The interpreter interacts with the graphical interface and the modeling language with the core of the system. The principles of organization of the NEST application contribute to the efficient use of computing resources of computers with a multi-core processor, multiprocessor computers and clusters. When modeling on a computing cluster or multiprocessor computers, each computer or processor recreates part of the network and stores information about the synaptic contacts of only its neurons. To distribute tasks to all computers (processors), NEST uses the Instant Messaging Interface (MPI) and POSIX threads (pthreads).

Figure 7. Illustrates modeling of a neural network in NEST. (a) absolute runtime; and (b) acceleration as a function of the number of processors involved.

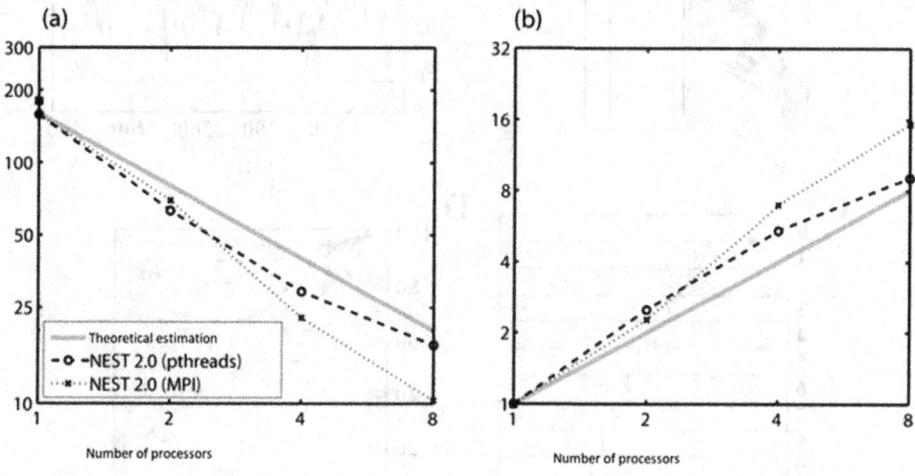

The graphs shown in Figure 7 demonstrate the dependence of NEST performance on the number of processors involved in calculating the network consisting of 12,500 elements described by the integrate-and-fire model (80% of the total number relates to exciting neurons and 20% inhibitory) each of which receives a signal from 10% of all neurons. The total number of synapses in the moody is 1.56×107. Neurons are initialized by random membrane potentials and are further stimulated by irregular exposure.

Examples of using NEST:

- Models for processing sensory information (visual, auditory, etc.)
- Models of the dynamics of neural network activity

- Training and plasticity in models of processing sensory information.
- Spike synchronization models in Synfire Chains networks.

The results of the study of spike synchronization in Synfire Chains networks using the NEST simulator are presented in Figure 8.

Figure 8.

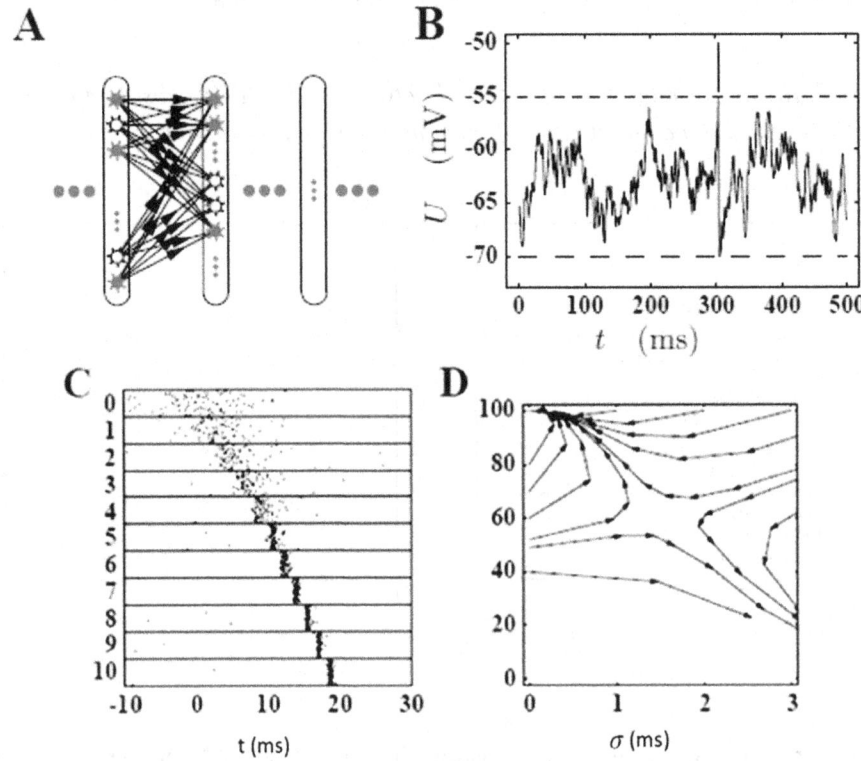

Figure 8 Illustrates:

(A) a diagram of a neural network known as a synfire chain. Groups of neurons are connected in a chain structure. Each neuron of the layer forms contacts (most of them exciting) with all neurons of the next layer, and receives a signal from all neurons of the previous layer.

(B) Fluctuations in the membrane potential of an individual neuron obtained by simulation.

(C) The result of neural network modeling is shown, where the first group of the chain is stimulated by a wide packet of incoming spikes. Subsequently, the phenomenon of spike synchronization is observed.(D) Phase portrait synchronization in synfire chain.

Neural Constructor of Rinat Galiautdinov for Simulating The Neural Circuits

The neural constractor of Rinat Galiautdinov for simulating the neural circuits is fully based on simulation of biological neurons, including all the processes running inside of the neuron, which includes but not limited to: simulation of the AP(action potential), opening of the calcium channels, moving of the calcium ions inside of the nerve cell, catching the calcium ions by the vesicles, movement of the vesicles with the neurotransmitters, calcium pump, effect of the neurotransmitters on the receptors, different kind of the protein receptors(NMDA, Non-NMDA), ability to simulate the processes such as Summation, Long-Term Potentiation, ability to construct the neural circuits based on the different types of the neurons such as: sensor neurons, inter-neurons, motor-neurons, ability to connect sensor neurons to the sensors and motor-neurons to the simulated muscles. Such the constructor is surely an innovative approach in science and allows both: research of the work of the constructed nervous system and simulation of behavior of the different kind of creatures. Eventually such the neural constructor could be the basis for the next generation computing systems.

The Main Modern Works in The Field Of Modeling Brain Functions

The approach to the study of the brain of mammals using computer simulation is one of the most promising today. The following are the most significant results:

- A model of visual attention (Silvia Corchs, Gustavo Deco). The model consists of interconnected modules that can be connected by different areas of the dorsal and ventral paths of the visual cortex.
- Model II / III layers of the neocortex (Mikael Djurfeldt, Mikael Lundqvist et al.)
- This model was implemented on a Blue Gene / L supercomputer. It includes 22 million neurons, 11 billion synapses and corresponds to the cerebral cortex of a small mammal.
- Self-sustaining irregular activity in a large-scale model of the hippocampal region (Ruggero Scorcioni, David J. Hamilton and Giorgio A.Ascoli). The

model consists of 16 types of neurons and 200,000 neurons. The number of neurons and their connections correspond to the anatomy of the rat brain. In the project, the authors analyze the emerging activity of the network and the effect on it of a decrease in the size or relationships of the network model.

- Blue Brain Project (Markram H. et al.)
- Model of the mammalian thalamocortical system (E.M. Izhikevich and G.M. Edelman, 2007).
- Neural constructor of Rinat Galiautdinov.

MATHEMATICAL MODEL OF BIOLOGICAL NEURON

Figure 9.

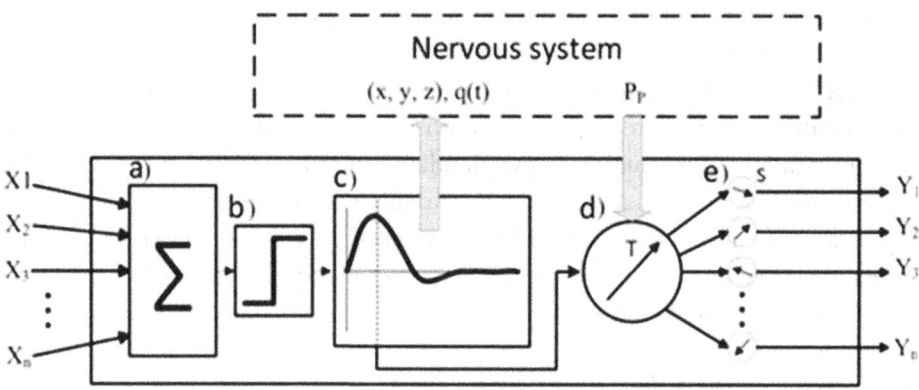

Imagine the beginning vector, which is located in the center of the active stand, and the end is directed to the pattern point defined for a given neuron. Denote as the vector of the preferred direction of propagation of the excitation (T, trend). In the biological neuron, the vector T can manifest itself in the structure of the neuroplasm itself, perhaps these are the channels for the movement of ions into the body of the cell, or other changes in the structure of the neuron. A neuron has the property of memory, it can memorize the vector T, the direction of this vector, can change and overwrite depending on external factors. The degree to which the vector T can undergo changes is called neuroplasticity (Choi Y.B. et al., 2014).

This vector, in turn, affects the functioning of the neuron synapses. For each synapse, we define the vector S beginning, which is located in the center of the cell, and the end is directed to the center of the target neuron with which the synapse is connected. Now the degree of influence for each synapse can be determined as follows: the smaller the angle between the vector T and S is, the more the synapse will be amplified (Boyen X. & Koller D., 1998); the smaller the angle, the stronger the synapse will weaken and may possibly stop the transmission of excitation. Each synapse has an independent memory property; it remembers the meaning of its strength. The indicated values change with each activation of the neuron, under the influence of the vector T, they either increase or decrease by a certain value.

The input signals $(x_1, x_2, \ldots x_n)$ of the neuron are real numbers that characterize the strength of the synapses of the neurons that affect the neuron.

A positive value of the input means a stimulating effect on the neuron, and a negative value means an inhibitory effect.

For a biological neuron, it does not matter where the signal exciting it came from, the result of its activity will be identical. A neuron will be activated when the sum of the effects on it exceeds a certain threshold value. Therefore, all signals pass through adder (a), and since neurons and the nervous system work in real time, therefore, the effect of the inputs should be evaluated in a short period of time, that is, the effect of the synapse is temporary. The result of the adder passes the threshold function (b), if the sum exceeds the threshold value, then this leads to neuron activity. When activated, a neuron signals its activity to the system, advanced information about its position in the space of the nervous system and the charge that changes over time (c). After a certain time, after activation, the neuron transmits excitation along all the available synapses, previously recounting their strength. The entire activation period of the neuron ceases to respond to external stimuli, that is, all the effects of synapses of other neurons are ignored. The activation period also includes the recovery period of the neuron.

The vector T (d) is adjusted taking into account the value of the pattern point Pp and the level of neuroplasticity. Next, there is a reassessment of the values of all synapse forces in the neuron (e).

Note that blocks (d) and (e) run in parallel with block (c).

The next simplification of the Hodgkin-Huxley model is the MorrisLecar model, proposed in 1981. This system of equations describes the complex relationship between the membrane potential and the activation of ion channels in the membrane. Mathematically, the model is written as follows:

The open state probability functions, MSS (V) and WSS (V), are obtained from the assumption that in equilibrium the open and closed states of the channels are delimited, according to the Boltzmann distribution. Changes in the external current, I, are accompanied by a saddle-node bifurcation, leading to the birth of a limit cycle. In the field of theoretical modeling of neural oscillators the author as an independent researcher developed the number of new math models of neural dynamics (Cheng, J. and Druzdzel M., 2000).

Figure 10.

$$\begin{cases} C\dfrac{dV}{dt} = I - g_L(V - V_L) - g_{Ca}M_{ss}(V - V_{Ca}) - g_K N(V - V_K) \\[2mm] \dfrac{dN}{dt} = \dfrac{N - N_{ss}}{\tau_N} \end{cases},$$

$$M_{ss} = 0.5(1 + \tanh[\frac{V - V_1}{V_2}]),$$

$$N_{ss} = 0.5(1 + \tanh[\frac{V - V_3}{V_4}]),$$

$$\tau_N = 1/(\phi \cosh[\frac{V - V_3}{2V_4}]),$$

One of the most interesting developments is the model of the modified FitzHugh-Nagumo generator, which is a simplified version of the Hodgkin-Huxley model. This model has a separatrix threshold manifold that separates signals into subthreshold oscillations and suprathreshold excitation pulses, which are further used for communication between neurons. In addition, the model simultaneously possesses the properties of an integrative response typical of threshold systems and resonance characteristics similar to oscillatory systems. In other words, there is a fundamental possibility of simultaneously performing both frequency and phase encoding and decoding of information.

Previously, the authors of the model of the modified FitzHugh-Nagumo generator proposed a model of a neuron with spontaneous periodic oscillations below the excitation threshold. Such neurons, in particular, play a crucial role in the problem of coordination of movements of the brain, setting the universal rhythm of muscle contractions. The model is based on well-known dynamic systems and is described by a system of fourth-order differential equations.

Figure 11. Phase plane of the FitzHugh-Nagumo model with a threshold manifold

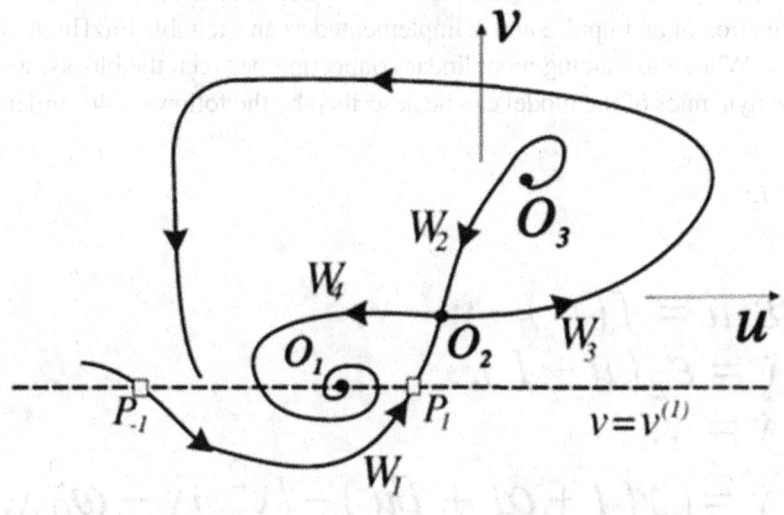

Figure 12. Model of an excitable element with subthreshold oscillations. Functional diagram

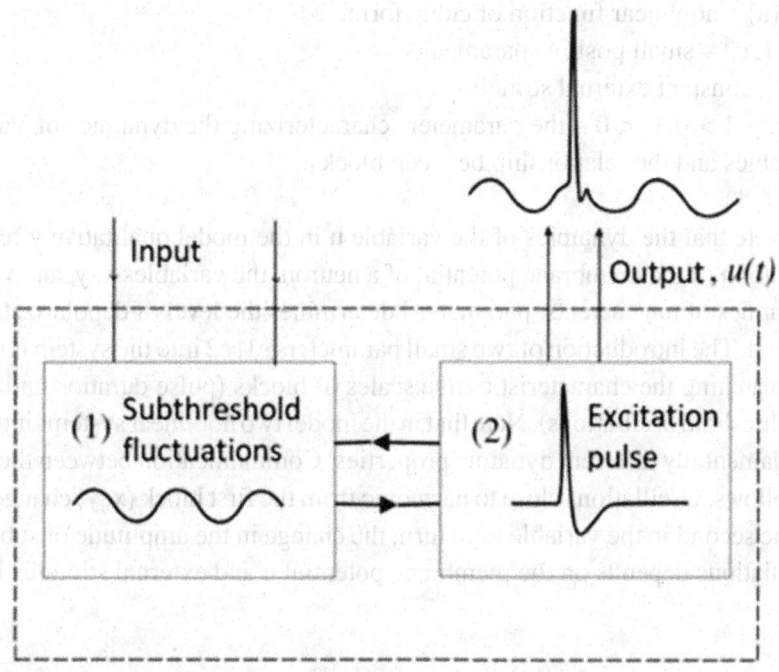

The first block describes subthreshold oscillations and can be implemented as a Van der Pol generator in a soft excitation mode. The second block is responsible for the formation of an impulse and is implemented as an excitable FitzHugh-Nagumo element. When introducing a nonlinear connection between the blocks, we obtain that the dynamics of the model can be described by the following 4th-order system

Figure 13.

$$\begin{cases} \varepsilon_1 \dot{u} = f(u) - v - y; \\ \dot{v} = \varepsilon_2(u + I), \\ \dot{x} = y, \\ \dot{y} = (\gamma(1 + \alpha I + \beta u) - lx^2)y - \omega_0^2 x, \end{cases}$$

Where:
  the variables x and y describe the dynamics of the first block,
  u, v – of the second one,
  f(u) – nonlinear function of cubic form,
  $\varepsilon 1, \varepsilon 2$ – small positive parameters
  I – constant external stimulus
  $\gamma, \beta, l > 0$, $\alpha < 0$ – the parameters characterizing the dynamics of Van der Pol variables and the relationship between blocks.

Note that the dynamics of the variable u in the model qualitatively reflects the evolution of the membrane potential of a neuron, the variables x, y, and v show the dynamics of ion currents, parameter I determines the level of depolarization of the neuron. The introduction of two small parameters $\varepsilon 1, \varepsilon 2$ into the system is necessary for matching the characteristic time scales of blocks (pulse duration and period of subthreshold oscillations). Note that in the model two nonlinear systems interact with fundamentally different dynamic properties. Communication between the blocks is as follows. Oscillations close to harmonic from the first block (x, y) change the state of the second in the variable u. In turn, the change in the amplitude of subthreshold oscillations depends on the membrane potential u and external stimulus I.

Figure 14.

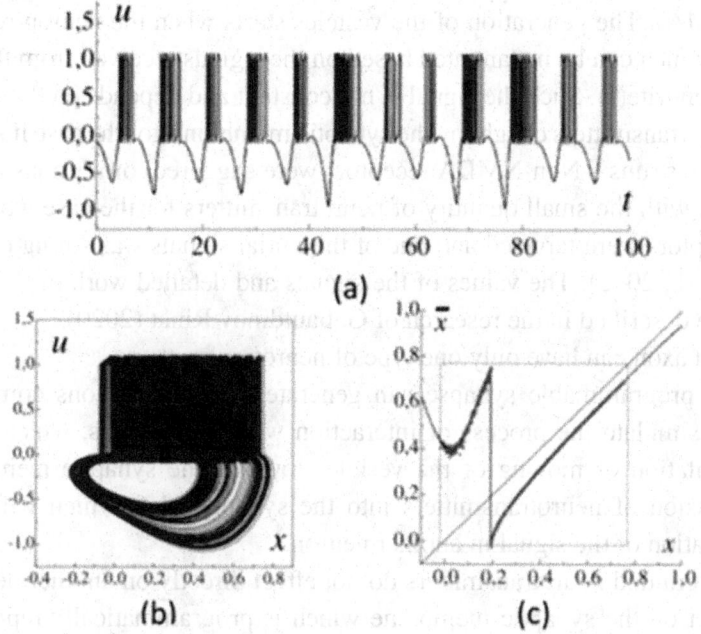

(a)

(b)   (c)

Chaotic Burst Vibrations in a Model I=-0.027. (a) Membrane potential evolution. (b) Phase trajectory in the projection onto the plane (x,u). (c) Poincare map corresponding to a chaotic attractor.

Note that for a certain ratio of the characteristic time scales of the model blocks, the oscillations are in the form of bursts, which can be both regular and chaotic. An example of a chaotic temporal realization and the Poincaré map corresponding to a chaotic attractor are presented in the figure.

EXPERIMENTS

Programming Implementation and AI Based Constructor's Experiments

The programming model of the artificial neuron which is based on the biological model contains the following key features:

- Each artificial biological neuron has generic list of the Queues for dendrites(input signals) and axons(output signals)

- Each synapse can generate the vesicles containing different type o neurotransmitters(in my research I used only 2 neurotransmitters: Glu, GAMA). The generation of the vesicles starts when the neuron receives the AP which can be instantiated based on the signals received from the Qeue(s) of dendrite(s). Such the signal is not constant and depends on the quantity of neurotransmitters caught by the synaptic membrane(for the case if summation process runs – Non-NMDA receptors were triggered) or it's constant enough even with the small quantity of neurotransmitters for the case when NMDA receptors were targeted and one of the initial signals was strong enough (Jin I. et al., 2012). The values of the signals and detailed work of the processes were described in the research of Galiautdinov Rinat (2020)
- Each axon can have only one type of neurotransmitters.
- The programmable synapse can generate the calcium ions coming inside and simulate the process of interaction with the vesicles, which effects on simulation of moving of the vesicles towards the synaptic membrane and emission of neurotransmitters into the synaptic cleft which will effect on initiation of the signal in another neuron.
- The emitted neurotransmitters do not effect directly on another neuron, they effect on the synaptic membrane which is programmatically represented as an input object of the dendrite connected with another neuron. The emission result fully depends on the type of the receptors and the whether NMDA receptor(if this is a case) is "turned on"(what could be caused by the initial strong signal)

Such the approach was used by the author in creation of the neuron based constructor in the beginning of 2016. This neuro-constructor allows to construct the neural circuits including the virtual muscles and virtual sensors (Galiautdinov Rinat et al., 2019A; Galiautdinov Rinat et al., 2019B). The virtual sensors serve as a triggering mechanism effecting on the neural circuit and the virtual muscles serve as outcome. Each neuron and each synapse generate the logs, which includes the data of the APs, the number of emitted calcium ions, the number of instantiated vesicles with neurotransmitters, the number of emitted neurotransmitters, the type of the receptors and the data related to the newly generated signals. This constructor allows to move the experiments on biological objects (such as Aplysia – the mullusc) into the virtual sphere where no animal is necessary for exploring of the work of the nervous system. More important, it allows to construct extremely complex virtual neural circuits and research its behavior. Such the approach allows to simulate the nervous system even of the complex creatures. During the experiment, with the help of the author's neuro-constructor, the author virtually created the nervous system of Aplysia and Planarian(Tricladida). The generated neural circuit was able to simulate the work and behavior of the natural creatures.

The usage of such the approach could be considered as revolutionary as it fully simulates all the processes of the nervous system and inside of each neuron. It fully simulates the behavior of a biological creature. And eventually this could be used not only in simulation of the nervous systems.

The major idea and solution in the frame of the title of the article is that every device can be used as a sensor or "muscle" connected to the "live" constructed neural circuit which will behave as a single live creature providing the owner of the neural circuit with all the benefits as if it was a live loving creature.

SUMMARY

Artificial intelligence has already become the norm in our world, and as its role in the Internet of Things is becoming increasingly important, we need to look for "smart" solutions that will facilitate their merger. In addition, AI will soon be ready to oust the human operator from many areas of activity, which further emphasizes the need for reliable systems that can cope with any problem corresponding to this ecosystem.

The use of AI platforms along with solutions for storing data and memory of an industrial class is a way to ensure that the equipment is ready to fulfill its tasks, and this is one of the key points in creating the "Internet of things" of the future.

However the most important thing in the context of usage of the AI along with the IoT is enclosed in the usage of the simulated biological neurons and neural circuits which allow to convert the area with the things into a single live creature having its own nervous system and using any device either as a sensor of the neural circuit or as a "muscle".

In the article the author described the math model of the biological neuron and suggested the new approach in Artificial Intelligence and the next generation of the AI. The author described created by him the neuro-constructor and the received results of simulation of the work of the virtually created neural circuits of Aplysia(the mollusc) and Planarian(Tricladida) showing that the same solution can be applied to the IoT. The author showed that AIoT based on simulation of the biological nervous systems could be more effective in all the tasks related to the sphere of Artificial Intelligence and simulation of the biological processes.

REFERENCES

Abbott, L. F. (1997). Synaptic depression and cortical gain control. *Science*, *275*(5297), 220–224. doi:10.1126cience.275.5297.221 PMID:8985017

Abbott, L. F., & Kepler, T. B. (1990). Model neurons: From hodgkin-huxley to Hopfield. *Statistical Mechanics of Neural Networks*, *18*, 5–18.

Abdelbar, A. M., & Hedetniemi, S. M. (1998). The complexity of approximating MAP explanation. *Artificial Intelligence*, *102*, 21–38. doi:10.1016/S0004-3702(98)00043-5

Boyen, X., & Koller, D. (1998). Tractable inference for complex stochastic processes. *UAI98 – Proceedings of the Fourteenth Conference on Uncertainty in Uncertainty in Artificial Intelligence*, 33–42.

Cheng, J., & Druzdzel, M. (2000). AIS-BN: An adaptive importance sampling algorithm for evidential reasoning in large Bayesian networks. *Journal of Artificial Intelligence Research*, *13*, 155–188. doi:10.1613/jair.764

Choi, Y. B., Kadakkuzha, B. M., Liu, X. A., Akhmedov, K., Kandel, E. R., & Puthanveettil, S. V. (2014). Huntingtin is critical both presynaptically and postsynaptically for long-term learning-related synaptic plasticity in Aplysia. *PLoS One*, *9*(7), e103004. doi:10.1371/journal.pone.0103004 PMID:25054562

Dayan, P., & Abbott, L. F. (2001). Theoretical Neuroscience: Computational and Mathematical Modeling of Neural Systems. *Neuroscience*, *39*(3), 460.

Gong, P., & Van Leeuwen, C. (2007). Dynamically maintained spike timing sequences in networks of pulse-coupled oscillators with delays. *Physical Review Letters. APS.*, *98*(4), 048104. doi:10.1103/PhysRevLett.98.048104 PMID:17358818

Izhikevich, E. M., Gally, J. A., & Edelman, G. M. (2004). Spike-timing dynamics of neuronal groups. Cerebral Cortex, 14(8), 933–944. doi:10.1093/cercor/bhh053

Jin, I., Udo, H., Rayman, J. B., Puthanveettil, S., Kandel, E. R., & Hawkins, R. D. (2012). Spontaneous transmitter release recruits postsynaptic mechanisms of long-term and intermediate-term facilitation in Aplysia. *Proceedings of the National Academy of Sciences of the United States of America*, *109*(23), 9137–9142. doi:10.1073/pnas.1206846109 PMID:22619333

Kempter, R., Gerstner, W., & Van Hemmen, J. (1999). Hebbian learning and spiking neurons. *Physical Review E. APS*, *59*(4), 4498–4514. doi:10.1103/PhysRevE.59.4498

Markram, H. (1997). Regulation of synaptic efficacy by coincidence of postsynaptic APs and EPSPs. *Science. AAAS, 275*(5297), 213–215. doi:10.1126cience.275.5297.213 PMID:8985014

Michie, D., Spiegelhalter, D., Taylor, C., & Campbell, J. (1994). *Machine Learning, Neural and Statistical Classification*. Ellis Horwood.

Morrison, A., Diesmann, M., & Gerstner, W. (2008). Phenomenological models of synaptic plasticity based on spike timing. *Biological Cybernetics, 98*(6), 459–478. doi:10.100700422-008-0233-1 PMID:18491160

Nicholls, J.G. (2001). *From Neuron to Brain*. Academic Press.

Pfister, J. P., & Gerstner, W. (2006). Triplets of spikes in a model of spike timing-dependent plasticity. *Journal of Neuroscience. Social Neuroscience, 26*(38), 9673–9682. PMID:16988038

Rinat, G. (2020). Brain machine interface: The accurate interpretation of neurotransmitters' signals targeting the muscles. *International Journal of Applied Research in Bioinformatics, 0102*. doi:10.4018/IJARB.2020

Rinat, G., & Vardan, M. (2019A). Math model of neuron and nervous system research, based on AI constructor creating virtual neural circuits: Theoretical and Methodological Aspects. In V. Mkrttchian, E. Aleshina, & L. Gamidullaeva (Eds.), *Avatar-Based Control, Estimation, Communications, and Development of Neuron Multi-Functional Technology Platforms* (pp. 320–344). Hershey, PA: IGI Global. doi:10.4018/978-1-7998-1581-5.ch015

Rinat, G., & Vardan, M. (2019B). Brain machine interface – for Avatar Control & Estimation in Educational purposes Based on Neural AI plugs: Theoretical and Methodological Aspects. In V. Mkrttchian, E. Aleshina, & L. Gamidullaeva (Eds.), *Avatar-Based Control, Estimation, Communications, and Development of Neuron Multi-Functional Technology Platforms* (pp. 345–360). Hershey, PA: IGI Global. doi:10.4018/978-1-7998-1581-5.ch016

Stent, G. S. (1973). A physiological mechanism for Hebb's postulate of learning. *Proceedings of the National Academy of Sciences of the United States of America, 70*(4), 997–1001. 10.1073/pnas.70.4.997

Tsodyks, M., Uziel, A., & Markram, H. (2000). Synchrony generation in recurrent networks with frequency-dependent synapses. The Journal of Neuroscience, 20(1).

Chapter 3
Hybrid Genetic Approach for Solving Fuzzy Graph Coloring Problem

Mohamed Amine Basmassi
ISO Laboratory, Faculty of Sciences, Ibn Tofail University, Morocco

Sidina Boudaakat
SSDIA Laboratory, ENSET Mohammedia, Hassan II University of Casablanca, Morocco

Lamia Benameur
LIROSA Laboratory, Faculty of Sciences, Abdelmalik Essaadi University, Morocco

Omar Bouattane
SSDIA Laboratory, ENSET Mohammedia, Hassan II University of Casablanca, Morocco

Ahmed Rebbani
SSDIA Laboratory, ENSET Mohammedia, Hassan II University of Casablanca, Morocco

Jihane Alami Chentoufi
(iD) https://orcid.org/0000-0002-7167-7620
ISO Laboratory, Faculty of Sciences, Ibn Tofail University, Morocco

DOI: 10.4018/978-1-7998-4381-8.ch003

ABSTRACT

A hybrid genetic approach (HGA) is proposed to solve the fuzzy graph coloring problem. The proposed approach integrates a number of new features, such as an adapted greedy sequential algorithm, which is integrated in genetic algorithm to increase the quality of chromosomes and improve the rate of convergence toward the chromatic number. Moreover, an upper bound is used to generate the initial population in order to reduce the search space. Experiments on a set of five well-known DIMACS benchmark instances show that the proposed approach achieves competitive results and succeeds in finding the global optimal solution rapidly for complex fuzzy graph.

INTRODUCTION AND BACKGROUND

The Graph coloring is highly studied as a combinatorial optimization problem (Pardalos, 1998). Several practical problem can be modeled by graph coloring such as traffic light signal, frequency assignment problem (Roberts, 1979), register allocation (Chaitin, 1981), etc. It includes both vertex coloring and edge coloring. However, the term graph coloring usually refers to vertex coloring rather than edge coloring (Jensen, 2011).

The objective of the graph coloring problem (GCP) is to find minimum number of vertices clusters with respect to the adjacency constraint in such a way that two connected vertices cannot be in the same cluster, each cluster use a color to mark his vertices.

A graph is called a k-colored graph, if accept a k-coloring, and k is called the chromatic number χ when it is the minimum possible color for coloring the graph.

The chromatic number is given by the following formula:

$$\varsigma(G) = \min\left(k : P(G,k) > 0\right)$$

The graph coloring problems are very interesting from the theoretical standpoint since they are a class of NP-complete problems that also belong to constraint satisfaction problems (Garey, 2002).

The coloring problem in the real world applications are not always made of sure relation between items so connection between vertices should not be defined in connected or not connected but can be presented in a certain degree of connection.

The fuzzy graph coloring problem (FGC) is an extension of the GCP introduced for the first time by Kaufmann (Kaufmann, 1976), while Rosenfeld (Rosenfeld,

1975) presented another developed definition, including fuzzy vertices and fuzzy edges. Generally, Fuzzy graph can be defined by three ways fuzzy set of vertices with crisp edges or fuzzy edges with crisp vertices set or fuzzy vertices and fuzzy edges (Chentoufi, 2007). In this paper we deal with crisp vertices and fuzzy edges.

In past years, many works deal with the fuzzy graph coloring in the literature such as (Munoz, 2005), where The concept of chromatic number of fuzzy graph was introduced in two different approaches, the first one is based on the successive coloring functions of crisp graphs, the second approach is based on an extension of the concept of coloring function by means of a distance defined between colors. In addition (Eslahchi, 2006) defined the chromatic fuzzy sum and strength of fuzzy graph and to color the fuzzy graph they separate the vertices into different classes and the number of distinct color classes is the fuzzy chromatic number. The authors in (Meirong, 2015) designed a new algorithm using a semi-tensor product and α-cuts with two conditions for the fuzzy graph to find all the feasible coloring schemes. Furthermore, (Gómez, 2006) deal with the image classification by conceiving it as a fuzzy graph problem and define it on a set of pixels where fuzzy edges represent the distance between pixels to get a more flexible hierarchical structure of colors. Moreover, (Keshavarz, 2016) worked on fuzzy graphs coloring problem, with crisp vertices and fuzzy edges, they formulated a binary programming problem and a hybrid local search genetic algorithm to solve the binary programming. In this work a hybrid solution is proposed to solve the fuzzy graph coloring using the genetic algorithm and the greedy sequential concept.

Genetic algorithm is a metaheuristic that belongs to a large class of evolutionary algorithms, inspired by the process of natural selection according to Darwin's evolution theory, for solving both constrained and unconstrained optimization problems (Mitchell 1998). The genetic algorithm is often used to generate high-quality solutions. The process of this evolutionary algorithm start with generating a random initial generation made of suggested solution called chromosomes, to measure the quality of these solutions a fitness function is used to evaluate chromosomes. In each generation, parent selection methods provide chromosomes for crossover to create off-springs for the next generation. Next a mutation function modifies randomly a minority of off-springs. The next generation will be composed of the new off-springs and an elite selection of the parent generation. Commonly, the algorithm is completed when a maximum number of generations is produced, or a satisfactory fitness level is reached.

The greedy sequential algorithm (GSA) is a heuristic, where the vertices of the graph are labeled according to some specific order and assigns the smallest possible color to a vertex that has not been assigned to its neighbors. Such ordering is called a coloring heuristic (Erciyes, 2013).

The greedy algorithm focus on searching the local optimal solution to prioritization, and thus their prioritization results may not be the optimal solution. The proposed approach uses the concept of GSA to improve the chromosomes quality.

MAIN FOCUS OF THE CHAPTER

The proposed approach combines Genetic Algorithm (GA) and Greedy Sequential Algorithm (GSA) to solve the FGC problem.

In order to solve fuzzy graph coloring $\tilde{G} = (V, \tilde{E}, \mu)$, its chromatic number is a fuzzy set $\tilde{\chi}(G) = \{(\alpha, \chi\alpha)\}$ where α are the different membership values of edges, and $\chi\alpha$ is the chromatic number of crisp graph. For reach this purpose, we start with the lowest α-cut values which engender the complex crisp graph of the fuzzy graph. Then, the genetic algorithm functions are applied until mutation function, the GS function come to improve the fitness and the quality of chromosomes. When the optimal solution of the specified α-cut reached, the numbers of used colors become the upper bound chromatic number of the next α-cut crisp graph. We follow the same manner until obtain the chromatic fuzzy set $Ç(G)$.

The pseudo-code of the proposed approach is as follow:

Hybrid Genetic Algorithm

Generate α-cut set in ascending order
    UpperColorBound ← Max degree of G
    Repeat
    UsedAlpha ← Select α-cut$_i$ value
    Generate crisp graph (UsedAlpha)
    Generate initial POP$_0$ (UpperColorBound)
    Evaluation of POP$_0$ function
    Repeat
    Parent selection function
    Crossover function
    Mutation function
    Apply the adapted greedy sequential function (POP$_{j+1}$)
    Evaluation of POP$_{j+1}$ function
    Until (Terminating condition is reached)
    EChrom←Get elite chromosome from final POP
    UpperColorBound ← get used colors number of EChrom
    Until (All Alpha-Cut are treated)

Search Space

A search space contain all possible solutions to solve a problem which are divided in feasible and non-feasible solutions, to find the best feasible solution there are different ways, such as, an exact method which tests all possible solutions, build the best solution by applying a specific functions, or a random methods that work on a part of solutions, which follows a heuristic or meta-heuristic method to guide search. In the genetic algorithm each point in the search space represent a proposed solution called chromosome and characterized by a fitness value.

The explorations of the search space depend on the size and dimensions. Wherefore it is complex needs more time to solving problem.

The genetic algorithm use a few points from the search space to initialize the population and guide the search to other points as the process of finding a solution continues until locating a global solution. The solution found by the genetic algorithm is often considered as a good solution, but it is not always the global optimum.

Generate α-Cut Set and Crisp Graph

The proposed approach begins with generating all α-cuts from the fuzzy graph $\tilde{G}$ and putting them on ascending ordered set. A crisp graph is created for every α-cut in our set, for the first crisp graph which is the complex one. To create the chromosomes of the initial population an upper bound of the chromatic number is used which gets the maximum degree of the graph. After starting the process, the upper bound value changed to the used colors number of the best optimal solution found for the previous α-cut.

Initial Population

The function begins with a population P of n feasible or non-feasible coloring graphs. The population can be obtained by any graph-coloring algorithm that is able to generate different proper colorings for a graph. In this case, the upper bound is used to generate different chromosomes with random number of colors.

Fitness Score

Each chromosome has a fitness score that show the ability of the chromosome to survive. A population of n chromosomes is maintained with their fitness scores by the genetic algorithm. The chromosomes with optimal fitness score have more chance to mate and transfer their genes to produce new offspring.

The population size is stable. So, the parent chromosomes (the old population) and the new arrivals compete to pass in the next generation. The chromosomes with the best fitness score move to next generation while least fit die.

Parent Selection

The parent selection is one of the GA outline, chromosomes are selected from the population to be parents and mate to create offspring for the next generation. The problem is how to select these chromosomes. Parent selection is very sensitive process that impacts the convergence rate of the GA and helps to avoid being stacked in an optimal solution. There are several methods used for the parent selection, for example steady state selection, rank selection, tournament selection, roulette wheel selection, and some others.

Tournament Selection is a Strategy used for selecting the fittest candidates from the current population, these selected candidates are then moved on to crossover. In a K-way tournament selection, a K -chromosomes are selected and run a tournament among them. The weak candidates have a smaller chance of getting chosen when the tournament size is larger, due to that the competition become stronger between chromosomes. In this work, four chromosomes are randomly selected from the population, and then the best of them is selected to become a parent.

Crossover

In this function, two parents are selected and two off-springs are produced using the genetic material of the parents. In this approach, a two random points of crossover are used.

Mutation

The uniform mutation is used to maintain and introduce diversity in the genetic population and it is applied with a small probability. The applied mechanism work with inversion or insertion. Also, the position of the mutation is random.

The Adapted Sequential Algorithm

This algorithm is mainly based on GSA proposed in (Erciyes, 2013) and it used in this work to allow the correction and upgrade chromosomes from a non-realizable solution to a realizable or almost a realizable one. As showed in pseudo-code HGA, for each chromosome all genes that do not verify the constraints of the problem are checked. In this case, a different color from its neighbor is affected, but it should be one of the used colors in the chromosome. Otherwise, no modification is performed. The use of this algorithm gives an impulse to increase the convergence of the algorithm.

Evaluation

All the generated chromosomes are exposed to the objective function to calculate their fitness based on graph coloring constraints. The evaluation process goes on two phases. The number of used colors in every chromosome is reviewed. Then, the fitness is calculated based on the adjacency matrix.

SOLUTIONS AND RECOMMENDATIONS

Data

The hybrid genetic approach is tested on different fuzzy graphs. No benchmark set of fuzzy graphs is found in the literature in our known. A set of fuzzy graphs are generated from a set crisp graphs from DIMACS benchmarks (Dataset, 2002).

The adjacency matrices of those crisp graphs are modified in such way that all connection value of edges gets a membership value. The membership value of the edges is randomly generated with uniform distribution in an interval [0, 1].

For the hybrid genetic algorithm used at this work, the crossover probability is Pc=90%, and the mutation probability is Pm=0.6%. The algorithm is run for 50 generations and the population size used is between 100 and 500.

Results and Discussion

As we know for every graph there is a space (S) of solution that are categorized to realizable solution and non-realizable solution. On the other hand, the genetic algorithm tries to locate the global optimal solution through a random selection of a fixed number of S chromosomes. To reduce the searching space an upper bound of the chromatic number will help the initial population to select chromosomes near to the global optimal solution.

The objective of initializing the upper bound of the initial population with the number of used colors of the optimal solution is reducing the space searching of the next crisp graph and that help to fast convergence and saving time.

Using only the greedy sequential algorithm to solve the global problem will provide only a local realizable solution; by using it for a specific proposed solution it will increase the quality of chromosomes which help to increase the rate of convergence.

To check validity of the proposed HGA we tested five DIMACS benchmark graph (Dataset, 2002), that were modified by adding to them edges membership value. After generating the α-cut set, is observable that the case where α=0, the associated crisp graph represent the original graph. HGA was successful to reach the chromatic number of the original graph.

It is shown in all graphs, that number of used color obtained by HGA decreases when the value of α-cut increases, and the proposed HGA needs very low CPU time to solve the FGP even for large fuzzy graphs see Table I. We could not compare the proposed approach with other works. A result of that, there is no paper work on FGP with benchmarks and concrete results.

Table 1. Experimental results

Graph	|V|	|E|	χ(G)	α-cut								
				0,1	0,2	0,3	0,4	0,5	0,6	0,7	0,8	0,9
				K-coloring								
Queen7_7	49	476	7	7	7	6	6	4	4	4	3	2
Jean	80	254	10	10	6	6	6	5	4	4	3	3
2_Insertion_4	149	541	4	5	4	4	3	3	3	3	3	3
Zeroin.i.2	211	3541	30	30	30	27	18	16	13	12	6	4
Fpsol2.i.2	451	8691	30	30	30	25	21	19	18	16	10	4

FUTURE RESEARCH DIRECTIONS

This HGA show a performance in the localization of the global optimal solution on the majority of the studied cases. In future, a modelization of the traffic light junction problem in a fuzzy graph will be done due to the correspondence of this theoretical problem with all his constraints.

CONCLUSION

In this work, a Hybrid genetic algorithm used to solve the fuzzy graph coloring problem. It is an extension of the graph coloring problem in the graph theory.

The proposed approach integrates a number of features. First, the greedy sequential algorithm was adapted and integrated in genetic algorithm to increase the quality of chromosomes to growth the rate of convergence. Second, the upper bound is used in the generation of initial population to reduce the search space.

Figure 1. Variation of the optimal solution depending on the α-cut for every grap

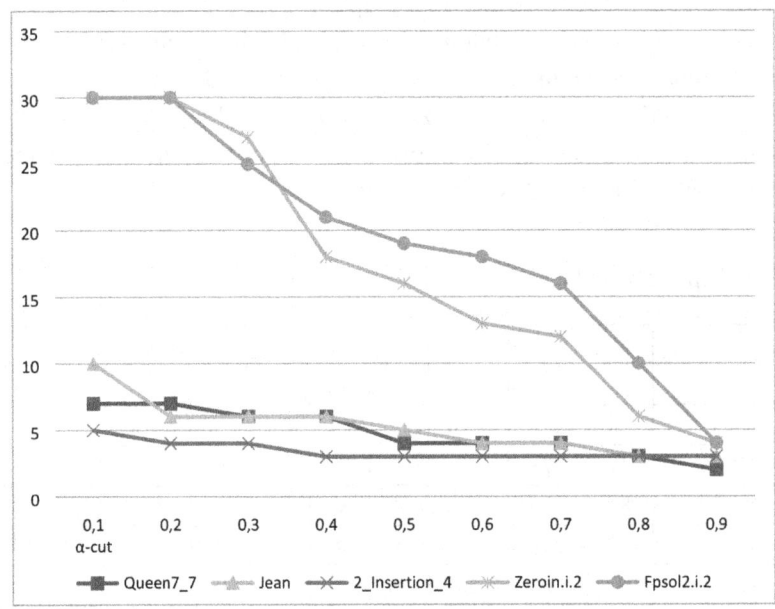

The main advantage of our proposed HGA is its simplicity and fast convergence as well as from its power to find the global optimal solution rapidly for complex fuzzy graph.

ACKNOWLEDGMENT

This research received no specific grant from any funding agency in the public, commercial, or not-for-profit sectors.

REFERENCES

Chaitin, G. J., Auslander, M. A., Chandra, A. K., Cocke, J., Hopkins, M. E., & Markstein, P. W. (1981). Register allocation via coloring. *Computer Languages*, 6(1), 47–57. doi:10.1016/0096-0551(81)90048-5

Chentoufi, J. A., El Imrani, A., & Bouroumi, A. (2007). A multipopulation cultural algorithm using fuzzy clustering. *Applied Soft Computing, 7*(2), 506–519. doi:10.1016/j.asoc.2006.10.010

Erciyes, K. (2013). *Distributed graph algorithms for computer networks*. Springer Science & Business Media. doi:10.1007/978-1-4471-5173-9

Eslahchi, C., & Onagh, B. N. (2006). Vertex-strength of fuzzy graphs. *International Journal of Mathematics and Mathematical Sciences, 2006*, 43614–1. doi:10.1155/IJMMS/2006/43614

Garey & Johnson. (1990). Computers and Intractability; a Guide to the Theory of Np-Completeness. W. H. Freeman & Co.

Gómez, D., Montero, J., & Yáñez, J. (2006). A coloring fuzzy graph approach for image classification. *Information Sciences, 176*(24), 3645–3657. doi:10.1016/j.ins.2006.01.006

Jensen, T. R., & Toft, B. (2011). *Graph Coloring Problems*. John Wiley & Sons.

Kaufmann, A. (1975). *Introduction à la théorie des sous-ensembles flous à l'usage des ingénieurs (fuzzy sets theory)*. Academic Press.

Keshavarz, E. (2016). Vertex-coloring of fuzzy graphs: A new approach. *Journal of Intelligent & Fuzzy Systems, 30*(2), 883–893. doi:10.3233/IFS-151810

Meirong, X., Yuzhen, W., & Ping, J. (2015, July). Fuzzy graph coloring via semi-tensor product method. In *2015 34th Chinese Control Conference (CCC)* (pp. 973-978). IEEE. 10.1109/ChiCC.2015.7259766

Muñoz, S., Ortuno, M. T., Ramírez, J., & Yáñez, J. (2005). Coloring fuzzy graphs. *Omega, Elsevier, 33*(3), 211–221. doi:10.1016/j.omega.2004.04.006

Pardalos, P. M., Mavridou, T., & Xue, J. (1998). The graph coloring problem: a bibliographic survey. In D. Z. Du & P. M. Pardalos (Eds.), *Handbook of combinatorial optimization* (Vol. 2, pp. 331–395). Boston: Kluwer Academic Publishers. doi:10.1007/978-1-4613-0303-9_16

Roberts, F. S. (1979). On the mobile radio frequency assignment problem and the traffic light phasing problem. *Annals of the New York Academy of Sciences, 319*(1), 466–483. doi:10.1111/j.1749-6632.1979.tb32824.x

Rosenfeld, A. (1975). Fuzzy graphs. In *Fuzzy sets and their applications to cognitive and decision processes* (pp. 77–95). Academic Press. doi:10.1016/B978-0-12-775260-0.50008-6

Trick, M. (2002). *Computational Series: Graph Coloring and its Generalizations*. Retrieved November 02, 2019, from https://mat.gsia.cmu.edu/COLOR02/

KEY TERMS AND DEFINITIONS

Chromatic Number: The chromatic number of a graph G is the smallest number of colors needed to color the vertices of G so that no two adjacent vertices share the same color. Calculating the chromatic number of a graph is an NP-complete problem and no convenient method is known for determining the chromatic number of an arbitrary graph.

Evolutionary Algorithm: Is any type of learning method motivated by their obvious and intentional parallels to biological evolution, including, but not limited to, genetic algorithms, evolutionary strategies, and genetic programming.

Fuzzy Graph: Is asymmetric binary fuzzy relation on a fuzzy subset. Pair of functions represents the object and the relation between them where the first one is a fuzzy subset of a non-empty set and the second one is a symmetric fuzzy relation on the first function.

Genetic Algorithm: Characterizes potential problem hypotheses using a binary string representation, and iterates a search space of potential hypotheses in an attempt to identify the best hypothesis, which is that which optimizes a predefined numerical measure, or fitness. GAs is, collectively, a subset of evolutionary algorithms.

Greedy Algorithm: The greedy approach is an algorithm strategy in which a set of resources are recursively divided based on the maximum, immediate availability of that resource at any given stage of execution. To solve a problem based on the greedy approach, there are two stages: scanning the list of items, optimization.

Chapter 4
IOT Technology, Applications, and Challenges:
A Contemporary Survey and Classification

Asha Gowda Karegowda
Siddaganga Institute of Technology, Tumakuru, India

Devika G.
ⓘD https://orcid.org/0000-0002-2509-2867
Government Engineering College, KRPET, Mandya, India

Ramya Shree T. P.
Siddaganga Institute of Technology, Tumakuru, India

ABSTRACT

The world we in is virtually becoming smaller since living and nonliving things are connected to the internet. Internet of things, or IoT, is a system of interconnected things, each with unique identifiers (UIDs) and the ability to exchange data without the need of human intervention. The rapid growth of IoT is considered the next wave for enhancing services in almost all sectors of life, at low cost and time. This chapter presents IoT in a broader context, in terms of its growth, IoT operating systems, architecture, and future trends of IoT. The major contribution is detailed information of umpteen IoT applications. The various benefits of IoT, matter of concerns with respect to IoT, scope of research work are also discussed. The integration of various technologies is the main enabling factor of IoT, yielding more benefits to society as a whole. Also, supports in understanding implementation technologies and the major applications of their domain where IoT plays a vital role and future problems for next 20 years are also explicated.

DOI: 10.4018/978-1-7998-4381-8.ch004

INTRODUCTION

The fact that technology in various fields will evolve through the years, is the reason why we observe a rapid change in the shape, size and capacity of various instruments, components and the products used in daily life. Currently available networks have made a world more connected nowadays, but still opportunities to enhance are endless. These networks can install or can connect devices with out or with minimum interference of humans. But, the current network standards and technologies landscape is highly fragmented from machine-to-machine communication (Elijah, 2018). This is because of different applied domains or of technologies being designed beyond basic communications and networking standards. Additionally, there is dire need for global standardization at inter-operatability level(Porkodi, 2014; Gubbi, 2013; Amendola, 2014).

IoT is the future of our increasingly digitized world in umpteen number of domains. IoT is a choice for intelligent systems as nowadays more machines are being connected to the Internet than people, likewise many more connected products are entering our lives over the next few years as embedded micro-electronics promulgate in day to day via usable things(Xu. 2014). IoT is called the world's fourth wave of information industry after computers and computer networks. IoT is the pivotal component of a new generation information technology.

SCOPE OF THE WORK

IoT can be everywhere and it can accommodate as well suit for various domain of application as well as integrates with different domains to sole real-time problems in an easier manner. The Internet of Things gives us an opportunity to construct effective administrations, applications for manufacturing, lifesaving solutions, proper cultivation and more. According to Forbes survey the top IoT segments in 2018 based on 1600 real IoT projects is presented in figure 1. The figure 2 depicts the IoT's applicability in various applications in this decade.

Figure 1. Number of articles published different areas of IoT during 2013-2018

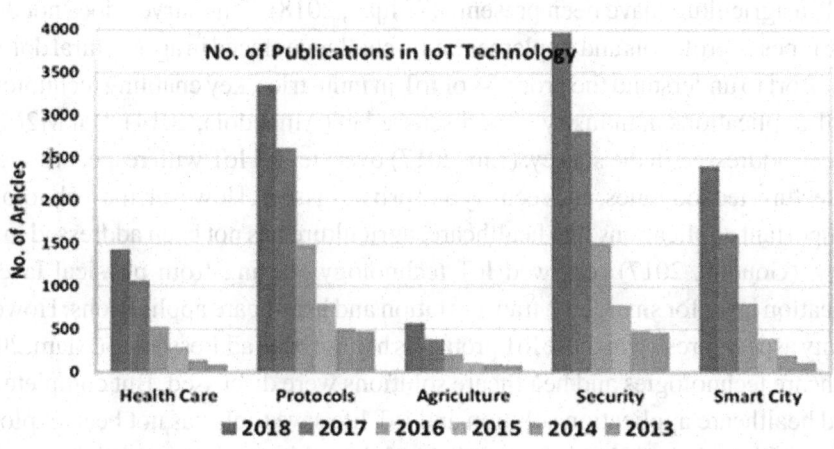

Figure 2. IOT Project investments under different domain in 2018

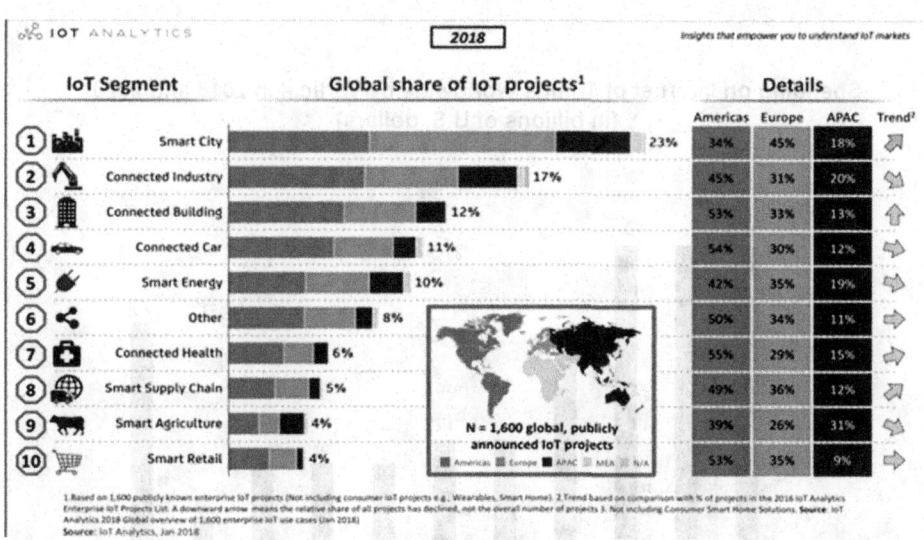

Related Work

In the past, many studies were conducted to review the IoT. Benefits and challenges of IoT in agriculture have been presented (Elijah, 2018). This survey does not detail on the types of protocols and implementation challenges faced in agricultural domain. In an effort to understand the progress of IoT in industries, key enabling technologies for IoT applications in industry were discussed in (Amendola, 2014). But M2M has not been addressed in the survey. (Lin, 2017) overviewed IoT with respect to system architecture, technologies, and security and privacy issues. However, the technologies for important applications like healthcare, agriculture has not been addressed in this survey. (Goudos, 2017) reviewed IoT technology starting from physical layer to application layer for smart city, transportation and healthcare applications. However security aspects present in these IoT protocols has not been addressed. In (Alam, 2018) healthcare technologies and healthcare solutions were discussed. But complete end to end healthcare application solution and IoT life save tools has not been explored. Authors (Gharaibeh, 2015) have highlighted the achievements in realizing various aspects of smart cities. Moreover this survey addresses smart health management in smart cities but smart structural health management which equally important in smart cities has not been addressed. From the source of Statista figure 3 compares

Figure 3. IoT Project investments between 2015-20

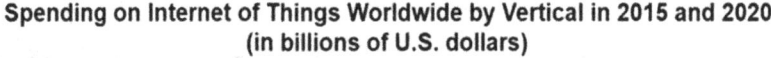

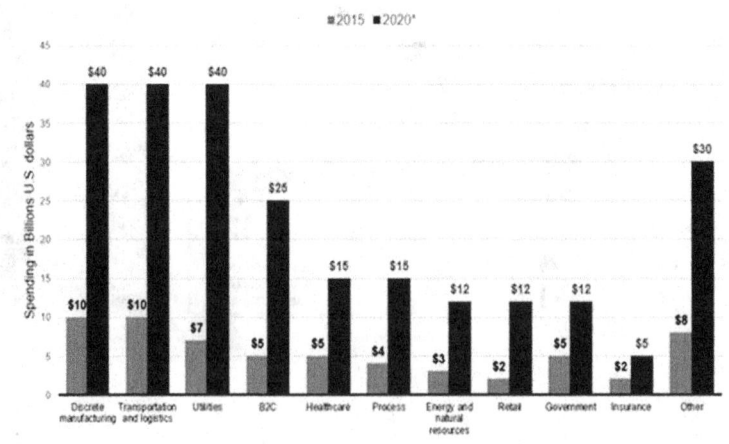

projected growth in spending by vertical globally from 2015-2020 and accordingly smart cities (23%), connected industries (17%) and connected buildings (12%).

Contributions

In order to adopt IoT approaches in different applications, firstly, an identification of the key characteristics and issues of IoT are considered. The review of such work will be highlighted. The various technologies of IoT will be covered. The enabling factor of IoT integration with contemporary domains to yield more benefits in applications will be presented. The challenges, advantages, disadvantages and future scope of the technologies will be discussed.

INTERNET OF THINGS (IOT)

The term 'IoTs' is not new; it is attributed to Kevin Ashton of Google in 1999, but has captured people's imagination and supported lots of new conferences in 2014. The explosion of the RFID market in 2005 marked the dawn of the thinking about the IoTs. IoT is synonymously also used in terms as in practice machine-to-machine (M2M) communications, Web of Things, Industry 4.0, Industrial Internet (of Things), smart systems, pervasive computing and intelligent systems. The IoT is the network of physical objects or things, wherein each thing is uniquely identified through its embedded computing system and is able to interoperate within the existing Internet infrastructure. Experts estimate that the IoT will consists of almost 50 billion objects by 2020.[wiki2015]

IoT is embedded with electronics, software, sensors, and network connectivity, which enables to collect and exchange data. It integrates the ubiquitous communications, pervasive computing, and ambient intelligence. In other words, IOT is a vision, where "things", especially everyday objects, such as all home appliances, furniture, clothes, vehicles, roads and smart materials, etc. are readable, recognizable, locatable, addressable and/or controllable via the Internet. This will provide the basis for many new applications, such as energy monitoring, transport safety systems, building security etc. The basic level and the development direction of IoT's system are based on wireless sensor network structure. The IoT allow objects to be sensed and controlled remotely across existing network infrastructure, creating opportunities for more direct integration between the physical world and computer-based systems, and resulting in improved efficiency, accuracy and economics benefits (Stankovic, 2014).

There are many projections for the potential value generated by IoT. One of the more extreme, from US IT company Cisco, predicts a value of US$14.4 trillion for companies and industries globally over the next decade. This leads to a new hope for technologies development in near future which will lead to potential economic and social benefit arising from new connected applications and new service-orientated business models. Certainly, it will change completely the applications, businesses and industry sectors, which were not included till date, henceforth will bring benefit to all real time applications. This vision will surely change with time, especially as synergies between Identification Technologies, Wireless Sensor Networks, Intelligent Devices and Nanotechnology will enable a number of advanced applications. Innovative use of technologies such as RFID, NFC, ZigBee and Bluetooth, are contributing to create a value proposition for stakeholders of IOT as depicted in Figure 4. IoTs will connect the world's objects in both a sensory and intelligent manner through combining technological developments in item identification ("tagging things"), sensors and wireless sensor networks ("feeling things"), embedded systems ("thinking things") and nanotechnology ("shrinking things") (Porkodi, 2014; Gubbi, 2013; Amendola, 2014).

Figure 4. Tracking of IoT](postscapes)

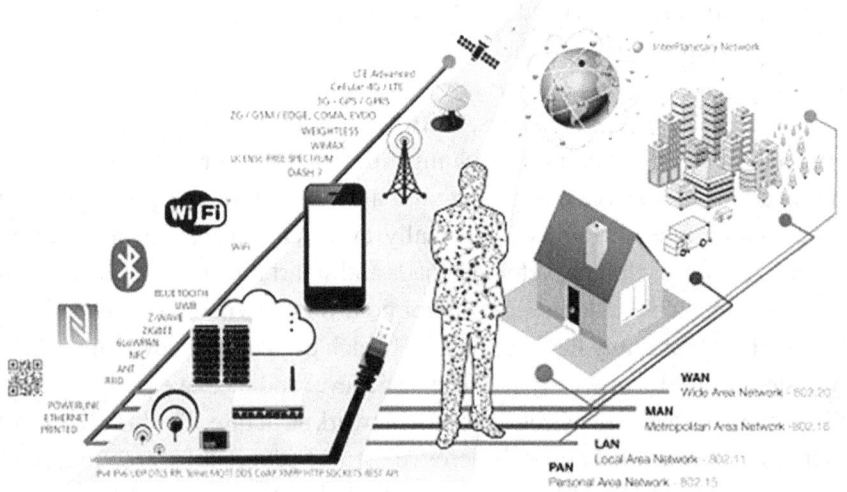

Features of IOT

IoT includes some basic features which includes: intelligence, expressing, safety and sensing. In addition to these features it needs to have following features: (Ovidiu, 2013)

- **Self-adaptation**: Communicating nodes, as well as services using them to react in a timely manner to continuously changing context i.e. accordance with instance, business policies or performances objectives.
- **Self-optimisation**: Efficient usage of resources such as processor, memory, bandwidth etc.
- **Self-configurable**: Since IoT includes thousands of devices, configuration of devices has to be done in remote manner or self-management applications automatically needs be configured.
- **Self-protection:** System has to be adjusted autonomously by itself for changing levels of security and privacy issues.
- **Self-healing:** The problems need to be diagnosed when ever occur and needs be fixed in an autonomous way.
- **Self-description:** The things and resources should be able to describe their characteristics and capabilities in an expressive manner in order to allow other objects to interact with them.
- **Self-discovery: The** devices and services have to be dynamically discovered and used by others.
- **Self-matching:** The devices must provide reliable service, have capability to provide service enhancements, future services while exchanging of information among objects.
- **Self-energy supplying:** This feature supports to realize and deploy sustainable solutions.

Architecture of IoT

The basic architectural requirement of IoT is to connect billions or trillions of homogeneous or heterogeneous devices/objects through Internet with flexibility among layer operations. Even though with increase in number of proposed architectures there is no single standard IoT architectural model that is agreed on but there is a need for few end-to-end architectural need such as concurrent data collection, efficient data handling, connectivity, communication, scalability, security, availability, quality of service, modularity, flexibility, platform independence and device management. Common and basic models of IoT through review of (khan, 2010; Wu, 2010; Yang, 2011) which includes 3 to 6 layer architectures are illustrated in figure 5. The perception layer represents the physical sensors of IoT to represent, collect and process information. Network layer includes functionality for connectivity and communication. The service layer manages layer as pair with its request based on address and names. The business layer regulates and manages entire activities of IoT system and services. Application layer provides services required by customers. The edge layer manages activities of data collection in and from data centres or in clouds.

Figure 5. Three to Six Layers architecture of IoT (source: smart science)

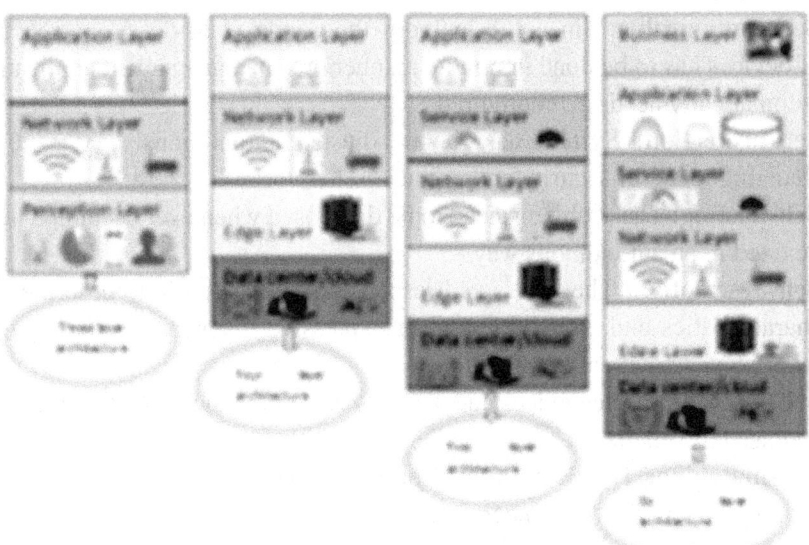

The basis for the IoT Reference Architecture structuring approach of different layers are briefed below:

- **Functional** – Describes the system's runtime functional elements and their responsibilities, interfaces, and primary interactions
- **Information** – Describes the way that the architecture stores, manipulates, manages, and distributes data and information.
- **Deployment** – Describes the environment into which the system will be deployed, including the dependencies the system has, in its runtime environment

- **Operational** – Describes how the system will be operated, administered, and supported when it is running in its production environment

Operating Systems for IoT

IoT devices are integrated through the interaction of software at a dynamic level along with the use of wireless sensor network and RFID technologies using the internet infrastructure (Kopetz, 2011; Zikria,2019). The few of the operating systems for IoTs are listed as follows:

i.   ARM Mbed IoT: It is a device platform which provides the operating system, cloud services, tools, and developer's ecosystem to make the creation and deployment of commercial, standards-based IoT solutions possible at scale.

ii. Snappy Ubuntu core: A new, transitionally updated Ubuntu for clouds and devices. Snappy Ubuntu Core is a new rendition of Ubuntu with transactional updates - a minimal server image with the same libraries as today's Ubuntu, but applications are provided through a simpler mechanism. The snappy approach is faster, more reliable, and provides stronger security guarantees for apps and users — that's why we call them "snappy" applications. Snappy apps and Ubuntu Core itself can be upgraded atomically and rolled back if needed. It's called "transactional" or "image-based" systems management, and it will be available on every Ubuntu certified cloud.

iii. Contiki OS: Contiki is an open source operating system for the IoTs. Contiki connects tiny low-cost, low-power microcontroller to the Internet. It supports IPV6, and IPV4 Internet protocols.

iv. Raspbian: Raspbian is a free operating system based on Debian optimized for the Raspberry Pi hardware. An operating system is the set of basic programs and utilities that make Raspberry Pi run. Raspbian is still under active development with an emphasis on improving the stability and performance of as many Debian packages as possible.

v.   RIOT: RIOT is the direct successor of μkleos. It is re-branded to avoid problems with spelling and pronouncing the name of the operating system. Support for 6LoWPAN, RPL, and TCP was integrated.

vi. Spark: Particle (formally Spark) is a complete, open source, full-stack solution for cloud-connected devices.

vii. Webinos: A web based application platform that allows a developer to access native like capabilities through JavaScript APIs. A set of interoperable protocols allows a device to remotely use the services of another device. A security framework that protects applications running on all device types, and designed for cross device, inter person sharing.

viii. Android things: This OS is been introduced by Google which can run on low power Bluetooth and WIFI technology. It aims at removing all obstacles and simplify IoT development.

ix. Huawei Agile IoT: The Huawei Agile IoT Solution consists of three core components: (1) LiteOS, an IoT operating system (OS); (2) Agile IoT gateways; (3) the Agile Controller.

x. LiteOS: LiteOS is a lightweight, open source IoT OS. Compared with other IoT OS, LiteOS is three times smaller in size and consumes four times less power. Its microsecond response speed is also 20 percent faster. In addition, the LiteOS is an open source system that provides a unified development platform. The system fosters the development of the IoT system as partners can obtain codes from the open source community free of charge and quickly build their own IoT products. Agile IoT Gateways: The Agile IoT Gateway is designed to bridge sensor and IP networks.

xi. Zephyr: It is a real time OS built for IoT applications to support under LINUX platforms. It supports for library and has memory protection system.

Table 1. Summary of Operating Systems supporting IoT.

OS	Product	Programming language	Protocols supported	Open source/ not	Features
Mbed	ARM	C++	Bluetooth, Wi-Fi, Zigbee, Ethernet, 6LoWPAN	Not open-source	Common OS for IoT devices Open standards for connectivity. Updatable and secure device
RIOT	GitHub	C, C++	IPV6, TCP, UDP, 6Lowpan, NHDP	Open-source	Low memory footprint, high energy efficiency, real-time capabilities, a modular and configurable communication stack, and support for a wide range of low-power devices.
Contiki OS	Contiki	JAVA	CoAP, 6Lowpan, RPL	Open-source	Hardware portability Low memory usage, and compatible with well-known standards
FreeRTOS	FreeRTOS	C	IPV6, IPV4, TCP, UDP, 6LowPAN	Open-source	With millions of deployments in all market sectors, blue chip companies trust FreeRTOS because it is professionally developed, strictly quality controlled, robust, supported, free to use in commercial products without a requirement to expose proprietary source code, and has no IP infringement risk.
Nano-RK	Nano-RK	C	RT-Link, PCF-TDMA, b-mac, U-Connect, WiDom	Open-source	C GNU tool-chain, Classical Preemptive Operating System Multitasking Abstractions, Real-Time Priority Based Scheduling, Built-in Fault Handling
Snappy Ubuntu core	Ubuntu	C, C++, Python, node JS	IPV6, TCP, UDP, 6Lowpan,	Open-source	· Faster, more reliable and stronger security guarantees for apps and users. · Atomic transaction upgrades for apps and the Ubuntu Core software itself, all of which can be rolled back if needed, for simple maintenance and upgrades. · Separation of OS and application files as a set of distinct read-only images, to

IoT Platforms

The main task of IoT is connectivity to optimize the application requirements, in such cases devices under application cannot be expected to reside in same environment. This gap between device sensors and network of data is filled or junctured by IoT platforms. It gives a platform to use and connect required devices effectively. Currently many IoT platforms exist, among them few common platforms are considered by Cowen software developer's survey on key tech trends as depicted in figure 6.

Figure 6. IoT platform % usage under different domains

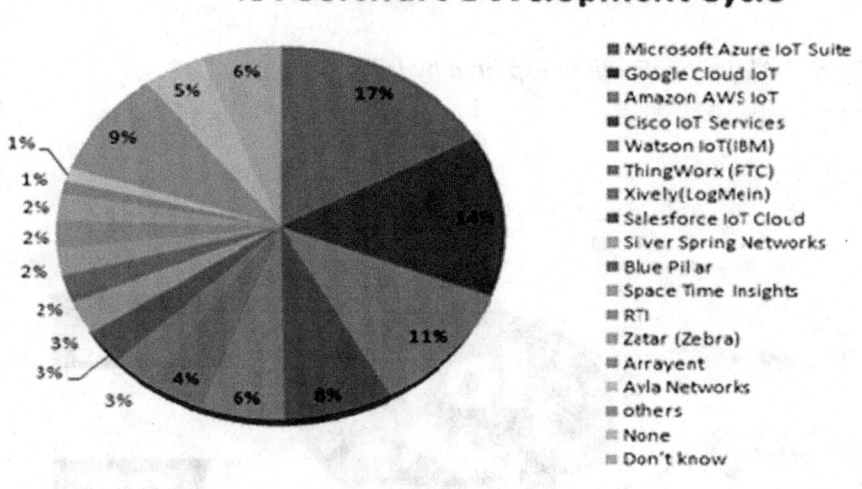

IOT APPLICATIONS

The Internet of Things (IoT) has emerged as a combination of multiple technologies with different applications. The broad theme that emerges in IoT is the use of devices that are connected over a network to provide a specific functionality. The information gathered by the devices is fed into other systems that act on it. The devices could be commonly used or purpose built.

Currently, the IoT solution has been deployed in many areas. IoT spans many vertical markets. The main sectors are: consumer goods, industrial equipment and processes, vehicle telematics and smart infrastructure. The IoT will deliver practical, real-time asset management, providing the ability to optimise use of abstraction licences, boreholes, pumps, reservoirs, pipe networks, sewers, treatment plant and discharge licences to ensure lowest cost and most effective service to customers. It will allow systems and plant to be run closer to their limits of capacity, condition and energy efficiency. The real values can be collected by deployed devices. IoT applications are shown in figure 7.

Cities are becoming smart by introducing IoT for smart parking, structural health, noise urban maps, traffic congestion, smart lighting, intelligent transportation systems etc,. IoT is contributing for environment by deploying IoT for forest fire detection, air pollution, land slide and avalanche prevention, earthquakes early detection, rare species tracking, theft detection in forest, etc. It can also be useful in fields like industry, logistics, agriculture, health, retail management, sports, and others.

Figure 7. Major applications explored by IoT

Smart Cities and Smart Homes

The smart home is the fastest growing field of IoT technology and it provides innovative, intelligent, ubiquitous and interactive services to users using different operations but with security as the major challenge. Smarts are the most popular application of IoT from all measured channels inclusive of Internet search results. Currently the total amount of its funding exceeds $2.5 billion and the list includes prominent startup names as Nest or AlertMe as well as a number of multinational corporations like Philips, Haier or Belkin (https://iot-analytics.com). It is future of residential technology designed to deliver and distribute number of services inside and outside house via networked devices. It creates smart home environment in which physical objects like home appliances, mobile devices, etc., connect to Internet and provide innovative and smart services to human (). The smart home appliances include smart ACs, heating, refrigerator, TV, others which are connectable via Internet to make peoples life more opportune and alluring (Fernandes, 2016). Like-wise home can be controlled and monitored in smarter way such as lighting, temperature, the climate of home, doors and windows (Jacobsson, 2016). Typical smart home architecture is shown in figure 8(a), which contain devices and appliances connected to tablet are controller to provide services in smarter way (https://doi.org/10.23919/IConAC.2017.8082057). The architecture of smart home is divided into internal and external wherein, the external environment contains all the entities of smart grid such as smart meters. The internal environment has appliances which belong to home. The internal and external environment of smart home is represented in figure 8(b) (Siano, 2016). The number of components is reduced in this architecture, management done with entity energy management system for internal environment and energy management interface for external environment of smart home. Energy management interface used as an interface between smart grid and home (Rahim, 2016). In (Gelazanskas, 2014) has still improved model with adaptation of load management with smart grid to perform better in terms of reliability and flexibility by an efficient management of energy (Siano, 2016). The energy management interface with different entities shown in dotted green line, energy management systems in solid orange line and blue dotted line shows communication with external environment.

Figure 8. (a) Typical smart home environment (b) Architecture of a smart home with Internal and external environment

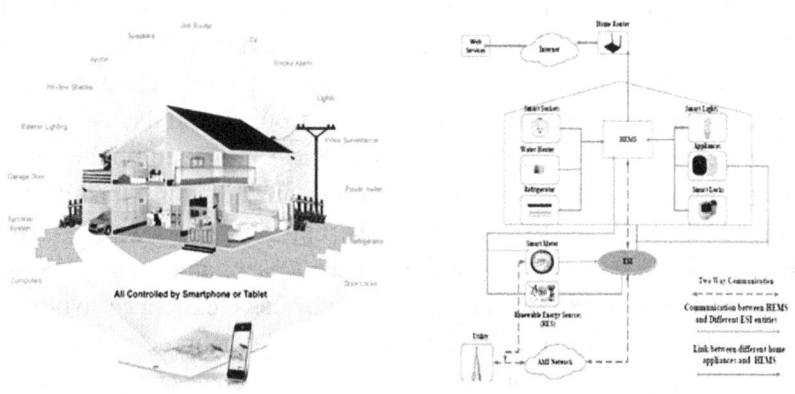

Although smart home in use is more convenient and controls all home appliances on the other side, due to internet connected, dynamic and heterogeneous nature of the smart home has challenges in security issues (Gubbi, 2013). The smart home environment contains more critical and private information of the user, so demands cyber-security requirements. Internet connected with hundreds of devices, data exchange among these devices and required software update facilitated system and such system should be deployed in a smart home (Jacobsson, 2016). That's why; there is huge need of strong authentication mechanism to prevent the attackers (Muñoz, 2011) security issues in the smart home environment using several scenarios. We investigated the security threats, classified these threats according to security objectives and evaluated their impact on the overall system. We identified security requirements and their solutions in the smart home environment. Based on several scenarios, we set security goals for the smart home. Based on historical data, we forecast of security attacks (like malware, virus, etc.) that how many attacks are expected to be launched in coming five years. The approach of layering (five layers) is applied (https://doi.org/10.23919/IConAC.2017.8082057) as in figure 9(a) with five layers as: resource, interface, agent, kernel and application layers. The household articles are included in resource layer. The second layer provides required suitable interfaces for all kind of resources. The third layer consist of different agents; producers, consumers and brokers of services. The next layer manages agents for authentication, state data transportation and management called kernel layer. The security issues of smart home are discussed in (Elmaghraby, 2014; Aravinthan, 2011;

Figure 9. (a) Layers of Smart Home Architecture (b) Smart city architecture

(a) Layers of Smart Home Architecture

User Application Layer		
Kernel Layer	Agent Management	Data Transportation
	Authentication and Authorization	State Monitor
Agent Layer	Agents	
Interface Layer	Appliance Specific Service Interface	
Resource Layer (Household appliances)		

(b) Smart city architecture

GOVERNMENT MANAGED	SUSTAINABILITY & ENVIRONMENTAL PROTECTION	CITIZEN RELATED DOMAINS	Education	Entertainment
			Tourism	Healthcare
			Entrepreneurship	...
		NATURAL RESOURCES RELATED DOMAINS	Agriculture	Renewable energies
			Smart grid	Waste management
			Water management	...
		INFRASTRUCTURE RELATED DOMAINS	Advertisement	Building and Housing
			City monitoring	Logistics
			Public safety	Public transport
			Traffic	...

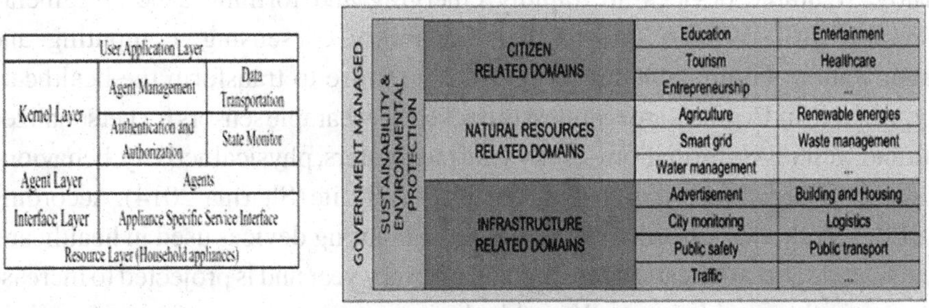

Cintuglu,2017; Gaikwad, 2015) with aim to include authentication, authorization, confidentiality, integration and availability.

The IoT smart city relay on components such as electric grids, healthcare or home automation, are connected to a single and global network. The resilience of interdependent infrastructure has to be modeled in IoT smart city architectures. The main challenges of smart city modeling are abstracting complexity and availability of model simulations includes all protocols (Sterbenz, 2017). Challenges for inclusion of big data, compatibility, investment, precision and security metrics to smart cities presented in (Ahmed, 2016) and in addition to these metrics risk management, low power and cost communication, trust and connectivity issues are described by (Mehmood, 2017). (Balsamo, 2017) focused on devices based on wearable's in cities. The communication, reliability, energy management, fabrication, interconnection individual participation is included with awareness situations consideration. The typical smart city has been described in (Corcuera, 2019). The essential atomic domains with multiple assignments including various categories are included in figure 9(b). The requirements of sustainability and environmental protection are included in orange box, and the governmental managing identities in yellow box. The players of smart city are included in inside boxes.

From review the key components of smart cities are:

- Inner operability of systems
- Energy and water efficiency sustainability
- Connectivity city-wide
- Security
- Effective transportation
- Development of private/public partnership

Wearable's

Recently, wearable devices are rapidly emerging and forming a new segment––"Wearable IoT (WIoT)" due to their capability of sensing, computing and communication. Future generations of WIoT promise to transform the healthcare sector, wherein individuals are seamlessly tracked by wearable sensors for personalized health and wellness information—body vital parameters, physical activity, behaviors, and other critical parameters impacting quality of daily life (Sharma, 2014). According to ABI Research, the consumers of wearable computing devices used in healthcare, wellness, and sport applications are increasing every year and is projected to increase even more in the coming years. Wearable devices encompass a variety of functions including data collection from on-body sensors, preprocessing the data, momentary data storage, and data transfer to internet-connected immediate neighbors such as mobile phones or to a remote server. It is the characteristic of wearability that adds value to these devices and allows customizing the collection of body's physiological or motion data depending upon the end-user application. While wearable sensors offer significant advantages to healthcare by automating remote healthcare interventions that include diagnostic monitoring, treatments and interoperability between patients and physicians, they face barriers such as the requirement to work in close proximity to other computing devices to compensate for low computing power, short battery life, and short communication bandwidth. The demand for interconnecting smart wearables is increasing over the years. IoT has moved from basic internet services to social networks to wearable web. This emergence of wearable devices is giving a new dimension to IoT by creating an intelligent fabric of body-worn or near-body sensors communicating with each other or with the internet. In other words, WIoT can be defined as a technological infrastructure that interconnects wearable sensors to enable monitoring human factors including health, wellness, behaviors and other data useful in enhancing individuals' everyday quality of life(). WIoT aims at connecting body-worn sensors to the medical infrastructure such that physicians can perform longitudinal assessment of their patients when they are at home(https://internetofthingsagenda.techtarget.com/definition/smartwatch). WIoT is still in its infancy period and therefore, demands a chain of developments in order to boost its successful evolution and to enable its widespread adoption in the healthcare industry. It is not sufficient to design standalone wearable devices but it becomes vital to create a WIoT eco system in which body-worn sensors seamlessly synchronize data to the cloud services through the IoT infrastructure.

Figure 10. Architectural Elements of Wearable IoTs

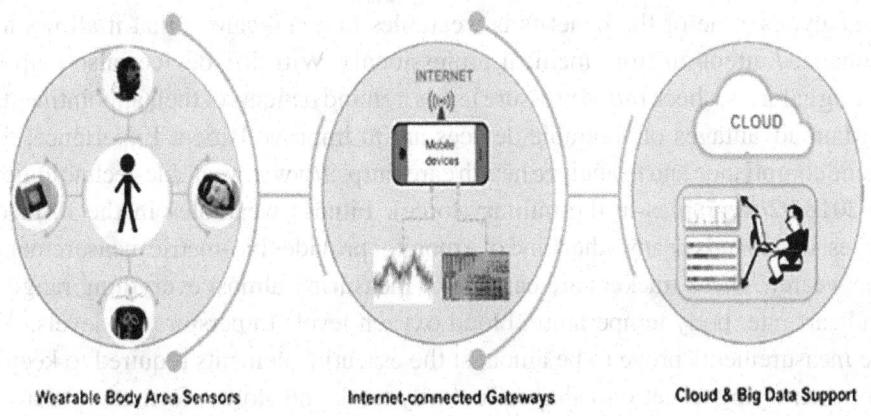

| Wearable Body Area Sensors | Internet-connected Gateways | Cloud & Big Data Support |

Wearable body area sensors architecture is presented in figure 10. The elements included are wearable body area sensors, Internet connected gateways and cloud& bigdata support. The wearable body area sensors (WBAS) are frontend components of WIoT and unobtrusively envelop the body to capture health-centric data. WBAS are primarily responsible for 1) collecting the data either directly from the body through contact sensors or from peripheral sensors providing indirect information of body and its behaviors preparing the data for either on-board analysis for close-loop feedback or remote transmission for comprehensive analysis and decision support. Internet-connected Gateways are rarely standalone systems due to their limited computing power and communication bandwidth. Therefore, they need to transmit data to potential computing resources that are either companion devices such as smart phones, tablets, and laptop PCs, or remotely-located cloud computing servers. In either case of data communication, companion devices are used as gateway devices, thus representing an important class of WIoT that enables the information to flow from the sensors to the cloud or server centers for storage and further analysis. The Gateway devices comprise of short-range communication technology such as Bluetooth, used to exchange data with wearable sensors, and of heterogeneous networks, such as WIFI and GSM, used to send the data to the cloud. Some Gateway devices have the capacity to store data, to run some pre-processing algorithms evaluating whether the data is clinically relevant, and to send the data intermittently to remote servers. The congruence of close companions wearable sensors and smart phones will flood the cloud centers with medical data at an unprecedented rate. Gaining knowledge from this data is as important as acquiring the information from the body. A cloud computing infrastructure can facilitate the management of wearable data and can support advanced functionalities of data mining, machine learning, and medical big data analytics. One of the benefits is wearables in healthcare is that it allows for personalized attention from medical professionals. With IoT devices also help in counting calories, check blood pressure levels, remind patients of their appointments, etc. Main advantages of wearable devices are to improve Patient Experience, for hygiene compliance and to analyze health care (https://www.wearable-technologies. com/2018/02/wearables-in-the-military-force). Fitness wearables in the form of watches, wristbands or any other kind of a monitor provides bio-metric measurements to the wearer. These trackers are capable of measuring almost everything ranging from heart rate, body temperature, blood oxygen levels to perspiration levels. All these measurements prove to be amongst the essential elements required to keep a track when trying to get into shape. To easily track and store information to track daily activity and even compare it with the previous days. Thus, it gets easy to track one's daily progress and even necessary changes in the exercise regimen can be implemented, if required. Fitness trackers also amp up one's fitness goals as the progress reports can easily be tracked. With increased global awareness of and dedication to health and fitness, a multitude of apps and tech devices help consumers track their progress. Health and fitness wearables are top amongst the ladder of IoT wearables (Mardonova, 2018). Google Glass is a type of wearable technology with an optical head-mounted display (OHMD). It displays information in a handsfree format. Wearers of the glass can communicate using the natural language voice command using the internet. A smart watch is a wearable computing device that closely resembles a wristwatch or other time-keeping device. In addition to telling time, many smart watches are Bluetooth-capable. The watch becomes, in effect, a wireless Bluetooth adaptor capable of extending the capabilities of the wearer's smart phone to the watch. In such a case, the wearer can use the watch's interface to initiate and answer phone calls from their mobile phone, read email and text messages, get a weather report, listen to music, dictate email or text messages or ask a digital assistant a question (Ching, 2016). Vandrico Solutions Inc. in Canada is developing a head-mounted device similar to Google Glass that is described as the first Smart Glass application intended for the mining sector. The project is expected to be used in 50 mines around the world where the Metrics Manager by Motion Metrics International Corp. is currently being used (Botta, 2016). Wearable technologies will enable soldiers to either track or be tracked in real-time and great precision which will ensure safety of these soldiers as well as to diminish risk of failure in high-risked operations. Norinco, a Chinese enterprise, is developing a new generation of military exoskeletons. These will enable soldiers to carry over 100kg of supplies without much struggle since the soldier would have its strength powered up. A system called Arielle, by ST Engineering, has been combining elements such as helmets embedded with smart-glasses, smart-clothing, bracelets, different sensors, cameras, smart watches, you name it. These gears will change the daily-life of thousands of soldiers when it comes to: providing soldiers with AR images that will ease communication and coordinate tracking, body temperature tracking, long-lasting batteries for the most various devices, and notifying colleagues with the probability of armed conflicts as well as day-night vision (http://www.egr.msu. edu; Gubbi, 2016).

SMART GRID

The IoT is being used in smart grid (SG) to speed up the information of power grid system and benefits effective management of power grid infrastructure. The disasters prevention and reduction is another application fields because of IoTs of advanced sensing and communication technologies to reduce economic loss (Kaur, 2016). The IoT technology, SG is an instance of cyber-physical systems. The development of most parts of SG can be enhanced by applying IoT. Through IoT, the whole power grid chain, from electricity generation to consumption, will enable intelligence and two-way communication capabilities to monitor and control the power grid anywhere and anytime (Bekara, 2016). The pervasive deployment of smart metering in IoT-based SG will generate energy big data in terms of its huge volume, large scale, and structural variety. The three categories of Grids business data is described as: (i) Grid operation and equipment testing or monitoring data, such as supervisory control and data acquisition (SCADA) data and sampling data of smart meters, (ii) Electric power marketing data, such as transaction price and electricity sales data, and (iii)Electric power management data, such as internal grid data (Cinar, 2018). There are many potential advantages to be derived from energy data for the goal of optimal operation, including real-time monitoring of energy consumption data generated by advanced metering infrastructure (AMI) and smart meters, detection of energy losses by fault or fraud, early blackout warning, fast detection of disturbances in energy supply, and intelligent energy generation, planning, and pricing. Huge data generated at the second level and concurrent peak demands from different homes may cause blackouts at some substations due to power imbalance introduced by inaccurate energy forecast. Energy big data are also very useful for realizing situational awareness. Based on long-term monitoring, security- related information can also be characterized (Bekara, 2016). The grid representation is presented in figure 11(a). There is increasing public awareness about the changing paradigm of our policy in energy supply, consumption and infrastructure. For several reasons future energy supply should no longer be based on fossil resources, neither nuclear energy a future proof option. Inconsequence future energy supply needs to be based largely on various renewable resources. Increasingly focus must be directed to our energy consumption behaviour. Because of its volatile nature such supply demands an intelligent and flexible electrical grid which is able to react to power fluctuations by controlling electrical energy sources (generation, storage) and sinks (load, storage) and by suitable reconfiguration. Such functions will be based on networked intelligent devices (appliances, micro-generation equipment, infrastructure, consumer products) and grid infrastructure elements, largely based on IoT concepts. The IoT smart grid connectivity for house hold appliances is displayed in figure 11(b). A smart grid is an energy delivery system that moves from a centrally controlled system, like we have today, to a more consumer driven, iterative system relying on bi-directional communication to constantly adapt and tune the delivery of energy.

Figure 11. (a) A Smart Grid Representation (b) Smart grid connectivity enabling smart home services

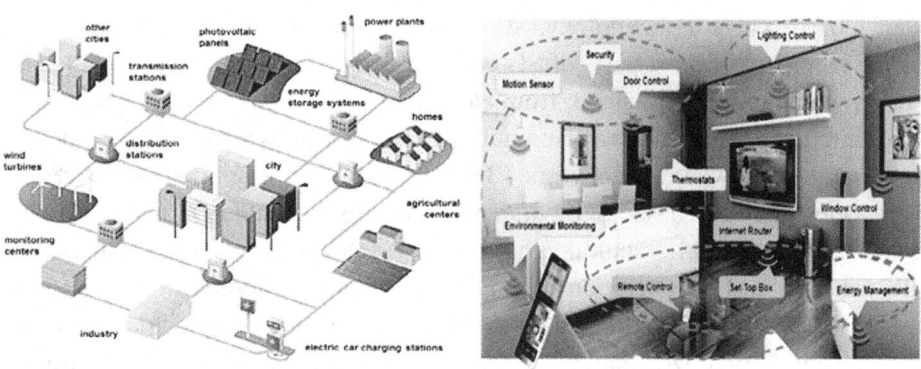

The five fundamental technologies of Smart Grid Technologie's (SGT) as propounded by the US Department of Energy is detailed in (Deshpande, 2016) which includes: integrated communications; Sensing & Measurement Technologies; Advanced Components; advanced Control Methods and Improved Interfaces & Decision Support. The cyber-physical system IoT-based smart grid system deals with issues related to impersonation/identity spoofing, eavesdropping, data tampering, authorization and control access, privacy, compromising and malicious code, availability and dos issues and cyber-attack. The few of major applications of IoT based smart grids system are briefed in (Xu, 2014) table 2.

Table 2. Applications of Smart Grid

Future Apps and Services	Real Time Energy Markets
Business and Customer Care	Application Data Flow to/from End-User Energy Management Systems
Smart Charging of PHEVs and V2G	Application Data Flow for PHEVs
Distributed Generation & Store	Monitoring of Distributed Assets
Grid Optimization	Self-healing Grid: Fault Protection, Outage Management, Remote Switching, Minimal Congestion, Dynamic Control of Voltage, Weather Data Integration, Centralized Capacitor Bank Control. Distribution and Substation Automation, Asset Protection, Advanced Sensing, Automated Feeder Reconfiguration.
Demand Response	Advanced Demand Maintenance and Demand Response; Load Forecasting and Shifting.
AMI	Provides Remote Meter Reading, Theft Detection, Customer Prepay, Mobile Workforce Management.

INDUSTRIAL IoT

Internet of Things (IoT) in industries has created a new revolution in industries. IoT in industry has given rise to the term "INDUSTRY 4.0" where systems are connected to each other over the internet and can communicate with each other to take necessary decisions (also called as M2M communication) through artificial intelligence. Controlling applications over Internet is one of the best way to deal with the industrial applications (Xu, 2014). Industrial Internet of Things (IoT) provides best and suitable ways of connecting industrial machineries and sensors, to each other, over the internet, allowing the authorized user of the industry to use information from these connected devices to process the obtained data in a useful way. IoT-connected applications typically support data acquisition, aggregation, analysis, and visualization. The IoT architecture includes latest technologies such as computers, intelligent devices, wired and wireless communication and cloud computing. Illustration of IoT based Industrial Automation is included in (Breivold, 2015). The IoT allows objects to be sensed and controlled remotely across existing network infrastructure. This is implemented as in figure 12. Different sensors such as temperature sensor, Pressure sensor, Humidity sensor, Vibration sensor, Intrusion sensor are used to percept the environment and object conditions in industrial application. Admin set threshold to every sensor placed in Industry and Android check this threshold against incoming analog signal. When it encounters an uneven condition, then d devices (Buzzer, Alarm, motor, fan) are used to take accurate measures such as Alarm/Alert and notified to Admin. Artificial Intelligence is adopted to take adequate steps to solve these problems which makes use of past experience and similar previous condition stored in database as training data. In this we use cloud as database for scalability. Design considerations for industrial IoT applications are energy, latency, throughput, scalability, topology and security and safety as mentioned in (Mahamune, 2017).

Figure 12. (a) Block diagram of IoT based Industrial Automation (b) IoT based Industrial Automation

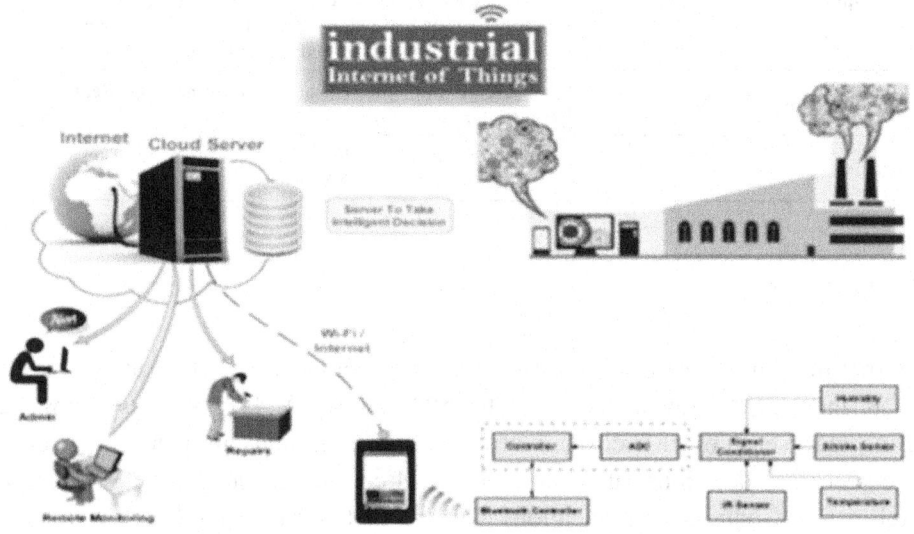

Transportation

The fundamental concept of the smart connected car is connectivity, and such connectivity can be provided by three aspects, such as Vehicle-to-Vehicle (V2V), Vehicle-to-Infrastructure (V2I),and Vehicle-to-Everything (V2X). To meet the aspects of V2V and V2I connectivity, we developed modules in accordance with international standards with respect to On-Board Diagnostics II (OBDII) and 4G Long Term Evolution (4G-LTE) to obtain and transmit vehicle information (Usha, 2016). Figure 13(a) shows Vehicles-To-X Connectivity.

Figure 13. (a) Overview of vehicles-to-x (b) Vehicle to Vehicle communicate (c) Vehicle to its Owner Communication (d) Vehicles to Centralized Server Communication

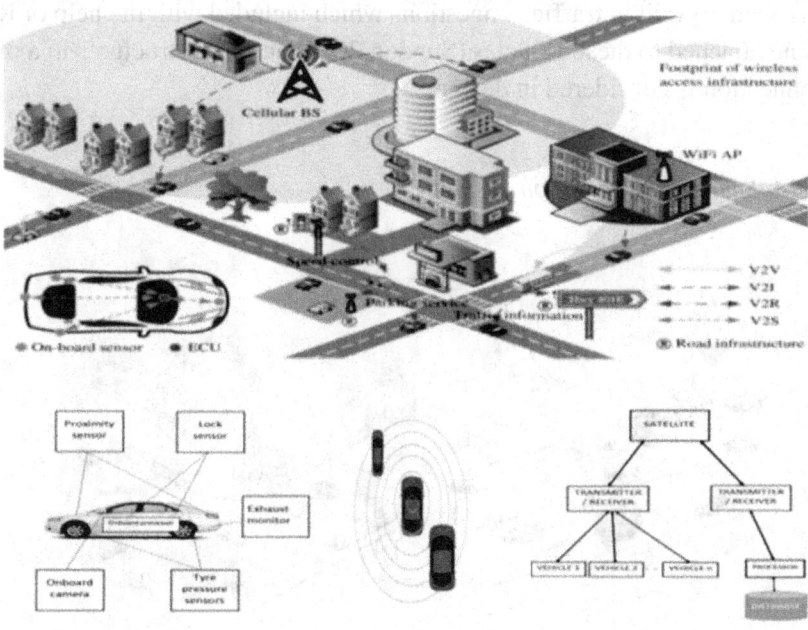

The prospects of Internet of Vehicles (IoV) to be a reality and must be able to communicate continuously. The possible vehicle communication are; between the vehicles and the vehicle owners (figure 13(b), between vehicles (figure 13(c), between vehicles and a centralized server and between server and third parties like police patrol, ambulance, fire-engine, etc (figure 13(d)). The metrics of the automobile that need to be monitored are; tyre pressure, fuel level, speed, velocity reading, exhaust gases content, vehicle lock and others (Abideen, 2017). The IoT has transformed industry by altering transportation systems to gather data and information by bringing together the major technical and business trends of mobility, automation and data analysis. It refers to use of embedded sensors, actuators, and other devices that can collect and transmit information about real-time activity in the network. The data gathered from these devices can then be analyzed by transportation authorities to improve the traveler experience, increase safety, reduce congestion and energy use, and improve operational performance. The clustering strategy is applied to generate vehicular networks with adaptive weighted clustering protocol (Mohamed, 2015). Numerous regulator constraints are included and considered data delivery rate for adaptive weighted clustering. The results, spacing, and opposite relating to distance were considered in design, but only problem is constant clustering architecture is not maintained. The emergency vehicles are transported with high priority, without allowing them to wait in traffic congestions which included with the help of RFID tags being attached to these vehicles (Sundar, 2015) and infrastructures in a multi-road connection is considered in design.

Figure 14. Smart transportation System

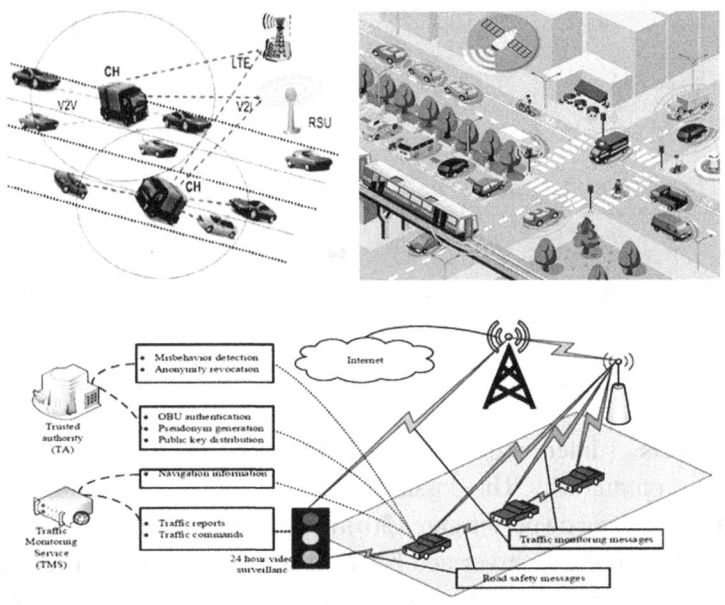

CONNECTED HEALTH

Health issues are rising progressively due to the ignorance of health monitoring on regular basis which can now be looked after by the fast growing communication technologies and smart devices. According to World Health Organization (WHO), every year 4.9 million people die from lung cancer due to the consumption of snuff, overweight causes 2.6 million deaths, 4.4 million deaths are caused by elevated cholesterol and 7.1 million people die from high blood pressure. It is said that in the next 10 years, deaths from chronic diseases will increase by 17%, which means in figures of about 64 million people. These chronic diseases are highly variable in their symptoms, evaluation and treatments which demand continuous monitoring otherwise can cause end of patient's life. A robust healthcare model is developed for continuous monitoring of the patient even when the patient is traveling. Sensitive data is collected from the Internet of Things (IoT) sensors connected to the patient's body and sent to the server through the smart phone of the patient (Selvi, 2018). New micro ovens like GE Brillion Micro oven, MAID micro oven already are connected to internet and can share data to smart phone. Also new IoT devices are coming up in connected kitchen space. New products like smart food scale allows user to place a food item on the scale and the smart sensors sending a full nutritional profile to your smart phone within seconds. The scale currently recognizes over 300,000 food items and 80,000 restaurant dishes. This Wi-Fi and Bluetooth compatible Food Scale is great for nutrition and dieting (Figure 28) (Selvi, 2018). Due to the advancement of modern technology, diagnosis of health issues is not that much difficult in this era as it was some decades ago. Now there is a wide variety of monitoring systems available in the market such as heart rate monitor, Glucometer, blood pressure cuff etc. which allow the patients to monitor their condition daily which communicate over some network without the collaboration of human beings (Selvi, 2018).

Figure 15. (a) Generic-Healthcare model (b) Architecture of IoT in Health Care

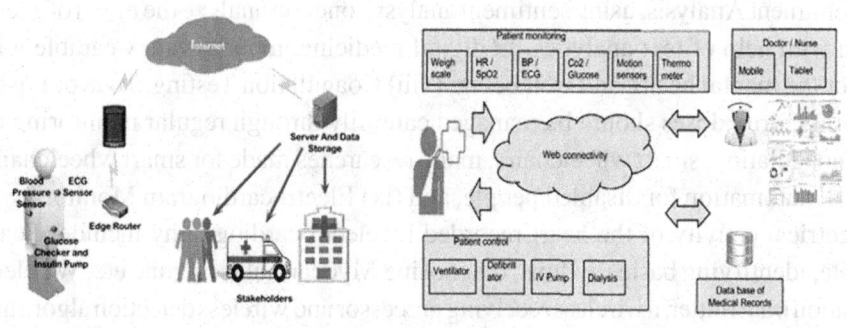

In Figure 15(a) a generic-healthcare model is visualized. However, in case of failure of paired smart phone with the sensor/smart device, it would not be able to send data, which can lead to disaster. On the other hand in case of robust-healthcare model, it will try to find next available optimal smart phone in vicinity. Optimal in the sense that device has active Internet connection and maximum battery standby (Yin, 2016). The architecture of IoT health care is shown in figure 15(b). (Hassan, 2015). The conversion of sensor data from physical world data to the digital world done for receiving signals from the environment for analysis or actuators for controlling based on inputs or both sensors and actuators. These devices connect with each other through Internet transfer and cloud storage for communication with similar devices and people, as shown in Figure 16. There are multiple studies from various research companies for the projected figures of these IoT devices ranging from 26 – 212 billion in 2020. The applications of IoT in health care are presented in(Sakr, 2016). They includes (i) open artificial pancreas system, open source software patient can continuously monitor the glucose level through connected device (ii) Hand Hygiene Compliance, inexpensive prevention measure to avoid increasing number of deaths, extra days of hospital stay and costs, (iii) Tracking during Cancer Treatment, trackers helps to measure the activity level, logging appetite level and all data will be saved to patients smart phones through specific app, (iv) Connected Inhalers, allows doctors to keep accurate track of whether patients are strictly adhering the treatment or not, (v) Real Time Location Services(RTL), the ability to track the specified location of people within a certain area through wireless network, (vi) Connected Contact Lenses, of communication called backscatter which allows devices to exchange information, (vii) Sentiment Analysis, using sentiment analysis, one can analyze the mood of every human with help of text analytics. In digital medicine, mood-aware wearable will monitor the mental health and well being, (viii) Coagulation Testing, To avoid risks of clotting, drug doses should be managed carefully through regular monitoring of blood coagulation, smart wheelchairs, many researches made for smart wheelchairs with full automation for disabled people, and (ix) Electrocardiogram Monitoring is the electrical activity of the heart recorded by electrocardiography includes heart beat rate, identifying basic rhythms, diagnosing Myocardial Ischemia, etc. Wireless acquisition transmitter, a wireless receiving processor and wireless detection algorithm can be used for monitoring ECG signals. The security requirements and challenges for IoT-based health care devices are summarized in (Moosavi, 2015). The security requirements for IoT based healthcare are; Confidentiality, Integrity, Authentication, Availability, Fault tolerance, and Data freshness. Challenges for secure IoT healthcare services covered are power consumption, Storage limitations, Mobility, Scalability, Communication media, Dynamic network, and Dynamic security updates.

Figure 16. (a) Connected Inhalers (b) Temperature Monitoring Sensors (c) Coagulation Testing (d) Automated Wheel-Chairs

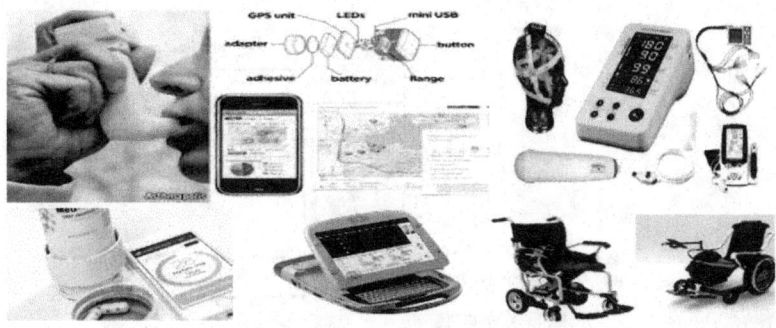

SMART RETAIL

Globally the retail environment is radically changing, due to the growth of the Internet. Retailers are developing new ways to interact with customers and customer channels. Literature indicates that retailers are often more adaptors of innovation, rather than innovators themselves. They are focused on creating new products, rather than innovating their services. The nature of service-based innovation is a rather under-searched area (Bajaj). It allows retailers to gather not only information about existing process other than existing tools in predicting and analyzing consumer market. The new data allows retailers to analyze needs of person in micro-segmentation. It enables real time analytics to create new level of flexibility and better chances to adequately adapt to changing market. The security and privacy aspects issues are included (Bajaj). The system architecture of smart retail is shown in figure 17(a). the architecture is for mobile based in-store mapping services. The Indoor Positioning System functions to "automatically" locate dynamic store items and send updated location information to the server for indexing. The Information Retrieval System maintains an inverted index and a database for store items containing catalog as well as location information. It functions as a server to receive query requests from clients (mobile devices) over the Internet, process them, and send ranked query results back to the clients. Customers interact with their smart mobile devices. With a mobile app (e.g., SmartMart) installed, such mobile devices function as clients, sending queries to the server and receiving query results through a friendly keyword search interface. The flow of smart retail operations are summarized in figure 17(b). The key actors are customers and retailers, customer is the mainly focused user in this system. The application is designed targeting the customer. The customer performs a list of item,

Figure 17. (a)System Architecture of Smart Retail as a Mobile Application (b) Flowchart for Smart Retail

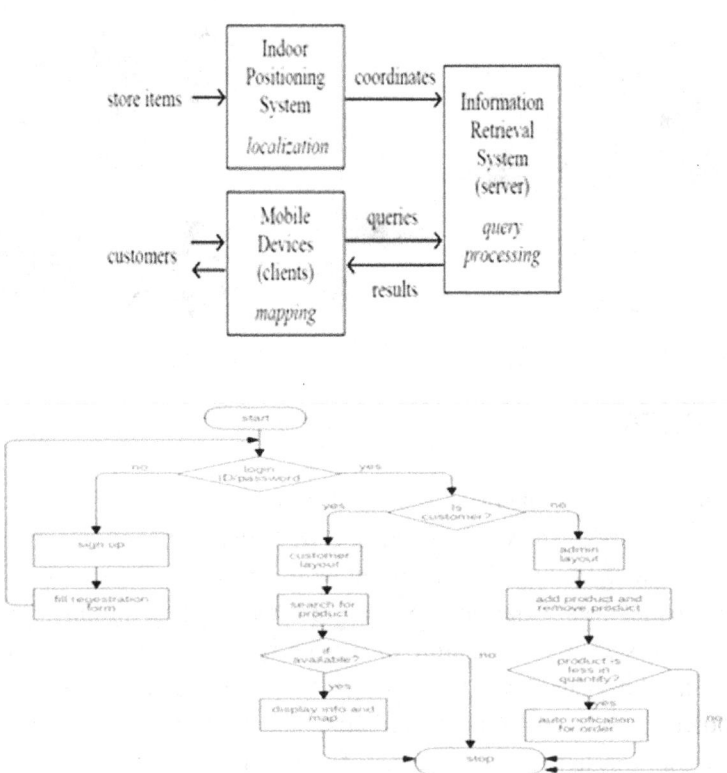

view the list, search the products in store and get the details of the product. The retailer shop owner is to access and manipulate administrator.

The applications of IoT based smart retail are listed in (Yuvaraj, 2016) which covers individual actions in IoT based retails such as mobile Phones, retailers are navigating customers inside their stores and at the same time collating information on customer buying behavior, Wearable Devices, dedicated efforts towards making this a reality by which customers will be able to make bill payments using these devices and that should replace the usage of cards and mobile applications. Beacons, that communicate with the smart devices such as a mobile phone through apps using the Bluetooth technology, Smart Mirrors, retailers selling apparels find this best to allure customers who prefer to try-on clothing before making a purchase, RFID Devices, Radio-frequency identification technology can significantly enhance key retail elements such as supply chain, inventory, logistics and fleet management. RFID scanners are already use for conducting faster audits by retailers, replenishing items on shelfs, etc. Even though more research taken in area of smart retail challenges still exist, they are (Machado, 2016), Tag Tamper-Proofing (Tag Security): The tag design must be resistant to the following misuses (a) Rewrites in order to pay less, (b) Obstructions and replacement by fake tags, (c)Swapping the tags of different items, (d) Breaking or tampering to avoid paying the price altogether, (e)Reading Collision: Intuitively, the reading range of the RFID reader should be carefully set to avoid collisions with other carts and (f) Communication Security: The communication in the smart shopping system needs to be protected.

SMART SUPPLY CHAIN

Supply Chain Management (SCM) is the central theme in today's global industries. SCM is the management of flow of goods and services. SCM includes movement and storage of raw materials and goods from one place to another place including design, planning, execution, control and monitoring of goods (Kumar, 2018). SCM methods are heavily followed from the areas such as operations management, logistics and information technology. SCM is the approach that manages the movement of raw materials into an organization and it is deployed in both indoor and outdoor environments and the information updates about the goods are uploaded in the server with the help of IoT. IoT refers to the wireless communication between the objects and it can be controlled and monitored from anywhere, any place and at any time. SCM keeps the record of the movement of goods from the supplier to the manufacturer which moves along with the wholesaler to the retailer and finally to the customer. It is used for the goods to be delivered to the customer without any damage and at specified time so that it satisfies the customer and also helps to increase the output efficiency for the organization. It also provides the detailed flow of information from the supplier to customer and also provides the time of arrival of goods. Ending Inventory is equal to the Beginning Inventory and Receipts of the goods (in) - Shipments (out). SCM provides the current information about the location, status and condition of your assets or goods. For indoor tracking of goods Radio Frequency Identification (RFID) is used. For outdoor tracking, Global positioning system (GPS) is used. Thus, from the process of SCM, Shipments of the Distribution Center equal to receipts of the Retail Location (Prathibha, 2017). The methodology for agricultural products in SCM provided in (Toma, 2018) covers process consists of maintaining the supply chain management website by taking input. Manager can have access to all the information about the products and the farmers can have access to their selling information of their product. For the purpose of tracking, RFID tags and readers are used, their entries are stored in the database as it will be needed further Figure 18 **(a-c)** The impact of supply chain management is more on inventory management which brings real-time visibility of the inventory, Real-Time SCM, **opti**mizes processes and collaboration with other companies in the supply chain (suppliers and customers) in order to create more value, and increases logistics transparency to embed load carried by a logistic operator with smart objects, which can make information about transport (destination, identification, transport conditions etc.) available to the entire supply chain, making the chain more transparent.

Figure 18. (a) Overall process of SCM (b) Overall Activity diagram of SCM for Agricultural products (c) Flow Goods in Supply Chain Management

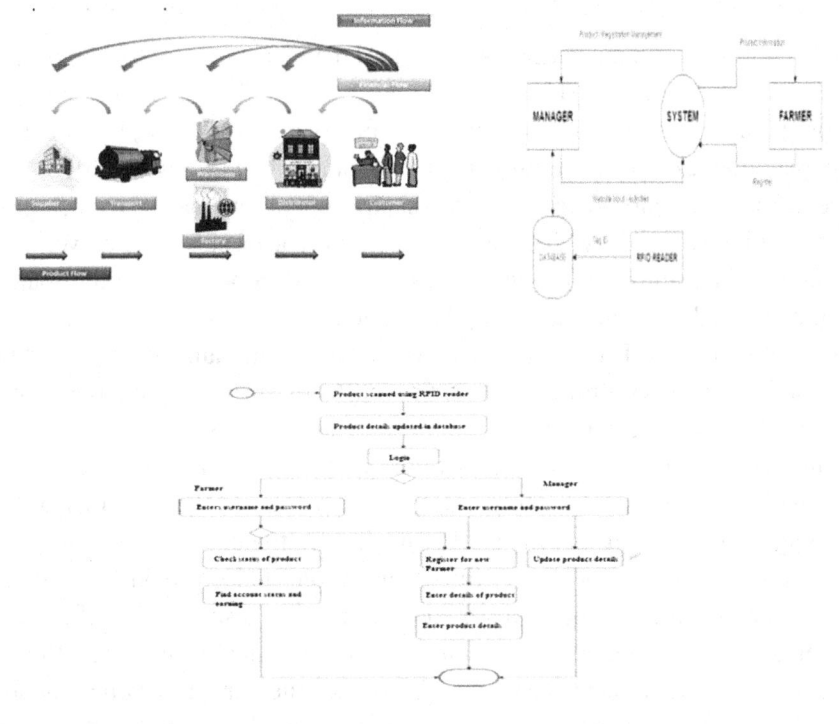

SMART FARMING

Climate changes and rainfall has been erratic over the past decade. Due to this in recent era, climate-smart methods called as smart agriculture is adopted by farmers. Smart agriculture is an automated and directed information technology implemented with the IOT (Internet of Things). Sensor technology and wireless networks integration of IOT technology has been studied and reviewed based on the actual situation of agricultural system. A combined approach with internet and wireless communications, Remote Monitoring System (RMS) is designed in (Ojha, 2016). By using IoT, we can expect the increase in production with low cost by monitoring the efficiency of the soil, temperature and humidity monitoring, rain fall monitoring, fertilizers efficiency, monitoring storage capacity of water tanks and also theft detection in agriculture areas. The combination of traditional methods with latest technologies as IoTs and Wireless Sensor Networks can lead to agricultural modernization. Moreover, smart agriculture will help in monitoring livestock productivity and health as well saving time, cost and manpower. IoT sensors are capable of providing farmers with information about crop yields, rainfall, pest infestation, and soil nutrition are invaluable to production and offer precise data which can be used to improve farming techniques over time. Internet of things, with its real-time, accurate and shared characteristics, will bring great changes to the agricultural supply chain and provide a critical technology for establishing a smooth flow of agricultural logistics. The role of IoT in agriculture (figure 19(a)) is of great help in enhancing the production and yield in the agriculture sector since these devices can be used to monitor soil acidity level, temperature, and other variables.

The IoT based smart agriculture architecture is shown in figure 19(b). The farm automation and monitoring typical architecture is showed in figure 19(c) & (d) It includes three modules; farm, server and client side. Farm side module includes operations related to crop and decision making, server side makes information storage and client side model includes communication to server and crop.

Smart Environment

The increasingly degrading quality of air in numerous regions of the world can be an unswerving repercussion of human hassles of increasing globalization and urbanization. The myriad consequences of the same can be regulated by keeping a periodic check on the quality of air and undertaking efforts to improve if it deteriorates beyond a certain extent. The foremost step here is the analysis and detection of the harmful gases present in the atmosphere yielding to poor quality of air which is a task that can be easily accomplished by the latest evolved high technology of 'IoT-Internet of Things'. Following models pose to be some of the

Figure 19. (a) Role of IoT in Agriculture (b) Proposed Smart Agriculture IoT model (c) Agricultural automation (d) Field monitoring

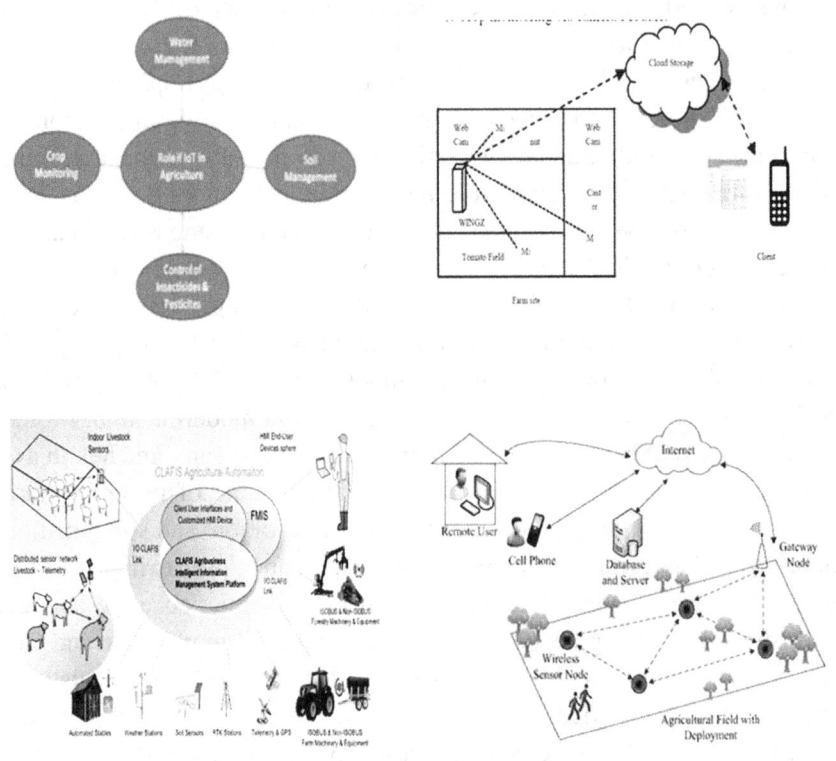

good solutions for the aforementioned problem. (Saha, 2017) extends its scope to detect intensity of UV rays and also takes in consideration the temperature and humidity of the area where it is deployed in. Extending the scope of (Saha, 2017) model proposed in (Joshi, 2017) progresses towards efficient and portable data extraction and accessibility. (Pal, 2017) proposed a year later, in 2018 describes an efficient deployment system to curb the harmful emissions from individual vehicles by putting restraint and generating an alert if any individual emission goes beyond tolerance value. Another model [5] proposed in same year, extends all models and instantiates a new approach, bringing in the factor of machine learning to predict and forecast the pollution after having monitored it for a certain duration. Similarly, water pollution is also of major concern. The rise in the value of water has been excruciating lately yet the clean amount of it is getting scarcer by the minute. This vital part of the existence of many life forms has been neglected for many decades over numerous hassles of human beings for unchecked development of mankind.

The water bodies are getting depraved with the harmful contaminants. To keep a control on the same, the extensively used technology of Internet of Things offers a good range of solutions. Some of which are surveyed herewith. The model portrayed in (Arys, 2017) describes an innovative way of reusing the wastewater released as effluents by using IoT to judge its potential and based upon the same it is converted to a renewable energy source. Model in (Myint, 2017) strives to keep a check on the water quality with efficient power consumption mechanism. The model in [8] proposes a solution to monitor the dumping lands by investigating the quality of water and certain other environment parameters to calculate the threat factor for surrounding areas. Model proposed in (Ramesh, 2017) moves towards the large water bodies monitoring by cumulative measurement of each region. Soil is the kitchen of mother Earth. No mouth will be able to feed, by the extension of food chain, if not for the healthy soil to reap a fine and abundant harvest. The class of this soil has been degrading overall while its value ever increasing, henceforth arising the need of soil pollution control mechanism. IoT is an excellent way to reduce the aforesaid pollution by monitoring soil environment and alerting the users in order to take necessary actions. Numerous ways have been discovered in order to maintain the soil quality by monitoring its parameters. (Brindha, 2016) describes a model which monitors the soil nutrient level at fixed intervals and aids the soil with nutrients from a nutrient tank on obtaining a lower value. (Vani,2017) presents an approach to keep an eye on soil moisture for cropping lands and efficiently updating the user on same and in (Patil, 2017) system is designed to measure the agricultural parameters and smartly deployed to analyse and act upon the readings obtained.

In the last few years, sensors have been widely used for fire detection. A work for fire detection in mines by using wireless sensor networks called WMSS used. For determining the hazardous factor in the mines, they used gas sensors and designed a wireless sensor network which collects and analyses the gas level in mines (Silva, 2015). The work with Zigbee-based wireless sensors for fire detection in forests applied in (Chiwewe, 2015) with temperature sensors to establish the intensity of fire in a forest using hardware devices, a similar model also designed a framework for forest fire detection(Buratti, 2009; Arrue, 2000)using clustering scheme and communication protocols with simulation work. The use of multi-sensor and wireless IP cameras to avoid false alarms applied in (Tan, 2007). Their system also connected to the internet via gateways for uploading the data to the cloud. A work proposed in (Zhang, 2008; Reheem, 2017) for forest fire detection was based on a ZigBee wireless sensor network designed in China. South Koreans (Son, 2006) also designed a system for fire detection in their mountains. They named their system FFSS (Forest fire Surveillance system). They developed their system by using WSN, middleware, and web applications. Network nodes (i.e., temperature sensors and humidity sensors) collect measurements and send them to the sink node. Afterwards, the sink node transmits that data to the cloud via a transceiver (gateway). Later, by using a formula, the fire risk level is determined in the middleware program. After detecting the fire, FFSS is activated automatically. TinyOS is used as an operating system for network nodes. Similarly, few other systems utilize the WSN for early fire detection. Few of them use IP-based cameras and mixed multi-sensors (Lloret,2009) on wireless mesh network to detect a fire efficiently. They use these three parameters to make an efficient system to identify and verify fire. A software application exists, which selects the closest IP base camera. When sensors sense a fire, they send the information to the central server where the software program selects the nearest camera. That camera takes pictures of the location and sends them back to the main sink. Alarm decision is made on the basis of the sensor's information and selected images. Also, a clustering-based forest fire detection system was proposed in (Aslan, 2012). The work consists of four major parts: (i) an approach for sensor deployment, (ii) the use of WSN for fire detection, (iii) intra-clustering protocols, and (iv) an inter-clustering communication protocol, similar work was proposed by (Wenning,) for disaster detection, including the fire event. This method can quickly adapt the routing work state on the basis of the threat of possible failure. In (Musolesi, 2005) proposed a protocol for the adaptation of the Context-Aware Routing Protocol (CAR) to WSN and named it SCAR. Energy, colocation with sink, and their connectivity was evaluated in SCAR. For the delivery of data packets to the sink, delivery probability and forecasted values were combined on the basis of previous knowledge of SCAR parameters. In this system, buffer space and delivery probability are exchanged periodically with neighbor nodes. An order list of neighbors is sorted by delivery probability, which is kept by each node. A work to create a model for fire detection. They performed a simulation by analyzing sensors data and geographic information. To differentiate from the existing work, they used topography of the environment under study (Garcia, 2008). The IoT based intelligent modeling of smart home is shown in figure 20.

Education

IoT is one of paradigms that contributes to such initiatives have transformed the learning fashion from traditional-based learning into digital-based learning(Dorobantu, 2016; ECDL,2017). A study to review the IoT smart campus model and applications concerning education is summarized in (Zhamanov, 2017; Dominguez, 2017; Ochoa, 2011). This study is an attempt to review the IoT-based education studies and provide an overview of the key challenges and open issues for the employment of IoT in education. An analysis for IoT conducted of IoT in education with merits and demerits (Pei, 2013). A detailed study with relationships between IoT, cloud computing and the emerging learning presented in (Sarıta, 2015). A learning kit to

Figure 20. IoT-based intelligent modeling of smart home for fire prevention

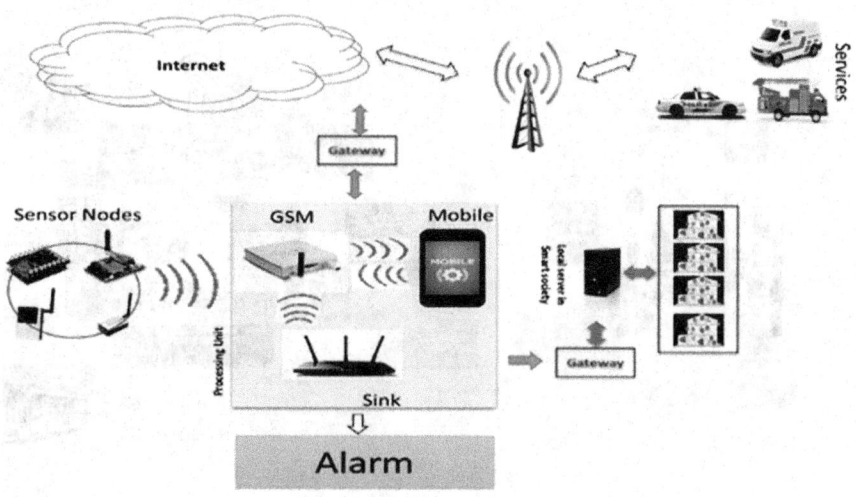

teach the basic concepts of IoT technology designed and implemented (Kamal, 2018). A model for making a university campus smart through the adoption of IoT technology presented in (Majeed, 2018), a 3D scheme IoT is extended to education by (Kassab, 2018). The adoption of IoT in vocational training yields qualified students, a safer educational environment, self-directed learning, and economical use of educational resources and enhanced learning (Vihervaara). Green IT as environmentally sound IT involves studying the design, manufacturing, use, and disposing of computers, servers, and associated subsystems such as monitors, printers, storage devices, and networking and communications systems efficiently and effectively with minimal or no impact on the environment (Murugesan,2008;Miao, 2018) through the use of screensavers, eco-friendly design, enabling power management, turning off systems when not in use, green data centers, energy conservation, and virtualization. The smart IoT in education is shown in figure 21.

Military

Soldiers equipped with a personal sensor system for health and equipment monitoring. It consists of three modules: tactical underwear, tactical vest and rugged Android device. The tactical underwear (Figure 22) is equipped with medical sensors: two pulseoximeters (one for each wrist), muscle activity sensors and heart rate sensor located on the chest. Sensory data is collected by the Arduino compatible micro board with Bluetooth Low Energy (BLE) module (Dyk, 2017).

Figure 21. Education IoT (https://www.pinterest.com/pin/420453315183452859/)

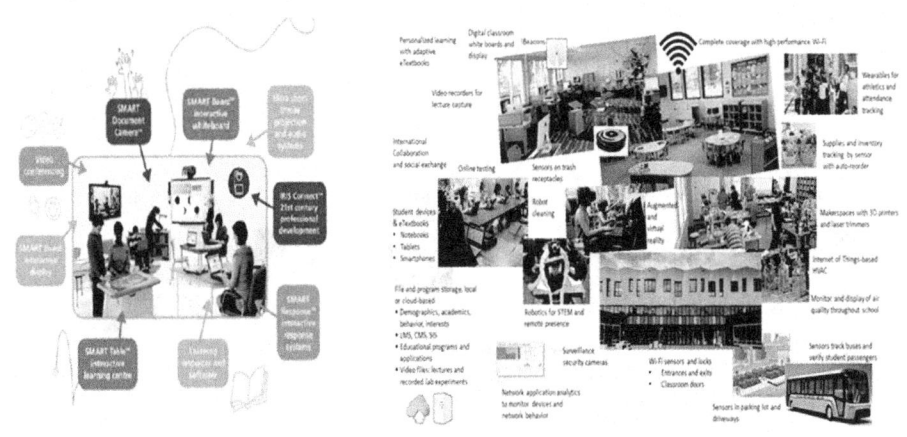

Figure 22. Smart IoT for military

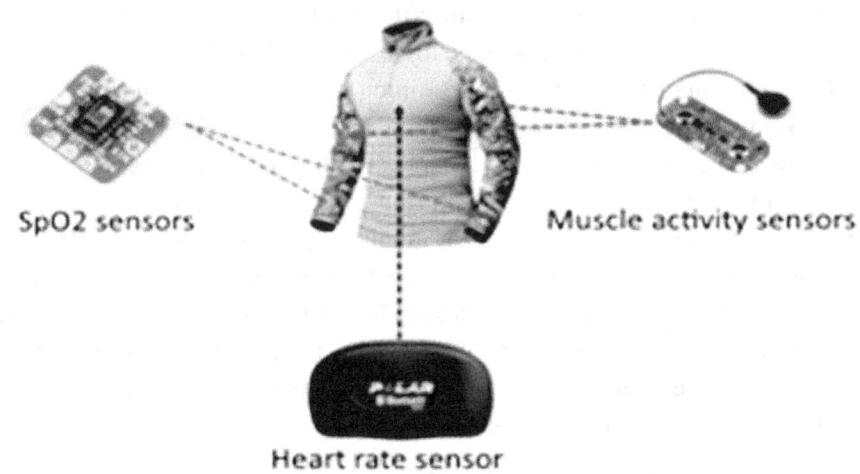

IOT RELATED FUTURE TECHNOLOGIES

Few of the future technologies related to IoT are briefed as follows:

- **Open, wireless communication networks**: Wireless communication in application like sensor networks is required. Data transfer protocols such as Bluetooth (low-power, short-range), Wi-Fi, Zigbee (medium –range) and GSM (medium-power, wide-area coverage) are becoming pervasive and low-cost. Security is a concern but is designed into the protocols and with multi-layer authentication at the edge device. Open standards are important to ensure multi-vendor interoperability – a term describing products from different manufacturers working together. Most open protocols will be based on IP (internet protocol), however some local networks, especially if ultra-low power is a requirement, as it is in water, will not use IP because it is not energy-efficient.
- **Cloud computing**: Cloud computing has drastically reduced the cost of data processing and storage due to the low cost of ownership and the lack of a need for server infrastructure or the additional cost of a data centre. A new term 'FOG computing' describes a more local distributed cloud resource which brings the processing to the data rather than pushing the data to the cloud. As the edge devices become smarter, the data that needs to be transmitted as messages will be less frequent and higher value.
- **IOT and Semantic technologies:** Future research on IoT is likely to hold the concept of Linked Open Data. This could build on the earlier integration of ontologies (e.g., sensor ontologies) into IoT infrastructures and applications. Semantic technologies will also have a key role in enabling sharing and re-use of virtual objects as a service through the cloud. The semantic enrichment of virtual object descriptions will realise for IoT. Associated semantic-based reasoning will assist IoT users to more independently find the relevant proven virtual objects to improve the performance or the effectiveness.
- **Big data analytics**: Big data analytics and machine learning, a vast array of techniques for extracting meaning and understanding from data, data-mining, self-learning algorithms and model-based reasoning, are a few of the mechanisms used for building automated decision support systems (DSS). When coupled with advanced data visualisation, actionable insight can be gained from real-time data coupled with new and legacy mathematical models. Integration with GIS and asset management and other enterprise business applications, such as energy optimisation, will also be critical business capabilities.

- **Mobile computing**: Smart phones and tablets deliver actionable insight to human users on the operational front-line. Mobile browsers and apps also provide a rich platform for user interaction and data collection, and become data processing Things in their own right.

RESEARCH DIRECTIONS FOR SELF MANAGING IOT SYSTEM

The following are the few research directions to progress towards self-manageable IoT systems:[8]

- Security and privacy issues: Large chunks of personal data relating to health, home, cars etc. are shared across varied devices and uploaded in the Cloud. The mismanagement and leakage of this data could make the security and privacy of its user vulnerable as well as lead to various cybercrimes.
- Excess power consumption: Since billions of devices are connected in IoT which are battery-operated and IoT demands huge data transfers between devices, it leads to an abundant consumption of power and energy which should be optimized for further expansion.
- Complexity: Any improper functioning or a bug in the IoT related software or hardware will create a lot of fuzz due to which its implementation, working and proper maintenance is quite toilsome
- Internet Bandwidth: IoT is nothing without a good internet bandwidth since, for proper functioning a SMART device would require fast internet connectivity and lack of bandwidth would hinder the working of the device and affect the Quality of Service
- Lack of flexibility and interoperability: Interoperability is characterized as the capacity of frameworks, applications, and services to work in a combined and predictable way but sometimes due to huge collection of data and sometimes due to devices are not able to interconnect with one another, it leads to inflexibility and flaws in the system.
- Work to done to develop:
- efficient and simple mechanisms for interaction with "things" Reliable and trustworthy participatory sensing
- foster Standardization for smart energy, grid and mobility
- Seamless integration of social and sensor networks
- Infrastructures for social interactions between Internet-connected objects
- Utility metrics and utility driven techniques for "Clouds of Things"
- Self-management systems should be designed with particular attention in contexts where the safety of the user can be impacted

- Novel methodologies, architectures, algorithms, technologies, and protocols should be developed taking into account IoT specific characteristics such as resource constraints, dynamic, un-predictive, error prone and lossy environments, distributed and real-time data handling and decision-making requirements, etc.
- Prototypes at early stages in order to validate the theoretical results by measuring the overhead that autonomy can bring to IoT systems.
- Scalability and Interoperability: IoT is expected to be composed of heterogeneous networks, thus standard interfaces should be defined for interoperability. In addition addressing scalability for a large scale IoT deployment is additional key issue.

Scope of Career in the IoT

Multiple surveys (including Cisco) in IoT have predicted there will between 25 billion to 50 billion connected devices by 2020. This significant number and with most of the countries vision in investing billions of dollars in IoT (in various domains including health, smart city, industry, agriculture, military etc.), itself indicates that there is umpteen number of job opportunity for IoT professions in near future. Construction of IoT requires a team of engineers from diverse branches which includes electronics, mechanical, computer science and many more. Since IoT is facing privacy issues as one the main challenge there is dire need of manpower who are well versed with vulnerability evaluation, authentication, hacking etc. People with knowledge of technologies for building user friendly interfaces between the device and the humans are in great demand. There is even more demand for android app developers since apps are the one means of communication between external hardware and sensors. In addition there is need of embedded engineers and Networking professionals knowing how to program interfaces with hardware (e.g. GPIO, I2C), aware of operating systems supporting IoTs, connectivity protocols, up-to-date standards in IoT communication etc. furthermore software engineers well versed in big data technologies (MongoDb,Cassandra etc) will be most sought after. Huge volume of data is garnered by sensors which demands for data analyst to make sense of data collected. Developers having awareness of Artificial intelligence (Deep learning, Tensorflow etc) will be in demand to develop apps to handle data to take autonomous decisions. People having skills in developing new breed of sensors also have bright future in IoT.

CONCLUSION

In a nutshell IoT helps in connecting all possible objects to communicate with each other via Internet with the objective of making life simpler for human. There is need for ensuring this communication among the objects in a secured way. The chapter highlights on the various applications which includes agriculture, supply chain management, health, education, smart home, environment etc., supported by IoT which ensure optimal cost and time reduction. A brief information about the various operating systems for IoT, architecture of IoT, features of IoT, carrier options in IoT, scope of future research in the field of IoT is also covered.

REFERENCES

Abideen. (2017). An IOT based robust healthcare model for continuous health monitoring. In *23rd International conference*. IEEE. doi:10.23919/IConAC.2017.8082012

Ahmed, E., Yaqoob, I., & Gani, A. (2016). Internet-of-thingsbased smart environments: state of the art, taxonomy, and open research challenges. *IEEE Wireless Commun, 23*(5), 10–16.

Al-Fuqaha, A., Guizani, M., Mohammadi, M., Aledhari, M., & Ayyash, M. (2015). Internet ofthings: A survey on enabling technologies, protocols, and applications. *IEEE Communications Surveys & Tutorials, 17*(4), 2347–2376, https://iot-analytics.com

Alam, M. M., Malik, H., Khan, M. I., Pardy, T., Kuusik, A., & Le Moullec, Y. (2018). A survey on the roles of communication technologies in IoT-based personalized healthcare applications. *IEEE Access, 6*, 36611–36631.

Amendola, S., Lodato, R., Manzari, S., Occhiuzzi, C., & Marrocco, G. (2014). RFID technology for IoT-based personal healthcare in smart spaces. *IEEE Internet of Things Journal, 1*(2), 144–152.

Antipolis. (2014). *New ETSI specification for Internet of Things and Machine to Machine Low Throughput Networks*. https://www.etsi.org/news-events/news/827-2014-09-news-etsi-new-specification-for-internet-of-things-and-machine-to-machine-low-throughput-networks

Aravinthan, Namboodiri, Sunku, & Jewell. (2011). Wireless AMI application and security for controlled home area networks. *2011 IEEE Power and Energy Society General Meeting*, 1–8.

Arrue, B.C., Ollero, A., & De Dios, J.M. (2000). An intelligent system for false alarm reduction in infrared forest-fire detection. *IEEE Intell. Syst. Appl.*, *15*, 64–73.

Aslan, Y.E., Korpeoglu, I., & Ulusoy, Ö. (2012). A framework for use of wireless sensor networks in forest fire detection and monitoring. *Computers, Environment and Urban Systems*, *36*, 614–625.

Atzori, Iera, & Morabito. (2010). The Internet of Things: A survey. *Computer Networks*, *54*(15), 2787–2805.

Bajaj. (2016). *IOT is the game changer fro retailer, aspire systems*. https://www.aspiresys.com/WhitePapers/IoT-is-the-game-changer-for-Retailers.pdf

Balsamo, D., Merrett, G.V., & Zaghari, B. (2017). Wearable and autonomous computing for future smart cities: open challenges. In *Proceedings of the 25th international conference on software, telecommunications and computer networks (SoftCOM)*. New York: IEEE.

Bandoyopadhayam. (2011). *Internet of Things - Applications and Challenges in Technology and Standardization*. arvix

Bekara. (2016). *Security issues and challenges for the IoT based smart grid*. https://www.sciencedirect.com/science/article/pii/S1877050914009193

Botta. (2016). Integration of cloud computing and internet of things: A survey. *Future Generation Computer Systems*, *56*, 684–700.

Breivold. (2015). IOT for industrial automation. *IEEE International conference on data science and data intensive systems*, 532-539. doi:10.1109/DSDIS.2015.11

Brindha. (2017). Involuntary Nutrients Dispense System for Soil Deficiency using IOT. *International Journal of ChemTech Research*.

Buratti, C., Conti, A., Dardari, D., & Verdone, R. (2009). An overview on wireless sensor networks technology and evolution. *Sensors (Basel)*, *9*, 6869–6896.

Ching & Singh. (2016). Wearable technology devices security and privacy vulnerability analysis. *International Journal of Network Security & Its Applications*, *8*(3), 19. doi:10.5121/ijnsa.2016.8302

Chiwewe, T.M., Mbuya, C.F., & Hancke, G.P. (2015). Using cognitive radio for interference-resistant industrial wireless sensor networks: An overview. *IEEE Transactions on Industrial Informatics*, *11*, 1466–1481.

Cinar. (2018). Optimization algorithm used in smart grids and comparison of results. *IOSR Journal of Electrical and Electronics Engineering*, *13*(6), 14-20.

Cintuglu, Mohammed, Akkaya, & Uluagac. (2017). A Survey on Smart Grid Cyber-Physical System Testbeds. *IEEE Communications Surveys and Tutorials, 19*(1), 446–464.

Cleveland. (2008). Cyber security issues for Advanced Metering Infrasttructure (AMI). *2008 IEEE Power and Energy Society General Meeting - Conversion and Delivery of Electrical Energy in the 21st Century,* 1–5.

Deshpande, Pitale, & Sanp. (2016). Industrial automation using IoT. *International Journal of Advanced Research in Computer Engineering and Technology, 5*(2).

Divya & Rao. (2016, August). Measurement and Monitoring of Soil Moisture using Cloud IoT and Android System. *Indian Journal of Science and Technology.*

Dominguez, F., & Ochoa, X. (2017). Smart objects in education: an early survey to assess opportunities and challenges. In *2017 Fourth International Conference on eDemocracy & eGovernment (ICEDEG),* (pp. 216–220). IEEE. doi:10.1109/ICEDEG.2017.7962537

Dorobantu, B. (2016, December 5). *Video ZF Live. Dana Războiu, preşedinta Asociaţiei Naţie prin Educaţie: 50 de echipe de elevi pasionaţi de robotică au primit kituri şi echipamente în valoare de 280.000 de euro pentru a construi roboţi.* Retrieved from https://www.zf.ro/business-hi-tech/video-zf-live-dana-razboiu-presedinta-asociatiei-natie-educatie-50-echipe-elevi-pasionati-robotica-au-primit-kituri-echipamente-valoare-280-000-euro-construi-roboti-16009941

Dyk, Chmielewski, & Najgebauer. (2017). Combat triage support using the Internet of Military Things. In Proc. of the 2017 Federated Conference on Computer Science and Information Systems Murugesan, S.: Harnessing green IT: principles and practices. *IT Professional, 10*(1), 24–33. doi:10.1109/MITP.2008.10

ECDL. (2017, October). *Lansare proiect imprimante 3D pentru şcolile din România. 3DUTECH - Modelează viitorul. Printează-l 3D!* Retrieved from https://www.ecdl.ro/en/news-article/lansare-proiect-imprimante-3d-pentru-scolile-din-romania.-3dutech-modeleaza-viitorul.-printeaza-l-3d_677.html

Elijah, O., Rahman, T. A., Orikumhi, I., Leow, C. Y., & Hindia, M. N. (2018). An overview of Internet of Things (IoT) and data analytics in agriculture: Benefits and challenges. *IEEE Internet of Things Journal, 5,* 3758–3773.

Elmaghraby & Losavio. (2014). Cyber security challenges in Smart Cities: Safety, security and privacy. *Journal of Advanced Research, 5*(4), 491–497. doi:10.23919/IConAC.2017.8082057 PMID:25685517

Fernandes, Jung, Prakash. (2016). Security Analysis of Emerging Smart Home Applications. *2016 IEEE Symposium on Security and Privacy (SP)*, 636–654.

Gaikwad, Gabhane, & Golait. (2015). A survey based on Smart Homes system using Internet-of-Things. *2015 International Conference on Computation of Power, Energy, Information and Communication (ICCPEIC)*, 330–335.

García, E.M., Serna, M.Á., Bermúdez, A., & Casado, R. (2008). Simulating a WSN-based wildfire fighting support system. *Proceedings of the International Symposium on Parallel and Distributed Processing with Applications*.

Gelazanskas & Gamage. (2014). Demand side management in smart grid: A review and proposals for future direction. *Sustainable Cities and Society, 11*, 22–30. doi:10.23919/IConAC.2017.8082057

Gharaibeh, A., Salahuddin, M. A., Hussini, S. J., Khreishah, A., Khalil, I., & Guizani, M. (2017). Smart cities: A survey on data management, security, and enabling technologies. *IEEE Communications Surveys & Tutorials, 19*(4), 2456–2501.

Goudos, S. K., Dallas, P. I., Chatziefthymiou, S., & Kyriazakos, S. (2017). A survey of IoT key enabling and future technologies: 5G, mobile IoT, sematic web and applications. *Wireless Personal Communications, 97*(2), 1645–1675.

Grandinetti, L. (2013). *Pervasive Cloud Computing Technologies: Future Outlooks and Interdisciplinary Perspectives*. IGI Global.

Gubbi, Buyya, Marusic, & Palaniswami. (2013). Internet of Things (IoT): A vision, architectural elements, and future directions. *Future Generation Computer Systems, 29*, 1645–1660.

Gubbi, J., Buyya, R., Marusic, S., & Palaniswami, M. (2013). Internet of Things (IoT): A vision, architectural elements, and future directions. *Future Generation Computer Systems, 29*(7),1645–1660.

Guillemin. (2015). *Internet of Things Position Paper on Standardization for IoT technologies*. European research cluster on the internet of things.

Hadded, Zagrouba, Laouiti, Muhlethaler, & Saidane. (2015). A multi-objective genetic algorithm-based adaptive weighted clustering protocol in vanet. *IEEE Congress on Evolutionary Computation (CEC)*, 994-1002.

Hassan, M. M., Lin, K., Yue, X., & Wan, J. (2015). A multimedia healthcare data sharing approach through cloud-based body area network. *Future Generation Computer Systems, 66*, 48–58. doi:10.1016/j.future.2015.12.016

Jacobsson, Boldt, & Carlsson. (2016). A risk analysis of a smart home automation system. *Future Generation Computer Systems, 56*, 719–733.

Joshi. (2017). IoT based Air and Sound Pollution Monitoring System. *International Journal of Computer Applications, 178*(7).

Kamal, N., Saad, M.H.M., Kok, C.S., & Hussain, A. (2018). Towards revolutionizing stem education via IoT and blockchain technology. *Int. J. Eng. Technol.* doi:10.14419/ijet.v7i4.11.20800

Kassab, M., DeFranco, J., & Voas, J. (2018). Smarter education. *IT Professional, 20*(5), 20–24. doi:10.1109/MITP.2018.053891333

Kaur & Kalra. (2016). A review on IoT based smart grid. *International Journal of Energy, Information and Communication, 7*(3), 11-22.

Khan, Khan, Zaheer, & Khan. (2012). Future Internet: The Internet of Things architecture, possible applications and key challenges. *Proc. 10th Int. Conf. FIT*, 257–260.

Khatoun & Zeadally. (2017). Cybersecurity and Privacy Solutions in Smart Cities. *IEEE Communications Magazine, 55*(3), 51–59.

Kopetz. (2011). Internet of things. In *Real-time systems*, (pp. 307-323). Springer US.

Kumar & Shoghli. (2018). A review of IOT applications in supply chain optimization of construction materials. *34th International symposium on automation and robotics in construction.*

Lee, Zappaterra, Choi, & Choi. (2014). Securing smart home: Technologies, security challenges, and security requirements. *2014 IEEE Conference on Communications and Network Security*, 67–72.

Lin, J., Yu, W., Zhang, N., Yang, X., Zhang, H., & Zhao, W. (2017). A survey on internet of things: Architecture, enabling technologies, security and privacy, and applications. *IEEE Internet of Things Journal, 4*(5), 1125–1142.

Liu, Xiao, Li, Liang, & Chen. (2012). Cyber Security and Privacy Issues in Smart Grids. *IEEE Communications Surveys and Tutorials, 14*(4), 981–997.

Lloret, J., Garcia, M., Bri, D., & Sendra, S. (2009). A wireless sensor network deployment for rural and forest fire detection and verification. *Sensors (Basel), 9*, 8722–8747. PMID:22291533

Machado. (2016). *Machado IOT impacts on supply chain saha machado second place grad, Scribd.* http://apicsterragrande.org/images/articles/Machado__Internet_of_Things_impacts_on_Supply_Chain_Shah_Machado_Second_Place_Grad.pdf

Mahamune & Amdani. (2017). IoT based connected vehiclein smart cities: review and research challenges. *IJERCSE, 4*(4). https://www.technoarete.org/common_abstract/pdf/IJERCSE/v4/i4/Ext_47935.pdf

Majeed, A., & Ali, M. (2018). How Internet-of-Things (IoT) making the university campuses smart? QA higher education (QAHE) perspective. In *2018 IEEE 8th Annual Computing and Communication Workshop and Conference (CCWC),* (pp. 646–648). IEEE. doi:10.1109/CCWC.2018.8301774

Mardonova & Choi. (2018). Review of wearable device technology and its application to the mining industry. *Energies, 11*(3), 547.

Market Study Report. (2013, Feb. 21). *Wearable Computing Devices, Like Apple's iWatch, Will Exceed 485 Million Annual Shipments by 2018.* ABI Research.

Mehmood, Y., Ahmad, F., & Yaqoob, I. (2017). Internet-of-things-based smart cities: recent advances and challenges. *IEEE Commun Mag, 55*(9), 16–24.

Miao, R., Dong, Q., Weng, W. Y., Yu, X.Y. (2018). The application model of wearable devices in physical education. *International Conference on Blended Learning,* 311–322.

Moosavi, S. R., Gia, T. N., Nigussie, E., Rahmani, A. M., Virtanen, S., Tenhunen, H., & Isoaho, J. (2015). End-to-end security scheme for mobility enabled healthcare Internet of Things. *Future Generation Computer Systems, 64,* 108–124. doi:10.1016/j.future.2016.02.020

Musolesi, M., Hailes, S., & Mascolo, C. (2005). Adaptive routing for intermittently connected mobile ad hoc networks. *Proceedings of the Sixth IEEE International Symposium on World of wireless mobile and multimedia networks.*

Myint, Gopal, & Aung. (2017). Reconfigurable Smart Water Quality Monitoring System in IoT Environment. *IEEE/ACIS 16th International Conference on Computer and Information Science (ICIS).*

Ojha, T., Misra, S., & Raghuwanshi, N. (2015). Wireless sensor networks for agriculture: The state-of-the-art in practice and future challenges. *Computers and Electronics in Agriculture, 118,* 66-84. Doi:10.1016/j.compag.2015.08.011

Ozturk, A., Umit, K., Medeni, I., Ucuncu, B., Caylan, M., Akba, F., & Medeni, D.T. (2011). Green ICT (Information and Communication Technologies): a review of academic and practitioner perspectives. *Int. J. EBusiness and EGovernment Stud.*

Pal, Gupta, Tiwari, & Sharma. (2017). IoT Based Air Pollution Monitoring System Using Arduino. *International Research Journal of Engineering and Technology, 4*(10).

Patil, Gawande, & Bag. (2017). Smart Agriculture System based on IoT and its Social Impact. *International Journal of Computers and Applications.*

Pei, X.L., Wang, X., Wang, Y.F., & Li, M.K. (2013). Internet of Things based education: definition, benefits, and challenges. *Applied Mechanics and Materials, 411–414,* 2947–2951. doi:10.4028/www.scientific.net/AMM.411-414.2947

Porkodi, R., & Bhuvaneswari, V. (2014). The Internet of Things (IoT) applications and communication enabling technology standards: An overview. In *2014 International conference on intelligent computing applications (ICICA).* IEEE.

Prathibha & Anupama. (2017). IOT based monitoring system in smart agriculture. *2017 International conference.* doi:10.1109/ICRAECT.2017.52

Rahim. (2016). Exploiting heuristic algorithms to efficiently utilize energy management controllers with renewable energy sources. Energy Build.

Ramesh, Nibi, Kurup, Mohan, Aiswarya, Arsha, … Guerrero-Barrantes. (2017). Characterization of biomass pellets from Chlorella vulgaris microalgal production using industrial wastewater. *International Conference in Energy and Sustainability in Small Developing Economies (ES2DE).*

Ramesh, Nibi, Kurup, Mohan, Aiswarya, Arsha, & Sarang. (2017). Water Quality Monitoring and Waste Management using IoT. *IEEE Global Humanitarian Technology Conference (GHTC).*

Rehman, A., Din, S., Paul, A., & Ahmad, W. (2017). An Algorithm for Alleviating the Effect of Hotspot on Throughput in Wireless Sensor Networks. *Proceedings of the IEEE 42nd Conference on Local Computer Networks Workshops (LCNWorkshops),* 170–174.

Sa'nchez-Corcuera, Nun˜ez-Marcos, Sesma-Solance, Bilbao-Jayo, Mulero, Zulaika, … Almeida. (2019). Smart cities survey: Technologies, application domains and challenges for the cities of the future. *International Journal of Distributed Sensor Networks, 15*(6). doi:10.1177/1550147719853984

Saha, Auddy, Chatterjee, Pal, Pandey, Singh, ... Maity. (2017). Pollution Control Using Internet of Things (IoT). *8th Annual Industrial Automation and Electromechanical Engineering Conference (IEMECON)*.

Sakr, S., & Elgammal, A. (2016). Towards a Comprehensive Data Analytics Framework for Smart Healthcare Services. *Big Data Research, 4,* 44–58. doi:10.1016/j.bdr.2016.05.002

Sarıta, M. (2015). The emergent technological and theoretical paradigms in education: the interrelations of cloud computing (CC), connectivism and Internet of Things (IoT). *Acta Polytechnica Hungarica, 12*(6), 161–179. doi:10.12700/APH.12.6.2015.6.10

Schiefer. (2015). Smart Home Definition and Security Threats. *2015 Ninth International Conference on IT Security Incident Management & IT Forensics,* 114–118.

Selvi. (2018). Difficulties and Data Mining Model For Internet of Things (IoT). *International Journal of Engineering Science Invention,* 44-48. http://www.ijesi.org/papers/NCIOT-2018/Volume-1/8.%2041-45.pdf

Sharma & Vinod. (2014). SPARK: Personalized Parkinson Disease Interventions through Synergy between a Smartphone and a Smartwatch. In *Design, User Experience, and Usability.* Springer International Publishing.

Siano. (2016). Demand response and smart grids—A survey. *Renewable & Sustainable Energy Reviews, 30,* 461–478.

Silva, B., Fisher, R.M., Kumar, A., & Hancke, G.P. (2015). Experimental link quality characterization of wireless sensor networks for underground monitoring. *IEEE Transactions on Industrial Informatics, 11,* 1099–1110.

Son, B., Her, Y.S., & Kim, J.G. (2006). A design and implementation of forest-fires surveillance system based on wireless sensor networks for South Korea Mountains. *Int. J. Comput. Sci. Netw. Secur., 6,* 124–130.

Stankovic. (2014). *Research directions for the Internet of Things.* IEEE.

Sterbenz, J.P. (2017). Smart city and IoT resilience, survivability, and disruption tolerance: challenges, modelling, and a survey of research opportunities. In *Proceedings of the 9th international workshop on resilient networks design and modeling (RNDM).* New York: IEEE.

Sundar, Hebbar, & Golla. (2015). Implementing intelligent traffic control system for congestion control, ambulance clearance, and stolen vehicle detection. *IEEE Sensors Journal, 15*(2), 1109–1113.

Tan, W., Wang, Q., Huang, H., Guo, Y., & Zhan, G. (2007). Mine Fire Detection System Based on Wireless Sensor Networks. *Proceedings of the Conference on Information Acquisition (ICIA'07).*

Thierer. (2015). The Internet of Things and Wearable Technology: Addressing Privacy and Security Concerns without Derailing Innovation. *Rich. J. L. & Tech., 21*(6).

Toma & Talpiga. (2018). Secure IOT supply chain management solution using blockchain and smart contrats technology. *International Conference on Security for Information Technology and Communication, SECITC.*

Usha & Rukmini. (2016). IOT in connected vehicles: challenges and issues-A review. *International Conference on Signal Processing, Communication, Power and Embedded Systems.* doi:10.1109/SCOPES.2016.7955769

Vermensan & Friess. (2013). *Internet-of-Things converging technologies for smart environment and integrated ecosystems.* River Publishers.

Vihervaara, J., & Alapaholuoma, T. (2016). Internet of Things: opportunities for vocational education and training—presentation of the pilot project. In *Proceedings of the 9th International Conference on Computer Supported Education*, (pp. 476–480). SCITEPRESS—Science and Technology Publications. doi:10.5220/0006353204760480

Wenning, B.L., Pesch, D., Timm-Giel, A., & Görg, C. (2010). Environmental monitoring aware routing: Making environmental sensor networks more robust. *Telecommun. Syst., 43*, 3–11. https://en.wikipedia.org/wiki/Internet_of_things

Wu, Lu, Ling, Sun, & Du. (2010). Research on the architecture of Internet of Things. *Proc. 3rd ICACTE, 5*, 484-487.

Xu, L., He, W., & Li, S. (2014). Internet of Things in industries: A survey. *IEEE Transactions on Industrial Informatics, 10*(4), 2233–2243.

Yang. (2011). Study and application on the architecture and key technologies for IOT. *Proc. ICMT*, 747–751.

Yin, Y., Zeng, Y., Chen, X., & Fan, Y. (2016). The internet of things in healthcare: An overview. *Journal of Industrial Information Integration, 1*, 3–13. doi:10.1016/j.jii.2016.03.004

Yuvaraj. (2016). *Smart supply chain management using IOT.* IEEE, WISPNET. doi:10.1109/WiSPNET.2016.7566196

Zhamanov, A., Sakhiyeva, Z., Suliyev, R., & Kaldykulova, Z. (2017). IoT smart campus review and implementation of IoT applications into education process of university. In *2017 13th International Conference on Electronics, Computer and Computation (ICECCO)*, (pp. 1–4). IEEE.

Zhang, J., Li, W., Han, N., & Kan, J. (2008). Forest fire detection system based on a ZigBee wireless sensor network. *Frontiers of Forestry in China, 3*, 369–374.

Zikria, Kim, Hahm, Afzal, & Aalsalem. (2019). Internet of Things (IoT) Operating Systems Management: Opportunities, Challenges,and Solution. *Sensors (Basel), 19*, 1793. doi:10.339019081793

Chapter 5

Nano–Biosensors Tech and IPM in Plant Protection to Respond to Climate Change Challenges in Morocco

Wafaa Mokhtari
Institut Agronomique et Vétérinaire Hassan II, Morocco & Complexe Horticole d'Agadir, Morocco

Mohamed Achouri
Institut Agronomique et Vétérinaire Hassan II, Morocco & Complexe horticole d'Agadir, Morocco

Abdellah Remah
Institut Agronomique et Vétérinaire Hassan II, Morocco & Complexe Horticole d'Agadir, Morocco

Noureddine Chtaina
Institut Agronomique et Vétérinaire Hassan II, Morocco

Hassan Boubaker
Ibn Zohr University, Morocco

DOI: 10.4018/978-1-7998-4381-8.ch005

ABSTRACT

In this chapter, the authors introduce two research axes: Part A, nano-biosensors as ad-hoc technologies designed to meet plant diagnostic sensitivity and specificity needs at point of care, and Part B, the study of the interaction of drought and infection stresses in crops investigating bio-control potential antagonists in developing integrated approach (IPM) for disease control measures in crops system. The first part will be revising most used nano-biosensors in plant pathogens detection using different platforms in greenhouses, on-field, and during postharvest. A special focus will be on optical and voltametric immuno/DNA sensors application in plant protection. The last part will present case studies of using nanoparticles functionalized with antibody/ DNA for detecting pathogenic Pseudomonas sp, mosaic viruses, Botrytis cinereal, and Fusarium mycotoxins (DON). The second part will be interpreting experimental results of a case study on evaluating bio-control efficacy of local Trichoderma spp. using root dips treatment in Fusarium solani-green beans pathosystem as a model.

INTRODUCTION

Different nano-biosensors have been developed and been commercialized for various Plant Protection applications. These new technologies have revolutionized the plant protection domain as it make the diagnostic more accessible to the stakeholder group diagnosticians as well as farmers. Furthermore, these technologies may save diagnosticians and farmers from the burdens in decision making of different disease management approaches (eradication, sanitation, protection etc.) affected by climate changes. The common definition of biosensors or nano-biosensor is a device that uses specific biochemical reactions mediated by isolated enzymes, immunosystems, tissues, organelles or whole cells to detect chemical compounds usually by electrical, thermal or optical signals (Khater et al., 2017). In another word, nano-biosensors are devices (most commonly bio-electronic chips) that detect an important analyte (pathogen, contaminant, and molecule) that is generally affinity triggered by a binding site called biological recognition element (antibodies, DNA, Aptamer). Simultaneously, this specific binding is readable as a signal by a transducer platforms or interface (Electrodes, arrays) layered on that chip. This platform incorporate nano-materials matrices to amplify the signal and, therefore, the out puts (Low Limit of Detection, low concentrations). In fact, analytes are detected, transduced to signals and measured using different measurement interfaces. These latter could be electrical, electrochemical, optical, magnetic or even vibrational signals depending on the nano-biosensor operating measurement approaches (figure 1).

Based on these bio-sensing approaches, tremendous nano-biosensors have been developed for detection of plant pathogens or their contaminants. For instance, gold-nanoparticles (AUNPs) have been utilized to "detect" gene/ nucleotides sequences or antigens of different phytopathogens that cause various blight, wilt and spot diseases in important vegetables and crops like *Pseudomonas syringae, P. argenus, Xanthomonas alfalfae subsp. Citrumeloni* and *X. axonopodis pv. citri* that causes the citrus cankers disease. Fluorescent DNA probes in microarrays is another nano-biosensor measurement approach utilized to detect causal agent of gray molds of grapes and berries, *Botrytis cinerea*. In the following discussion, we are going to present some case studies of nano-biosensors detection of phytopathogens and contaminants of food in postharvest (Vaseghi et al., 2013; Wang and Li, 2010; Yao et al, 2009).

Figure 1. different approach of nano-biosensors based on recognition element-analyte affinity, nature of detection measurement 'optical, electrochemical, acoustic" and the innovative Nano scale embodiment. CV: Cyclic Voltammetry, DPV: Differential Pulse Voltammetry, QCM: Quartz Cristal Microbalance. NT/AuP: Nanotubes Gold Particles

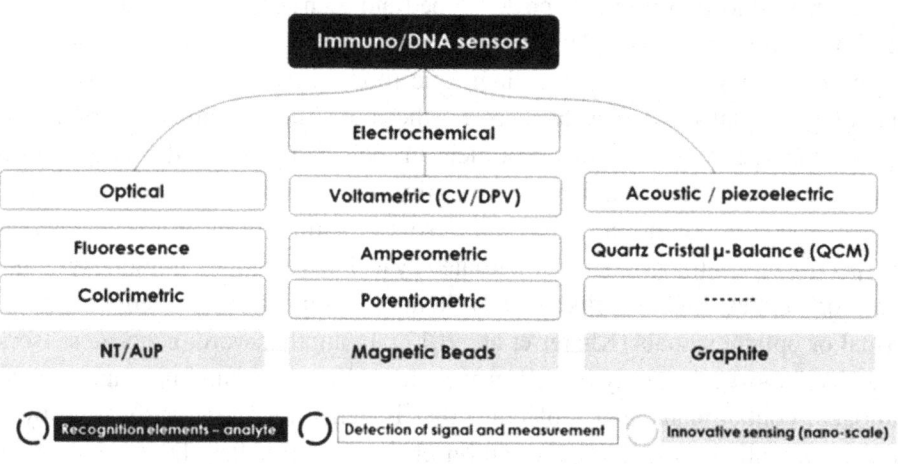

In the following paragraphs we will be discussing the most important nano-biosensors used in plant pathogens diagnosis, some of the elements used making them special for sensitive detection (surface interaction; geometric diffusion or fluidic diffusion properties, cost effective) and tricks in detection detailing four case studies where nano-biosensors are applied.

Microfluidic Nano-Chips to Detect *Botrytis cinerea* DNA

Recently, gold or silver Microfluidic nano-chips were fabricated to generate DNA micro-arrays platforms creating highly ordered DNA stretches based on the fluidic diffusion capacity of microfluids channels. Actually, microfluidic nano-chips have been used for rapid visual detection of plant fungal pathogens like *Phytopthora* species (Schwenkbier et al., 2015) and *Botrytis cinerea* (Wang and Li, 2010) based on colorimetric or fluorescent detection approaches. To detect *Botrytis cinerea*, a glass chip animated with polydimethylsiloxane (PDMS) polymer channels has been utilized as substrates. Simply, a PCR amplified genome or gene specific sequences of pathogens hybridize with complementary DNA sequence (oligonucleotides conjugated with fluorescent dye; 5Cyst dye) using contact printing microfluidic technique (figure2). The latter DNA sequences are immobilized in the already prepared 16 vertical PDMS (V-PDMS) channels. The microfluidic method allow optimizing DNA probe immobilization (recognition element) and *Botrytis* DNA (PCR products) hybridization by coating all the arrays of the micro-plate with Gold nanoparticles. Fluorescence signals of less than 0.5ng/µl of DNA are detected and the washing and moving out solution do not affect in any way the specificity of hybridizing *Botrytis* DNA pools (Wang and Li, 2010).

Figure 2.

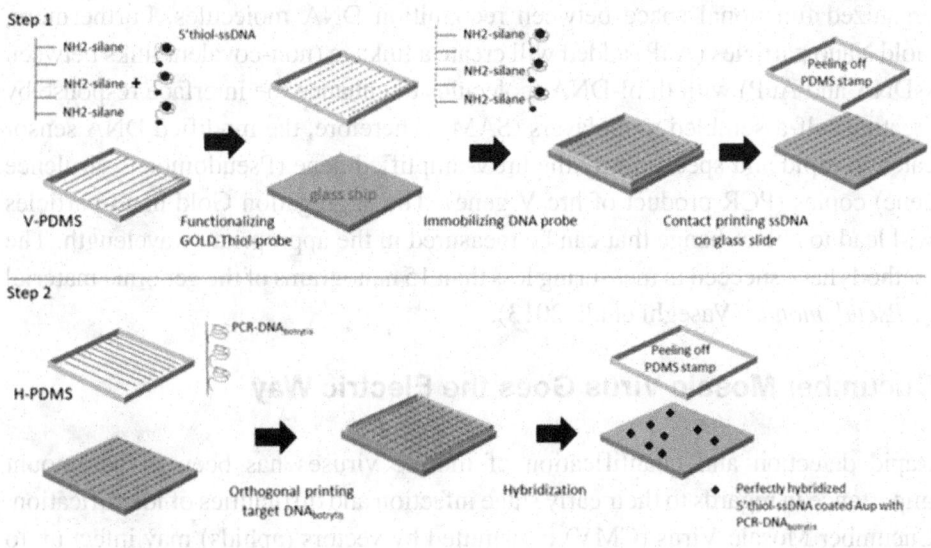

Figure 2: Microfluidic nano-chip on glass slide for DNA. The process consist of 2 major steps: 1st step is the formation of Self Assembled Monolayers (liking gold particles with thiol group and making interfacial space using PDMS functionalized with amine) and immobilizing recognition elements (DNA Probes). 2nd step involves the reaction between recognition element (DNA probes) and analyte (DNA$_{botrytis}$) based on their affinity, the intersection between probe lines (16 modified V-PDMS) and target DNA lines (16 horizontal lines in H-PDMS) will be visualized as fluorescent dots (fluorescent staining, Confocal microscopy) or simply by changing colors (colorimetric approach) when two strands are perfectly hybridizing.

Microfluidic nano-chips allow detecting and quantifying with high sensitivity and specificity *Botrytis cinerea* as well as other fungal plant pathogens spp based on DNA hybridization. Nanotubes, nanoparticles or nano-rods alone or in combination with enzymes and biomolecules able the early detection of *Botrytis* before any visual symptoms appeared. Therefore, it is a very successful early diagnostic and biotic asymptomatic on sites-application of plant pathogen detection.

Almost Same for *Pseudomonas* sp.

Similarly, a bacterial pathogen of different solanacea and crops called *Pseudomonas syringae* has been detected using portable DNA sensors based on the optical approach. At this time, different tricks on the immobilization procedures have been tempted. Thiol-DNA scaffolding have been used a functionalized material to create more organized functional space between recognition DNA molecules. Furthermore, Gold Nano particles (AuP) added will create a linkage (non-covalent links between ssDNA and AuP) with thiol-DNA molecules to enhance the interface response by creating self-assembled monolayers (SAM). Therefore, the modified DNA sensor catch in rapid and specific way the hrcV-amplified gene (Pseudomonas virulence gene) copies (PCR product of hrc V gene). The aggregation Gold nano-particles will lead to color change that can be measured in the appropriate wavelength. The methods have succeed in measuring less than 15 nanograms of the genomic material of *Pseudomonas* (Vaseghi et al., 2013).

Cucumber Mosaic Virus Goes the Electric Way

Rapid detection and quantification of mosaic viruses has been of paramount importance in regards to their early stage infection and difficulties of identification. Cucumber Mosaic Virus (CMV) transmitted by vectors (aphids) may infect up to 1000 plant species and cause damages of economic significance especially in cash crops. Rafidah el al. 2016 have developed an immune-sensor enzyme labeled and based on electrochemical measurement to detect and quantify CMV. Actually, the practical use of this approach is based on the protocols of immobilization, virus biding to the recognition element and finally signal transduction. In details, we can sketch the protocol of CMV detection as following; 1- directly immobilizing CMV polyclonal antibodies in the Screen-printed electrodes (SPE), we should notice that directly depositing the biological material on the working electrode surface would interfere in the sensing results and therefore the measurement. 2 – Rafidah et al have chosen an accessible screened (ordered layer) of carbon electrodes SCPE to have optimal working electrode sensitivity response. Moreover, to prevent bulky interaction or passivated electrode responses they have chosen to help the dispersion of CMV polyclonal on the SCPE surface using Gold-nanoparticles (AUP) 3- after immobilization, the research team have tempted to "sandwich" the CMV antigens by a secondary Horseradish Peroxidase (HPR) labeled antibodies. The incubation in the adequate substrate like 3,3, 5,5 -tertramethyl benzidin (TMB) will lead to the enzymatic catalysis of HPR (less than 10 minutes) following that a redox (TMB$_{reduction/oxidation}$) reaction in the presence of H2O2. This redox will change the electrolyte composition detected directly by the working electrode and measured here the potential of an ampere-meter. Chrono-amperimetry was used for measurement regarding the high performing measure of TMB$_{oxidized}$ released in the solute with 10 to 60 seconds detection time scale (Rafidah et al., 2016).

What About DON?

For Food-born toxic contaminants like deoxynivalenol (DON), Portable Voltametric immuno-sensors have been widely used for their cost effective and sensitive application. Actually, a screen printed electrodes (SPE) are widely used as nano-diagnostic kits. How it works a SPE? SPE technology cover the planar ink platform (Carbon, gold or graphite working Electrode) that can be modified implementing like conjugated antibodies as recognition element or diffusing Gold nanoparticles on the surface (see figure 3). The analysis of the food born toxin can be used in drop step, so adding little amount of solution (buffer chosen) of suspected food born toxin products can record a high sensitive Differential Pulsed Voltametric (DPV) or Cyclic Voltametric (CV) signals. The change on electrochemical behavior (redox complex on electrodes surface) recorded by potentiostat and the peaks will be readable in the connected computer. The following scheme showed use of portable SPE for on-site detection of Don toxins (Escriva et al., 2017). Figure 3 represents on site detection and quantification of DON toxins in wheat grains using SPE; 1: SPE, 2: SPE submerging the analytes (DON toxins suspected). 3: plane view of working electrode modified with, carbon Nanotube (CNT) and immobilized with DON antibodies (Ab). 4: electrochemical magnitude change recorded in peaks in case of absence (- Toxins) or presence (+ Toxins) of toxins.

Figure 3.

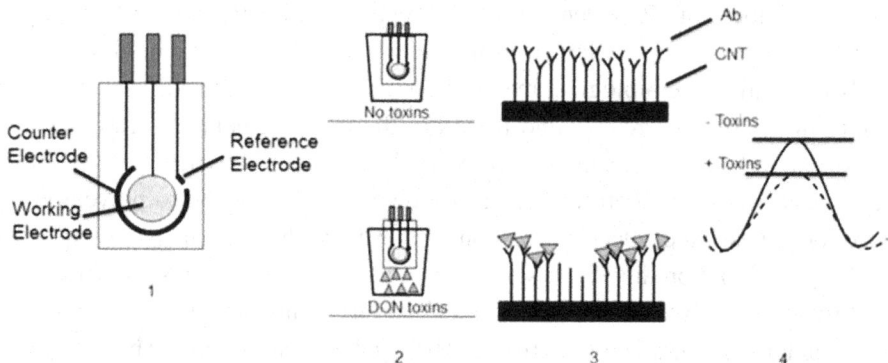

CONCLUSION

Biological biosensors or nano-sensors most often are affinity biosensors with a specific bio-recognition element that attach the analyte specifically in their binding sites (epitopes, receptors, nucleotides strands). Consequently, this affinity biosensors records sensing biological reaction changes when recognition element and the target analyte interact (optical, electric potential, or vibration). Bio-recognition molecules have other functionality in the biosensors such as detecting and transducing the signal. Probes can identify different plant pathogen target by transducing their signal through a plug transduction platform. Nano-scale embodiment help in amplifying the signal and therefore, increase the output response. Therefore, nanomaterial matrices play a role of transducers and amplifiers challenging the limit of detection and sensitivity of the devices use for measurement. These portable nano-sensors devices (immune-sensors, DNA sensors) showed effective, fast and on sites terms of their use. However, some of these nano-sensors composites are still expensive and not mass-producible. Therefore, there is still much research work to do taking into account the cost assessment. Saying this there is leader companies that offer nano-devices with expensive nano-materials in the market for the service of agriculture industry (Ricii et al., 2012; Khater et al.,2017; Escriva et al., 2017).

PART B: INTEGRATED BIO CONTROL APPROACH OF *TRICHODERMA* SPP AGAINST *FUSARIUM SOLANI* COMBINING DROUGHT AND INFECTION STRESSES

Introduction

Trichoderma spp biocontrol agents used for seed or seedling treatments often applied as "protectants" to reduce or suppress seed-borne or soil-borne diseases. In fact, at this level, *Trichoderma* species are considered an integrated bio-control approach to manage initial inoculum of pathogens (i.e. mycelium or sclerotia of *Rhizoctonia* and conidia of *Fusarium* sp.) at the site of infection. Hence, it is ultimately the effective approach in reducing disease development rate of different soil borne pathogens including *Fusarium solani*. *Fusarium solani f. sp phaseoli* has been identified as primary cause of root rot of common bean (*Phaseolus vulgaris L.*) that reduces beans yield (Singh, 2001). In Morocco, *Fusarium solani* has been reported as the causal agent of root diseases and causes different damages in different crops and fruits; banana, beans, olives and rice (Zehhar *et al.*, 2006; Meddah *et al.*, 2011; Chliyeh *et al.*, 2014). Moreover, it has been demonstrated that *F. solani* contributed to the synergistic interaction with other soil-borne phyto-pathogens and highly colonized different roots during the attack of crops in the country (i.e. Banana) (Loubane, 1991). The fungus is a mycotoxin producer that produces fusaric acid (Naiker *et al.*, 2004; Moretti *et al.*, 2010) and deoxynivalenol derivative of trichothecen family. These mycotoxins confer pathogenicity to *Fusarium* species and once they are present in food product may lead to severe infection in human body. In fact, *Fusarium solani* is considered as feared hazard mycotoxicoses inoculum source (Nucci and Anaissie, 2007). Green beans cultivar is a specific host of *Fusarium solani f. sp phaseoli*. However, it has been reported that green beans showed *Fusarium* symptoms only in wounded inoculated cultivars in greenhouses assays (Melgar and Roy, 1994; Gray *et al.*, 1999). Saying this, it is well known that *F. solani* as for other phyto-pathogenic *Fusarium* species perfectly attack stressed or injured plants. Furthermore, alleviation of abiotic stress like extended drought, flooding, pesticide and fertilizer injuries have been found to enhance *Fusarium* disease severity (Burke and Hall, 1991; Miller and Burke, 1977). Sinha and co-workers (2019) have demonstrated that combined drought and *F. solani* infection stresses lead to more yield and performance damage on chickpea compared to plants infected with *Fusarium* alone. Therefore, green beans-*Fusarium* would be one of the pathosystem model to study plant stress and *Fusarium* wilt disease biocontrol. Revealing suppressive interactions between *Trichoderma* spp and *Fusarium* in water stressed plant would tremendously help in designing the adequate and required disease management strategy of *Fusarium* wilt, canker or rot diseases depending on the environment of disease scouting. For instance, testing biocontrol potential of *Trichoderma* spp. using alike stressed host may allow identifying the way *Trichoderma* isolates can be applied against *Fusarium* disease (Huisman and Gerik, 1989; Blestos et al., 2003). In this part of chapter, we present *Fusarium* disease biocontrol mesures using *Trichoderma* spp. combining both water and infection stresses on green beans. In fact, in this work, we sought to establish root dipping of spores' suspension of three *Trichoderma* species to evaluate and screen biocontrol potential antagonists combining drought and *Fusarium solani* infection in potted green beans in green house.

MATERIALS AND METHODS

Obtaining Isolates of Fusarium Solani

Isolation and culturing of *Fusarium solani* was undertaken using isolation methods adopted by Lopez-Escudero and Mercado-Blanco (2011). Branches of olive trees diagnosed with *Fusarium* wilt symptoms were collected, washed with tap water and disinfected one minute in 10% sodium hypochlorite. Six small fragments from wilted branches were inoculated in PDA medium (PDA, Difco) and incubated at 21°C in the dark for 6 days.

Fusarium Solani Inoculum

Macroconidia and microconidia of *Fusarium solani* represent the inoculum source produced to inoculate green beans. To obtain *Fusarium* conidia we transfer plugs of *Fusarium* culture to grow in fresh PDA medium plates at 24°C for 7 to 10 days. Thereafter, conidia were harvested from the surface of *Fusarium* colonies by adding 10 ml of sterile di-ionized water, scraping the surface of culture and transferring it into beakers. Conidia suspension is adjusted to approximately final concentration of 10^6 to 10^7 conidia/ml (Romberg and Davis, 2007).

Green Beans Root- Dipping Inoculation with Trichoderma

For green beans inoculation with *Fusarium* suspension, plants were wounded-inoculated according to Atibalentja and Eastburn (1997). 2 to 3 centimeters (cm) cuts were performed in green beans roots prior the inoculation procedures. Plants were then dipped in *Trichoderma* spores' suspension (10^6 conidia/ml) then in *Fusarium* spores suspension of 10^6 to 10^7 conidia/ml (Atibalentja and Eastburn, 1997).

Water Stress Treatment on F. Solani Infected and Trichoderma root Dipped Green Beans

Prior to water stress treatment, all green beans were watered for seven days after pots transplanting for root and shoot establishment. Afterwards, imposition of drought stress was achieved applying no watering day interval of 5 days during the whole experiment. Soil water content was establishedusing a simple soil drainage test for potted plants based on determining the leachate volume after watering (Cottenie, 1980).

Measurement of Disease Incidence (DI) and Plant Development Parameters in Green Beans

Disease Incidence (DI) was measured as the percentage of number of plant units showing typical symptoms of the pathogenic fungus (cankers in stem and roots). Therefore, DI percentage was calculated as shown in the equation (1) given by Benson & Baker (1974):

$$\text{Disease Incidence (DI)} = \frac{\text{number of infected plant units}}{\text{number of plants}} * 100$$

Disease incidence was measured in aerial units of the plant (DI-AU) and root units (DI-RU). DI-AU was measured based on visible symptoms detected in the above-ground plant area. DI-RU was measured based on visible symptoms and signs in the diseased roots of related host plant. In fact, this method requires destructive sampling of plant, therefore, applied at the end of each experiment. Signs were identified based on the presence of pathogens' components like mycelium and propagules under microscopic observation (Campbell and Neher, 1994). In addition, plant parameters like plant height (PH) in centimeter (cm), leaves surface area (LSA) in centimeter square (cm^2) and green bean pods number (Pods N°) as well as roots dry weight (RDW) (Benson and Baker, 1974; Campbell and Neher, 1994).

Experimental Design

Three species of *Trichoderma; Trichoderma afro-harzianum* (T8A4), *Trichoderma reseei* (T9i12) and *Trichoderma guizouhense* (T4) were tested for their biocontrol efficacy against *Fusarium* causing roots and stem cankers on green beans. Green beans cultivars were grown for three to four weeks on 77 peat trays. Seedlings of different cultivars were then transplanted into pots after their inoculation with the fungus. Green beans seedlings were transplanted in 7 L pots filled with sterile substrate at 3:1 w/w peat to sand ratio. Substrate was fertilized using NPK and oligo-elements composition (20-20-20 hydrosoluble NPK plus Oligo-elements; Magenesium, Iron and Manganese) at 250 mg/l. Experimental design was organized in four randomized complete blocs with four replicates in each experimental unit. That is, four pots were used in each experimental unit.

RESULTS AND DISCUSSION

In the whole, *Trichoderma* spp. treatments reduced significantly *Fusarium* disease incidence in aerial plant units ($P = 0.000$). All DI-AU = 0.0% in treatment T1, T2, T3 and TC compared to DI-AU = 100% in infected controls Tm2. Disease incidence measures in root units were DI-RU = 100%, 25%, 18.8% and 6.3% in Tm2 controls and T3, T1 and TC treatment respectively. DI-RU recorded in healthy plants was 0.0% as indicated in figure 4.

Figure 4. Caption here

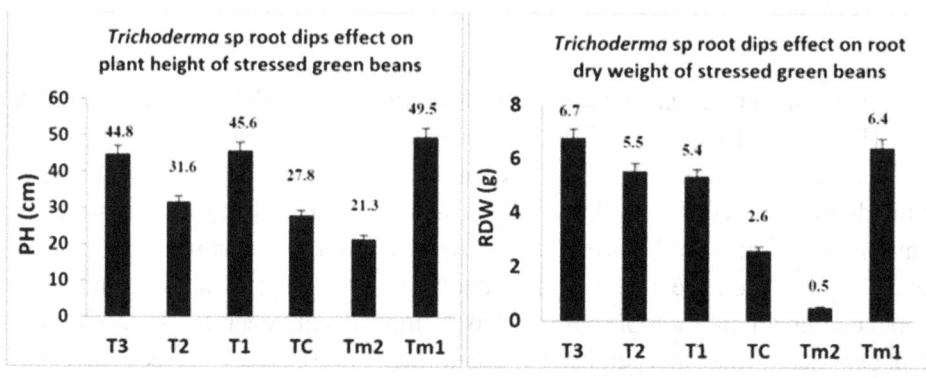

Figure 4: Trichoderma potential in disease incidence control on combining water and Fusarium infection stress in green beans with; Trichoderma afroharzianum T1, Trichoderma reseei T2 and Trichoderma guizouhense T3 and TC Trichoderma extracted from commercial product used as reference. Disease incidence assessed in above ground units DI-AU and in roots DI-RU. Tm1 and Tm2 are negative and positive controls respectively.

With respect to disease incidence measurement, symptoms like roots, crown and stem rots and lesions caused by *Fusarium solani* in green beans were observed. Symptoms assessed in positive controls Tm2 were creamy brown root, stem rots and lesions with root and stem discoloration. No wilting or stunting was observed in Tm2 green beans though. Root dry weight of green beans was RDW = 6.7 g, 5.5 g and 5.3 g when dipped in *T. guizouhense* spores suspension (T3), *T. reesei* (T2) and (*T. afro-harzianum* (T1) and were fairly equal to roots dry weight of healthy green beans in Tm1 controls with RDW = 6.4 g. However, root dry weight only of infected green beans with *F. solani* and treated with commercial *Trichoderma* (TC) showed the lowest RDW = 1.2 g. Similarly to RDW results, green beans heights in TC treatment were the lowest with PH = 29.3 cm while the highest green beans heights occurred in T2 treatment with PH = 52.5 cm compared to Tm1 with PH = 56.8 cm (see figure 5).

Figure 5.

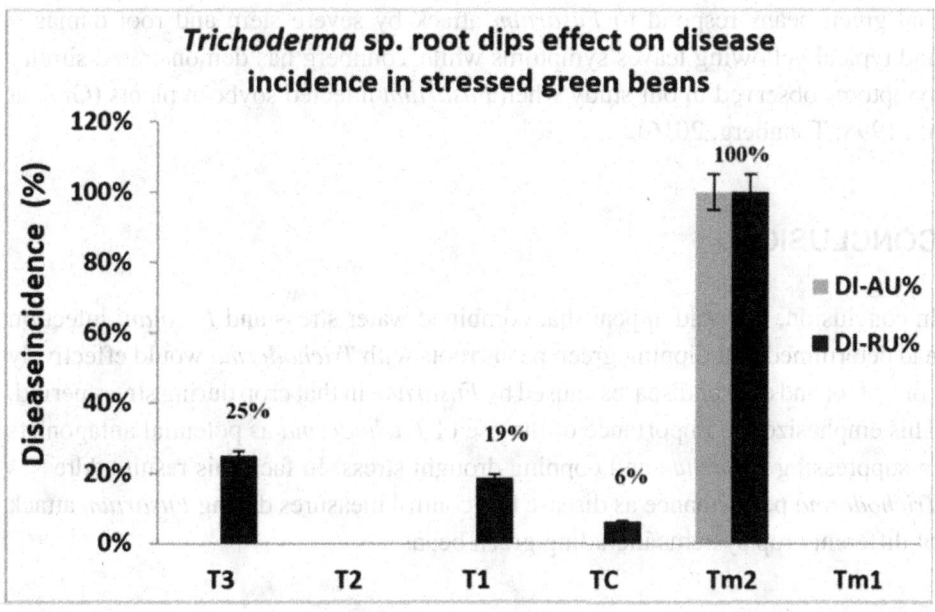

Figure 5: *Trichoderma* sp. effect on development parameters of plant combining water and *Fusarium* infection stress in green beans with; Plant height (PH) and Root dry weight (RDW).

From disease assessment and plant parameters development results, it can be inferred that *Trichoderma* spp and especially *T. reesei* (T2 dipped root treatment) remarkably alleviate disease responses, maintain and improve green beans development when infected with *F. solani*.

Previous study suggested the efficacy of *Trichoderma* treatments to control disease caused by *Fusarium solani* in different crop systems. For instance, Rojo *et al.* (2006) reported the efficient of *Trichoderma* to control *Fusarium* disease in peanut. In their work, Rojo et al. (2006) have assessed *Trichoderma* biocontrol against *Fusarium* estimating disease severity index decrease in host plants in greenhouse and field conditions (Rojo *et al.*, 2006). In earlier studies, attempts were made to inoculate *Fusarium solani* on green beans roots to assess *Fusarium* wilt or canker disease. Artificial inoculation succeeded in green beans' wounded roots (Melgar and Roy, 1994). Few research work have been investigated plant-pathogen interaction in stress conditions and really few researches have elaborated biocontrol of crops as abiotic control measures (Blanca et al., 2004; Eastburn *et al.*, 2010). Therefore, in this study we sought to establish wounded inoculation in green beans roots to evaluate *Fusarium* pathogenicity and examine antagonistic potential of *Trichoderma* spp in water stressed plants. For instance, research work of Gray at al., 1998 demonstrated that green beans respond to *Fusarium* attack by severe stem and root damages and typical yellowing leaves symptoms while Tonnberg has demonstrated similar symptoms observed in our study when *Fusarium* infected soybean plants (Gray at al., 1998; Tonnberg, 2016).

CONCLUSION

In conclusion, it would appear that combined water stress and *F. solani* infection had determined that dipping green beans roots with *Trichoderma* would effectively control rot and canker diseases caused by *Fusarium* in that crop during stress period. This emphasize the importance of the use of *Trichoderma* as potential antagonists in suppressing *Fusarium* and copping drought stress. In fact, this result addresses *Trichoderma* performance as disease bio-control measures during *Fusarium* attack of different crop systems including green beans.

REFERENCES

Atibalentja, N., & Eastburn, D. M. (1997). Evaluation of inoculation methods for screening horseradish cultivars for resistance to *Verticillium dahliae*. *American Phythopathology Society*, *81*(4), 356–362. doi:10.1094/PDIS.1997.81.4.356 PMID:30861815

Benson, D. M., & Baker, R. (1974). Epidemiology of *Rhizoctonia solani* pre-emergence damping-off of radish: Influence of pentachloro-nitrobenzene. *Phytopathology, 64*(1), 38–40. doi:10.1094/Phyto-64-38

Blanca, B. L., Juan, A., Navas-Cortés, R., & Jiménez-Díaz, M. (2004). Integrated management of Fusarium wilt of chickpea with sowing date, host resistance, and biological control. The American Phytopathological Society, 94, 946-960.

Blestos, F., Thanassoulopoulous, C., & Roupakias, D. (2003). Effect of grafting on growth yield and *Verticillium* wilt of eggplant. *Horticultural Science, 38*, 183–186.

Burke, D. W., & Hall, R. (1991). Fusarium root rot. In R. Hall (Ed.), Compendium of Bean Diseases (pp. 9-10), St. Paul, MN: The American Phytopathological Society.

Campbell, C. L., & Neher, D. A. (1994). Estimating disease severity and incidence. In C. L. Campbell & D. M. Benson (Eds.), Epidemiology and management of root diseases (pp. 117-147). Springer. doi:10.1007/978-3-642-85063-9_5

Chliyeh, M., Rhimini, Y., Selmaoui, K., Ouazzani Touhami, A., Filali-Maltouf, A., El Modafar, C., ... Douira, A. (2014). Survey of the fungal species associated to olive-tree (Olea europaea L.) in Morocco. *International Journal of Recent Biotechnology, 2*, 15–32.

Cottenie, A. (1980). Soil and Plant testing as a basis of fertilizer recommendations. *Food and Agriculture Organization Bulletin, 38*(2), 119.

Eastburn, D. M., Degennaro, M. M., Delucia, E. H., Dermody, O., Elrone, M., & Elevated, A. J. (2010). Atmospheric carbon dioxide and ozone alter soybean diseases at Soybean. *Global Change Biology, 16*(1), 320–330. doi:10.1111/j.1365-2486.2009.01978.x

Escrivá, L., Font, G., Manyes, L., & Berrada, H. (2017). Studies on the Presence of Mycotoxins in Biological Samples. *An Overview Toxins, 9*(8), 251. doi:10.3390/toxins9080251 PMID:28820481

Francesco Ricci, F., Adornettoa, G., & Palleschi, G. (2012). A review of experimental aspects of electrochemical immunosensors. ElectroChemica acta.

Gray, L. E., Achenbach, L. A., Duff, R. J., & Lightfoot. D. (1999). Pathogenicity of *Fusarium solani f. sp. glycines* isolates on soybean and green bean plants. *Journal of Phytopathology, 147*, 281-284. doi:10.1016/j.bios.2016.09.091

Huisman, O. C., & Gerik, J. S. (1989). *Dynamics o colonization of plant roots by Verticillium dahlia and other fungi in Vascular wilt disease of plants.* Berlin: Springer-Verlag.

Khater, M., Escosura-Muñiza, A., & Merkoçia, A. (2017). Biosensors for plant pathogen detection. *Biosensors & Bioelectronics*, *93*, 72–86. doi:10.1016/j.bios.2016.09.091 PMID:27818053

Loubane, M. (1991). Contribution à l'étude du complexe nématodes-champignons associés aux pourritures racinaires du bananier dans la région du souss-Massa. Engineering diploma Institut Agronomique Vétérinaire Hassan II, Complexe Horticole Agadir.

Meddah, N., Ouazzani, T. A., Benkirane, R., & Douira, A. (2011). Etude du pouvoir pathogène de quelques espèces de *Fusarium* sur le bananier sous serre au Maroc. *Bulletin de la Société Royale des Sciences de Liège*, *80*, 939–952.

Melgar, J., & Roy, K. W. (1994). Soybean sudden death syndrome cultivar reaction to inoculation in controlled environement and host range and virulence of causal agent. *American Phytopathology Society*, *78*, 265–268.

Miller, D. E., & Burke, D. W. (1977). Effect of temporary excessive wetting on soil aeration and *Fusarium* root rot of beans. *The Plant Disease Reporter*, *61*, 175–179.

Moretti, A., Ferracane, L., Somma, S., Ricci, V., Mulè, G., Susca, A., ... Logrieco, A. F. (2010). Identification, mycotoxin risk and pathogenicity of *Fusarium* species associated with fig endosepsis in Apulia, Italy. *Food Additives and Contaminants*, *27*(5), 718–728. doi:10.1080/19440040903573040 PMID:20352549

Naiker, S., & Odhav, B. (2004). Mycotic keratitis., profile of Fusarium species and their mycotoxins. Identification, mycotoxin risk and pathogenicity of Fusarium species associated with figendosepsis in Apulia, Italy. *Mycoses*, *47*, 50–56. doi:10.1046/j.0933-7407.2003.00936.x PMID:14998400

Nucci, M., & Anaissie, E. (2007). *Fusarium* infections in immune-compromised patients. *Clinical Microbiology Reviews*, *20*(4), 695–704. doi:10.1128/CMR.00014-07 PMID:17934079

Rafida, A. R., Faridah, S., Shahrul, A. A., Mazidah, M., & Zamri, I. (2016). Chronoamperometry Measurement For Rapid Cucumber Virus Detection in Plants. *Procedia Chemistry*, *20*, 25–28. doi:10.1016/j.proche.2016.07.003

Rojo, F. G., Reynoso, M. M., Sofia, M. F., & Torres, A. M. (2006). Biological control by *Trichoderma* species of *Fusarium solani* causing peanut brown root rot under field conditions. *Crop Protection (Guildford, Surrey)*, *26*(4), 549–555. doi:10.1016/j.cropro.2006.05.006

Romberg, M. K., & Davis, R. M. (2007). *Host range and phylogeny of Fusarium solani f. sp. eumartii from potato and tomato in California*. Academic Press.

Singh, J. (2001). Fusarium resistance via biotechnology. Final report for the Ontario research enhancement program.

Sinha, R., Vadivelmurugan, I., Basavaiah, M. R., Angappan, S., & Muthappa, S. K. (2019). Impact of drought stress on simultaneously occurring pathogen infection in field-grown chickpea. *Nature Report, 9*(1), 5577. doi:10.103841598-019-41463-z PMID:30944350

Tonnberg, V. (2016). *Evaluation of resistance against Fusarium root rot in Peas* (Master thesis). Faculty of Landscape Architecture, Horticulture and Crop protection science Swedish University of Agricultural Sciences, Anlarp, Sweden.

Vaseghi, A., Naser Safaie, N., Bakhshinejad, B., Mohsenifar, A., & Sadeghizadeh, M. (2013). Detection of *Pseudomonas syringae* pathovars by thiol-linked DNA–Gold nanoparticle probes. In G. Rivas & U. Weimar (Eds.), *Sensors and Actuators B: Chemical* (pp. 644–651). Tokyo, Japan: Elsiever. doi:10.1016/j.snb.2013.02.018

Wang, L., & Li, P. H. C. (2010). Gold nanoparticle-assisted single base-pair mismatch discrimination on a microfluidic microarray device. *Biomicroluidics, 4*(3), 032209. doi:10.1063/1.3463720 PMID:21045930

Yao, K. S., Li, S. J., Tzeng, K. C., Cheng, T. C., Chang, C. Y., Chiu, C. Y., ... Lin, Z. P. (2009). Fluorescence Silica Nanoprobe as a Biomarker for Rapid Detection of Plant Pathogens. *Advanced Materials Research, 79–82*, 513–516. . doi:10.4028/www.scientific.net/AMR.79-82.513

Zehhar, G., Ouazzani Touhami, A., Badoc, A., & Douira, A. (2006). Effet des *fusarium* des eaux de rizière sur la germination et la croissance des plantules de riz. *Bulletin Societé de Pharmacie de Bordeaux, 145*, 7-18.

Chapter 6
Reinstate Authentication of Nodes in Sensor Network

Ambika N.
(iD) https://orcid.org/0000-0003-4452-5514
Department of Computer Applications, SSMRV College, Bangalore, India

ABSTRACT

Transmitting data using the wireless medium provides a greater opportunity for the adversary to introduce different kinds of attacks into the network. Securing these monitoring devices deployed in unattended environments is a challenging task for the investigator. One of the preliminary practices adopted to ensure security is authentication. The chapter uses a new authentication procedure to evaluate the nodes. It uses sensors of a different caliber. Static and mobile nodes are used to bring the thought into play. The mobile nodes in the network demand vigorous defense mechanisms. The chapter defines a new authentication algorithm engaging itself to validate the communicating parties. The work is devised to substitute new nodes into the cluster to foster prolonged working of the network. By adopting the proposed protocol, backward and forward secrecy is preserved in the network. The work assures the security of the nodes by minimizing forge and replay attacks.

INTRODUCTION

The future demands machines to minimize human efforts and keep the work going at a faster pace. Sensors are a similar kind of device serving the purpose for quite a long time. These tiny devices (I.Akyildiz, Su, & Sankarasubramaniam, 2002) curtail the manual supervision done to monitor and track the objects of interest. These devices are available in variable sizes and hence are adaptable in many applications.

DOI: 10.4018/978-1-7998-4381-8.ch006

The applications range from military surveillance (Lee, Lee, Song, & Lee, 2009) to smart home monitoring systems (N.K.Suryadevara & Mukhopadhyay, 2012) (Zhang, Song, Wang, & Meng, 2011). Sensors were introduced into the real-time system as sensing equipment. They aid in the surveillance of the environment and report the monitored readings to the base station. Technology progress has added a wing making these devices more computational and smart. In turn, the demand has brought in a variety of new and sophisticated devices into existence. A node is also capable to move from one end of the network to another performing different task.

These tiny elements are not rechargeable. Hence to resolve this issue, new nodes are to be reinstated. Mobile nodes and static nodes are used in the proposed work. Static nodes engage themselves in sensing, processing and storing the data. The proposed work substitutes mobile nodes in place of inactive nodes to settle this issue. The mobile nodes act as collector nodes that can be replaced in place of surveillance nodes, which are either being discharged/ inactive or concluded as malicious nodes. The work also uses the benefits offered by today's technology to provide better security solutions. Previous practices use a multi-hop technique (Chatzigiannakis, Nikoletseas, & Strikos, 2006) (Isam.O & Hussain.S, 239-242) (Fabri.F, Buratti.C, & Verdone.R, 2008)to transmit the processed data to the base station. The same practice is used in this work. To preserve confidentiality (Baek, Foo, Tan, & Zhou, 2009), strong authentication algorithms are adopted.

Considering the different possibilities of node compromise authenticating (Ambika & Raju, 2014) both the communicating parties is the best available measure that can be adopted. Using this measure a new node can be entrusted into the cluster to gather data. A trusted entity has to certify the identity of new nodes before commencing any transmission. The proposed work adopts robust measures to provide a strong security solution to the same. The work assures better security to the nodes by tackling forge and replay attacks.

The work is divided into various segments. Section 2 details the previous work suggested by different authors. Section 3 summarizes different notations and the assumptions made in the study. Segment 4 details the proposed work. Section 5 analyses the security measures provided by the proposed protocol. Segment 6 provides a detailing description of the graphical representation of the work simulated in the NS2 environment. The work is concluded by highlighting the positive aspects of the study in section 7.

REVIEW OF THE PREVIOUS WORK

Verifying the identity of the communicating parties becomes essential before transmitting the confidential data. The previous works suggested by various authors focus on either one-way, mutual authentication or reliability by considering the acknowledgment sent by the sink node. This section details the work proposed by various authors.

In (Wong, Zheng, Cao, & Wang, 2006) the environment is divided into different zones. Any mobile user is an authorized user to access data from the sensors. The user is to register with the gateway by its name and password before communicating with the sensors. After registration, the user is provided a time to do his queries. During the procedure, the user is supposed to be stationary, login to the sensor login-node, query them and wait for the result. The procedure has to be repeated if the user has any more queries for the sensors. The scheme fails to provide security against replay and forgery attacks.

An energy-efficient communication paradigm is used in the work (Abraham & Ramanatha, 2006). The nodes receive the broadcasted messages of the base station through its neighbors. The nodes assume this to be a reliable path and utilize the same path to transmit their data. The base station gives flexibility for the new nodes to join the network by broadcasting *interest to join* message. One of the drawbacks of the work is that the new nodes joining the cluster are not authenticated before including them into the group.

In the study (Ibriq & Mahgoub, 2007) the nodes interested to join the clusters send their interest to the respective cluster head. On receiving the message the cluster forwards the same to the base station. The base verifies the legitimacy and reverts with a group ticket containing the identity of the nodes. On affirmation, the nodes are allowed to join the group. After authentication, the group head issues private key and group key to the base and the cluster members. The work falls short in providing better security as it considers using one-way authentication.

The study (Han, Shon, & Kim, 2010) uses the sink and other nodes in the network to maintain the node neighbor list. The nodes broadcast a nonce, using which the nodes not in the list acknowledge joining the list. The sink generates an authentication key by randomly choosing a nonce. A node willing to join the group chooses one of the nonces does a part of its processing and dispatches the same to the base station. A new authentication key is generated in case of connecting to a new sink. The work does not consider the possibility where the group head is compromised.

The study (Cheikhrouhou, Koubaa, Boujelbenl, & Abid, 2010)provides mutual authentication and session key agreement between the user and the server. The user is to be authenticated before gaining access to the nodes. The work (Butun & Sankar, 2011) is designed for the heterogeneous network which employs public and symmetric key cryptography approaches. The nodes of the cluster and the group head agree on a shared secret key generated by the Elliptic curve digital signature algorithm. The procedure proceeds with the registration phase where the user sends his ID encrypted with the public key. A certificate generated by the base station is transmitted to the user. The user uses this certificate to authenticate him for further communication. The user is provided with the flexibility to change his password. The user is not authenticated every time he tries to communicate. In both the suggested works the deployed node does not validate the user before transmitting the data.

The study (Chuchaisri & Newman, 2012) uses two new broadcast authentication schemes, key pool scheme, and a key chain scheme. Keys are stored in an access point while only a part of them are stored in the nodes. The scheme consists of 3 phases- the pre-deployment phase is where the local keys are embedded into the sensors. In the Signature phase, the second phase broadcast message and signature are generated. The last phase is known as message verification and forwarding, the node verification takes place. A key chain scheme is where a global key pool is used to store independent key chains without any partitioning. The work does not address the validation of newly deployed nodes in the environment.

In the study (Yao, Fukunaga, & Nakai, 2006), three phases are suggested- reliable broadcast of messages, legitimate acknowledgment from all the nodes in the network, disclosure of the message-authentication key. The server discloses the message authentication key after the server receives the positive acknowledgment from the nodes. The system specifies the number of messages to be authenticated per each of the keys in the one-way key chain. Making a comparison to the proposed work, the new nodes deployed do not hassle to validate its post-neighbors before joining the cluster.

Localized encryption and authentication protocol (Zhu, Setia, & Jajodia, 2004) are designed to support in-network processing tasks, restricting the nodes from getting compromised. Four sets of keys are generated, an individual key is shared with the base station, and a pairwise key is generated to share it with the other node in the network, a cluster key shared with multiple neighboring nodes and a group key shared with all the nodes in the network. The study uses one-way key chains for local broadcast authentication. The work secures the network from Hello flood attack, Sybil attack, and wormhole attack but not forge attack.

The protocol (Perrig, Szewczyk, Tygar, Wen, & Culler, 2002)is a security building blocks optimized for resource-constrained environments and wireless communications. The procedure consists of two secure building blocks. The *secure network encryption protocol* is designed to provide data confidentiality, two-party data authentication, and data freshness. The protocol adds 8 additional bytes per message giving low communication overhead, using cryptographic protocols, offers semantic security to the network. *µTESLA*, a protocol used for authenticated broadcast for severely resource-constrained environments. It is a timed, efficient, streaming, loss-tolerant authentication protocol. The protocol validates the initial packet with a digital signature using asymmetric schemes. The keys are disclosed once per epoch. It also restricts the number of authenticated senders. The issue of new nodes authentication is not dealt with in this work.

The study (P.Ning, Liu, & W.Du, 2008) is proposed to handle the Denial-of-Service attack against broadcast authentication. The work uses a message specific puzzle to defend the network from a DOS attack. The weak authenticator is verified, used as an additional layer of protection to filter out forged broadcast packets parallel reducing resource consumption. Strawman approach is used in the one-way key chain to provide weak authentication. The broadcast message along with the message index and the broadcast authenticator is considered as a puzzle. Undisclosed key in the one-way key chain is used to prevent forge attack and also at other end legitimate nodes will be able to verify the puzzling message. The work does not ensure protection against replay and forges attacks.

The combination of two kinds of sensors varying in their caliber is used in the work (Qiu, Zhou, Baek, & Lopez, 2010), dynamic and static nodes. When a mobile element moves to a new location, it sends a request to the base station. The sink makes a check into the revocation list, on affirmation transmits approval message with the session key to the cluster head. The head uses this message, recalculates the session key. Authenticating the cluster head is not given importance in the suggested work.

The algorithm (Wang & X.Liu, 2009) is an efficient short term public key broadcast authentication technique. Short length public/private keys are utilized, limiting the usage of keys to some minutes. There is a reduction in communication and computation, using one public key for a particular term. The work does not tackle forge and replay attacks.

The authors (Lu, Lin, Zhu, Liang, & Shen, 2012) have proposed work to reduce false data attacks into the network. A bandwidth-effective cooperative authentication scheme is based on random graph characteristics. Using this technique the network will be able to save a considerable amount of energy. The majority of false data with extra overheads at en-route nodes are eliminated.

A scalable authentication scheme (Li, Li, Ren, & Wu, 2012) is proposed enabling intermediate node authentication. An unlimited number of data can be transmitted eliminating the threshold problem. Message source privacy is enabled in the work.

The authors (Cho, Jo, Kwon, Chen, & Lee, 2013) have considered various to investigate – device types, detection methodologies, deployment strategies, and detection ranges. Based on the proposed criteria, the existing schemes are classified. The methodology aids in analyzing clone attacks in the sensor environment.

In the work, the authors (Wang, Liu, Zhang, & Zhang, 2014) collaborative false data injection attack is minimized. Two schemes are proposed in the work. The keys of the nodes are used to bind to the geographical location. Absolute locations of the sensors are used in the verification procedure. Relative positions are used in neighbor information based false data filtering scheme.

The authors (Jan, Nanda, Usman, & He, 2016) have designed a mutual authentication scheme for cluster-based networks. The methodology assists in deploying the optimal percentage of cluster heads to communicate with the neighboring nodes. The cluster head, in turn, authenticates the nearby nodes in forming the cluster.

A Symmetric key-based authentication scheme is proposed by the authors (Jaballah, Mosbah, Youssef, & Zemmari, 2013). Bloom filter mechanism is considered as the basis to design the proposed scheme. The binary tree analogy is used to distribute and authenticate keys.

A Reversible watermarking scheme was proposed by the authors (Shi & Xiao, 2013). A prediction error expansion analogy is used to design the work. The watermark bits are embedded by another group. The validation of data is done by the sink node. Table I provides the running time complexity of previous contributions made by various authors.

Table 1. Running Time Complexity of previous contributions

Previous contributions	Complexity	Previous contributions	complexity
(Wong, Zheng, Cao, & Wang, 2006)	$O(K^n)$	(P.Ning, Liu, & W.Du, 2008)	$O(n)$
(Abraham & Ramanatha, 2006)	$O(n)$	(Qiu, Zhou, Baek, & Lopez, 2010)	$O(n)$
(Ibriq & Mahgoub, 2007)	$\Theta(n\log n)$	(Wang & X.Liu, 2009)	$O(n)$
(Han, Shon, & Kim, 2010)	$\Omega(2^n)$	(Lu, Lin, Zhu, Liang, & Shen, 2012)	$O(n)$
(Cheikhrouhou, Koubaa, Boujelbenl, & Abid, 2010)	$O(n)$	(Li, Li, Ren, & Wu, 2012)	$\Theta(K^n)$
(Butun & Sankar, 2011)	$O(n\log n)$	(Cho, Jo, Kwon, Chen, & Lee, 2013)	$O(n)$
(Chuchaisri & Newman, 2012)	$\Omega(K^n)$	(Wang, Liu, Zhang, & Zhang, 2014)	$O(n)$
(Yao, Fukunaga, & Nakai, 2006)	$O(n\log n)$	(Jan, Nanda, Usman, & He, 2016)	$\Theta(n)$
(Zhu, Setia, & Jajodia, 2004)	$O(n\log n)$	(Jaballah, Mosbah, Youssef, & Zemmari, 2013)	$\Theta(2^n)$
(Perrig, Szewczyk, Tygar, Wen, & Culler, 2002)	$O(2^n)$	(Shi & Xiao, 2013)	$O(n)$

NOTATIONS AND ASSUMPTIONS

Notations

This section gives a description of the notations used in this work. Table II gives the notations used in the study with its description.

Table 2. Notations used in the work

Notation used	Description
N	Network under study
BS	Sink node/ base station
r	Transmission range of the communicating nodes
M_i	i^{th} dynamic node of the network
CH_i	Cluster head of ith cluster of the network N
Loc_i	Location details of node N_i of the network
S_i	Stored keys stored in the nodes
Id_i	Unique Identification number assigned to the node N_i by the base station
H_i	Hash message generated by node N_i
H_m	Hash message generated by mobile agent
Msg_{bs}	Evaluation message transmitted by the base station

Assumptions

The following assumptions are made in the proposed study-

- The base station is assumed to be trustworthy. It is the responsibility of the sink node to generate and embed keying credentials into the nodes. Keys, diverse algorithms and other keying material are embedded into the nodes before deployment.
- Static and mobile nodes are used in the work.
- The network is divided into regions. The mobile node is assigned to position itself in the required area and accomplish the task of the replaced node.
- All the nodes, mobile and static ones are assumed to contain a global positioning system component.
- All the nodes in the network are liable to get compromised.

PROPOSED SYSTEM

The network is divided into fixed regions. The nodes are embedded with the keying material inside them by the base station and deployed in the environment N under study. The static nodes are deployed randomly in the environment. The nodes self configure and broadcast *Hello* messages to form the clusters (Dasgupta & Namjoshi, 2003). In the notation (1), node N_i is broadcasting *Hello* message to the network N.

The sensors within the transmission range r, which can hear the broadcast message acknowledge with *Ack* message. In notation (2), the node N_j acknowledges with *Ack* message.

$$N_i \circledR N : Hello \tag{1}$$

$$N_j \circledR N_i : Ack \tag{2}$$

The nodes which can communicate form the cluster. The cluster members choose their cluster head based on the proximity from the next available hop and the residual energy. The cluster head is responsible for aggregate (Anitha & Sumathi, 2014) the cluster data.

Authentication Between Mobile Element and Base Station

When the clusters have a shortage of sensing nodes in their cluster they raise a request to the sink node. In equation (3) the cluster head CH_i is requesting Req_i to the base station BS to assign a new node to its cluster.

$$CH_i \circledR BS : Req_i \tag{3}$$

The base station assigns the job to one of its mobile nodes. The mobile node, in turn, positions itself in its assigned location. After taking its position in the allocated location, the mobile agent M_i transmits *Hello* message to the respective cluster head CH_i. The same is depicted in notation (4). If the transmission range is within r, the cluster head acknowledges the message.

$$M_i \circledR CH_i : Hello \tag{4}$$

In notation (5), the cluster head CH_i after the message acknowledges by transmitting *Ack* message to the respective communicating node.

$$CH_i \circledR M_i : Ack \tag{5}$$

The mobile agents in the study increase the lifespan of the network and reduce the effort of the investigator to install new nodes manually. Simultaneously it adds a threat caused due to unawareness of the agent. Hence the initial measure considered in the work is to check the legitimacy of the mobile node. The cluster head and mobile agent generates the respective hash message using its identification number id_i, randomly chosen stored key S_i and location details Loc_i. The same is depicted in equation (6). The nodes ensure to generate different hash messages for different authenticating parties. The mobile agent transmits the hash message to the respective cluster head. In equation (7), mobile node M_i is transmitting hashed message H_m to the cluster head CH_i.

$$H_i \circledR hash[id_i, S_i, Loc_i] \tag{6}$$

$$M_i \circledR CH_i: H_m \tag{7}$$

$$CH_i \circledR BS: H_i \| H_m \tag{8}$$

The cluster head CH_i, after receiving the hash message from the mobile node, transmits the received and generated hashed message to the base station BS. In equation (7) the cluster head CH_i is transmitting concatenated hash message H_i and H_m to the base station BS. Both the nodes delete the hash message after transmission to employ backward secrecy. The forward secrecy is maintained by choosing a random symmetric key.

After evaluating the legitimacy of the received message, the base station transmits the evaluated result (Lee, HangRok, Choi, & Kim, 2005) to both the communicating parties. In the equation (9) the base station BS is transmitting the evaluation message Msg_{bs} to the cluster head CH_i and the mobile node M_i.

$$BS_i \circledR CH_i, M_i: Msg_{bs} \tag{9}$$

After receiving the affirmation message from the base station the mobile agent is considered as one of the cluster members.

Algorithm: Authentication Procedure

Step 1: The cluster head and the mobile agent generate the respective hash message using its unique identification number, randomly chosen stored key and location information.

Step 2: The cluster head concatenates the generated and received hash message. The concatenated message is transmitted to the base station.

Step 3: The base station evaluates the information obtained by cluster head.

Step 4: The nodes receive the evaluation result from the base station. Based on the result, the mobile node is either included in the cluster or discarded followed by a message to the base station.

SECURITY ANALYSIS

Unsupervised environment carrying confidential information provides the adversary an opportunity to introduce different kinds of attacks into the network. Authentication is a general practice adopted to evaluate the communicating parties before commencing any transmission. Authenticating both parties ensures better security in the network.

The proposed work considers including a mobile agent as a cluster member. After receiving a node request, the mobile agent is assigned the task to replace the same. The agent positions itself in the assigned location. The mobile agent is programmed to transmit the hash message to the respective cluster head. The cluster head communicates the same along with its own generated hash message to the base station. The base station verifying the same sends an acknowledgment picturing the verification process. If the message turns to be positive, the mobile agent is included as a cluster member in the respective cluster. If the mobile agent is proved to be guilty, the cluster head discards the mobile agent. If the cluster head is proved guilty, other nodes of the cluster are notified to choose another node as its cluster head. The notification is transmitted to the base station.

Forge Attacks

The message transmitted from one entity to another can be replicated to introduce different kinds of attacks into the network. This kind of strike is known as Forge attack. The proposed work ensures a large amount of safety against this kind of attack (Karlof, Chris, & Wagner, 2003).

Static nodes are deployed in the environment under study to form the cluster. The cluster members are lead by the group head. The static nodes are responsible to sense the environment and the group head aggregates the data. The mobile agent is used to substitute the inactive/malicious nodes of the cluster. Both the communicating nodes are evaluated for their legitimacy using the hash message generated by the respective nodes. As cluster heads are the more trustable nodes among the two communicating parties, the group head collects the hash message from the mobile node. It concatenates the same with its generated hash message. The resultant message is dispatched to the base station. On affirmation, the mobile node is substituted for the inactive/malicious node of the group. As the nodes ensure to generate different hash message values for different communicating parties, forge attack is minimized.

Replay Attacks

Replay attack (Pathan, Khan, Lee, & Hong, 2006) is where the adversary replicates the data and retransmits them. The base station will get a false illusion of the environment and will not be able to take accurate action on time. To enhance security the generated authentication metric values calculated are erased after transmission, preserving backward secrecy of the network.

Simulation and Its Results

The work is simulated using NS2. The nodes of the network are deployed in the area of dimension 200 * 200 m. The area is divided into fixed regions of dimension 50m*50m. Tinynode 584 is used as the static node and Waspmote is used as the mobile node in the network. 63 static nodes are deployed in the environment randomly. 6 mobile nodes are used to evaluate the outcome of the proposed work. The hash message generated by the nodes deployed in the environment is of length 244bits and that of the base station is 120bits. The simulation is conducted for 240s. The experimental setup used to simulate the work is detailed in Table III.

Authentication Efficiency

Authentication is a process that aids in increasing the reliability of the data. To enhance security (Fatima, Ahmad, & Khan, 2014) in the network, both the communicating parties are evaluated for their legitimacy. The proposed work considers using a mobile agent to substitute the inactive/malicious nodes of the cluster. A mobile element is liable to get compromised likely than the static ones. Hence the mobile node is evaluated for its legitimacy as it takes its position in the assigned location. Both the communicating parties generate the hash message using location details, unique identification number and randomly chosen stored key. The generated hash message is transmitted to the base station to verify their legitimacy. The sink node after validating the received message transmits its acknowledgment to the cluster head and mobile agent. On receiving an affirmation message from the base station, the mobile agent is replaced in place of the inactive node.

Table 3. Experimental Setup Parameters

Description	Quantity
Area under surveillance	200m * 200m
Count of static nodes	63
Count of mobile agents	6
Number of clusters in the network	9
Region size (fixed)	50m*50m
Total number of regions assigned	16
Number of hops considered	0-3 hops
Time duration	240s
Length of hash message generated by the respective nodes	244 bits
Length of the hash message transmitted by the base station	120 bits
Waspmote configuration(mobile node)	
Data transmission speed	38,000bps
Data payload	59 bytes
Transmission time	1.89ms
Preamble	18 bytes
Number of keys stored in the node	300
Tinynode 584 configuration (static node)	
Data frame size	272 bits
Acknowledgement frame size	64 bits
Data bit rate	76kbps
Preamble	6 bytes
Number of keys stored in the node	75

Figure 1. Comparison of Authentication efficiency of proposed and previous models

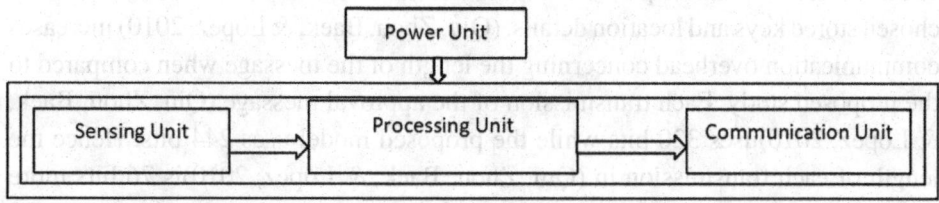

141

Fig 1, represents the comparison made between the proposed work and two previous works. In (Abraham & Ramanatha, 2006), the base station broadcasts messages to other nodes in the network through their neighbors. The nodes are given an assumption to use the path indicated by their neighbors as the best and reliable path. Hence the nodes use the same path to transmit their data.

In the study (Ibriq & Mahgoub, 2007)the nodes interested to join the clusters send their interest to the respective cluster head. On receiving the message the cluster forwards the same to the base station. The base verifies the legitimacy and reverts with a group ticket containing the identity of the nodes. On affirmation, the nodes are allowed to join the group. After authentication, the group head issues private key and group key to the base and the cluster members. While in the proposed model, an additional measure of security is taken. Apart from base station providing its confirmation on the validity of the mobile node, both the communicating parties are evaluated by the base station.

In (Qiu, Zhou, Baek, & Lopez, 2010)when a mobile element moves to a new location, it sends a request to the base station. The sink makes a check into the revocation list, on affirmation transmits approval message with the session key to the cluster head. The head uses this message, recalculates the session key. Comparing the work with the proposed study, the suggested works (Abraham & Ramanatha, 2006), (Qiu, Zhou, Baek, & Lopez, 2010)and (Ibriq & Mahgoub, 2007)does not provide a platform to evaluate the legitimacy of the cluster head before transmission. Hence the proposed work increases the lifespan of the network.

Communication Overhead

A Combination of two kinds of sensors varying in their caliber is used in (Qiu, Zhou, Baek, & Lopez, 2010). When a mobile element moves to a new location, the cluster head sends a concatenated hash message to the base station. The sink makes a check into the revocation list, on affirmation transmits acknowledgement to the cluster head and mobile agent. In the proposed work validation of the mobile node and cluster head is performed based on its identification number, randomly chosen stored keys and location details. (Qiu, Zhou, Baek, & Lopez, 2010) increases communication overhead concerning the length of the message when compared to the proposed study. Each transmission of the approval message (Qiu, Zhou, Baek, & Lopez, 2010)uses 320 bits while the proposed model uses 244 bits. Hence the length of each transmission in (Qiu, Zhou, Baek, & Lopez, 2010)is 76 bits more compared to the proposed work. The results obtained during the simulation are listed in Table IV.

Table 4. Results obtained during simulation

Weighed metrics	Proposed work
Reliability to data	>=0.99
Probability to detect compromised node	>=0.98
Probability of false alarm	<=0.001

CONCLUSION

Sensor nodes are liable to get compromised in the unsupervised environment. Hence better security measures are to be adopted. Authentication is the preliminary procedure adopted to ensure secure communication. The heterogeneity system is considered by including dynamic and static nodes in the network. To elongate the lifespan of the working nodes in the network, mobile nodes are replaced in place of inactive nodes. Since nodes are more liable to get compromised, they are evaluated for their legitimacy by the base station using the hash message generated by both the communicating parties. Forge attack and replay attacks are minimized by adopting this procedure. Hence overall the work assures better security in the network by fueling the network with active nodes.

The proposed work can be made more effective if energy consumption is minimized. The cluster head consumes a lot of energy to transmit the message to the base station. The future work can consider in minimizing energy to authenticate the new nodes simultaneously looking out for a reliable source for authentication.

REFERENCES

Abraham, J., & Ramanatha, K. S. (2006). An Efficient Protocol for Authentication and Initial Shared Key Establishment in Clustered Wireless Sensor Networks. In *Proc. Third International conference on Wireless and optical Communication networks* (pp. 5-10). Bangalore, India: IEEE. 10.1109/WOCN.2006.1666568

I.Akyildiz, Su, W., & Sankarasubramaniam, Y. (2002). A Survey On Sensor Networks. *IEEE Communications*, 102-114.

Ambika, N., & Raju, G. T. (2014). ECAWSN: Eliminating compromised node with the help of auxiliary nodes in wireless sensor network. *International Journal of Security and Networks*, 9(2), 78–84. doi:10.1504/IJSN.2014.060743

Anitha, C. L., & Sumathi, R. (2014). Comparative Analysis of Data Aggregation Algorithms Under Various Architectural Models in Wireless Sensor Networks. *International Journal of Information Technology*, 6(2), 757–763.

Baek, J., Foo, E., Tan, H., & Zhou, J. (2009). Securing wireless sensor networks— threats and Counter measures. In *Security and Privacy in Wireless and Mobile*. Troubador Publishing.

Butun, I., & Sankar, R. (2011). Advanced Two Tier User Authentication Scheme for Heterogeneous Wireless Sensor Networks. In *Proc. IEEE Consumer communications and networking conference* (pp. 169-171). Las Vegas, NV: IEEE. 10.1109/ CCNC.2011.5766446

Chatzigiannakis, I., Nikoletseas, S., & Strikos, A. (2006). Experimental evaluation of the performance of multi-hop wireless sensor networks. In Proc. 5th Communication Systems, Networks and Digital Signal Processing, (pp. 19-21). Patras, Greece: Academic Press.

Cheikhrouhou, O., Koubaa, A., Boujelbenl, M., & Abid, M. (2010). A lightweight user Authentication Scheme for Wireless Sensor Networks. In *International Conference on Computer Systems and Applications - AICCSA 2010* (pp. 1-7). Hammamet, Tunisia: IEEE. 10.1109/AICCSA.2010.5586995

Cho, K., Jo, M., Kwon, T., Chen, H.-H., & Lee, D. H. (2013). Classification and Experimental Analysis for Clone Detection Approaches in Wireless Sensor Networks. *IEEE Systems Journal*, 7(1), 26–35. doi:10.1109/JSYST.2012.2188689

Chuchaisri, P., & Newman, R. (2012). Fast response PKC-based broadcast authentication in wireless sensor network. *Mobile Networks and Applications*, 17(4), 508–525. doi:10.100711036-011-0349-8

Dasgupta, K. K., & Namjoshi, P. (2003). An efficient clustering-based heuristic for data gathering and aggregation in sensor networks. In *IEEE Wireless Communications and Networking Conference* (pp. 1948-1953). New Orleans, LA: IEEE. 10.1109/ WCNC.2003.1200685

Fabri, F., Buratti, C., & Verdone, R. (2008). A multi-sink multi-hop wireless sensor network over a square region: connectivity and energy consumption issues. In *GLOBECOM Workshop* (pp. 1-6). New Orleans, LA: IEEE. 10.1109/ GLOCOMW.2008.ECP.38

Fatima, S., Ahmad, S., & Khan, P. M. (2014). Certificate Based Security Services in Adhoc Sensor Network. *International Journal of Information Technology, 6*(2), 783–790.

Han, K., Shon, T., & Kim, K. (2010). Efficient Mobile Sensor Authentication in Smarth Home and WPAN. *IEEE Transactions on Consumer Electronics, 56*(2), 591–596. doi:10.1109/TCE.2010.5505975

Ibriq, J., & Mahgoub, I. (2007). A Hierarchical Key Establishment Scheme for Wireless Sensor Networks. In *21st International Conference on Advanced Networking and Applications* (pp. 210-219). Niagara Falls, Ontario, Canada: IEEE. 10.1109/AINA.2007.14

Isam, O., & Hussain, S. (239-242). An intelligent multi-hop routing for wireless sensor networks. In *IEEE/WIC/ACM International Conference on Web intelligence and Intelligent agent technology workshops* (pp. 239-242). Hong Kong, China: IEEE.

Jaballah, W. B., Mosbah, M., Youssef, H., & Zemmari, A. (2013). Lightweight Source Authentication Mechanisms for Group Communications in Wireless Sensor Networks. In *IEEE 27th International Conference on Advanced Information Networking and Applications (AINA)* (pp. 598-605). Barcelona, Spain: IEEE.

Jan, M., Nanda, P., Usman, M., & He, X. (2016). PAWN: a payload-based mutual authentication scheme for wireless sensor networks. *Concurrency and Computation Practice and Experience, 29*(17), 1-32.

Karlof, C., & Wagner, D. (2003). Secure routing in wireless sensor networks: Attacks and countermeasures. *Ad Hoc Networks, 1*(2-3), 293–315. doi:10.1016/S1570-8705(03)00008-8

Lee, S. H., Lee, S., Song, H., & Lee, H. S. (2009). Wireless sensor network design for tactical military applications: Remote largescale environments. In *Proc. IEEE conference in Military Communications* (pp. 1-7). Boston, MA: IEEE. 10.1109/MILCOM.2009.5379900

Lee, H., Choi, Y., & Kim, H. (2005). Implementation of tiny hash based on hash algorithm for sensor network. *Proceedings of World Academy of Science, Engineering and Technology, 10*, 135-139.

Li, Y., Li, J., Ren, J., & Wu, J. (2012). *Providing hop-by-hop authentication and source privacy in wireless sensor networks. In IEEE INFOCOM* (pp. 3071–3075). Orlando, FL: IEEE.

Lu, R., Lin, X., Zhu, H., Liang, X., & Shen, X. (2012). BECAN: A Bandwidth-Efficient Cooperative Authentication Scheme for Filtering Injected False Data in Wireless Sensor Networks. *IEEE Transactions on Parallel and Distributed Systems*, 32–43.

Ning, P., Liu, A., & Du, W. (2008). Mitigating DOS attacks against broadcast authentication in WSN. *ACM Transactions on Sensor Networks*, *4*, 1. doi:10.1145/1325651.1325652

Pathan, K. A.-S., Lee, H.-W., & Hong, C. S. (2006). Security in wireless sensor networks: issues and challenges. In *The 8th International Conference Advanced Communication Technology* (p. 6). Phoenix Park, South Korea: IEEE.

Perrig, A., Szewczyk, R., Tygar, J. D., Wen, V., & Culler, D. E. (2002). SPINS: Security Protocols For Sensor Networks. *Wireless Networks*, *8*(5), 521–534. doi:10.1023/A:1016598314198

Qiu, Y., Zhou, J., Baek, J., & Lopez, J. (2010). Authentication and key establishment in dynamic wireless sensor network. *Sensors (Basel)*, *10*(4), 3718–3731. doi:10.3390100403718 PMID:22319321

Shi, X., & Xiao, D. (2013). A reversible watermarking authentication scheme for wireless sensor networks. *Information Sciences*, *240*, 173–183. doi:10.1016/j.ins.2013.03.031

Suryadevara, N. K., & Mukhopadhyay, S. (2012). Wireless sensor network based home monitoring system for wellness determination of elderly. *IEEE Sensors Journal*, *12*(6), 1965–1972. doi:10.1109/JSEN.2011.2182341

Wang, J., Liu, Z., Zhang, S., & Zhang, X. (2014). Defending collaborative false data injection attacks in wireless sensor networks. *Information Sciences*, *254*, 39–53. doi:10.1016/j.ins.2013.08.019

Wang, W. D., & Liu, X. (2009). ShortPK: a short term public key scheme for broadcast authentication in WSN. *ACM Transactions on Sensor Networks*, *4*(1), 29.

Wong, K., Zheng, Y., Cao, J., & Wang, S. (2006). A dynamic user authentication scheme for wireless sensor networks. *Proc. IEEE International Conference on Sensor Networks, Ubiquitous, and Trustworthy Computing*, *1*, 8. 10.1109/SUTC.2006.1636182

Yao, T., Fukunaga, S., & Nakai, T. (2006). Reliable Broadcast authentication in Wireless sensor networks. In *Proc. EUC 2006 Workshops* (pp. 271-280). Seoul, South Korea: Springer. 10.1007/11807964_28

Zhang, J., Song, G., Wang, H., & Meng, T. (2011). Design of a wireless sensor network based monitoring system for home automation. *Proc. International conference on future computer sciences and applications*, 57-60. 10.1109/ICFCSA.2011.20

Zhu, S., Setia, S., & Jajodia, S. (2004). LEAP: efficient security mechanishms in large scale distributed networks. In *10th ACM conference on Computer and communications security* (pp. 62-72). Washington, DC: ACM.

Chapter 7
Comprehensive Ontological Model for Senior Wellness Activity Recognition in Smart Homes

Hajar Khallouki

(iD) https://orcid.org/0000-0002-1212-3740
Department of Software Engineering, Lakehead University, Canada

Rachid Benlamri
Department of Software Engineering, Lakehead University, Canada

Abdulsalalm Yassine

(iD) https://orcid.org/0000-0003-3539-0945
Department of Software Engineering, Lakehead University, Canada

ABSTRACT

There are several works in the field of smart homes for healthcare, with different types of sensors used to monitor medical, behavioral and environmental parameters for patients. In the context of smart home for the elderly, the use of sensors needs to be adapted to respect the privacy of elders and to work passively without the need for caregiver assistance. Most research in this area focused on activity recognition (e.g. eating, sleeping, watching TV, etc.) which may be defined as the identification of a sequence of actions (e.g. using microwave, lying down, etc.). In this chapter, we propose a comprehensive ontological model for well-being activity recognition in smart home. Our approach takes into account different aspects of the well-being context such as patient profile, object being used to perform the activity, the time of running the activity, its location, etc. In order to validate the proposed ontology and reason on it, we perform a set of queries and inference rules.

DOI: 10.4018/978-1-7998-4381-8.ch007

1. INTRODUCTION

In recent decades, the Western world has experienced many changes that could have a strong influence on our daily lives: the aging population and the development of advance technologies.

The aging population is a real problem for the elderly. They are most likely in need of particular help due to their loss of autonomy, sometimes forcing them to have a caregiver at home. However, the need to have a caregiver almost all the time with the elders depletes the human and financial resources of the health system. In order to reduce the impact of this issue, recent researches are oriented towards reducing the negative effects of traditional assistance. Availing the evolution of information technologies, electronics and especially the emergence of the field of ubiquitous intelligence (Ramos et al., 2008). This new field aims to improve people's life by introducing tools, transparently to the user miniaturized technologies in their daily routine. The vision of assisted technology is to create a smart environment (e.g. smart homes) filled with sensors and devices to monitor and track people's behavior in order to enhance their wellbeing (Icquebourg et al., 2006).

Research on smart homes has been the subject of great enthusiasm in recent decades because of the corresponding societal stake towards improving the quality of life. However, the numerous projects carried out in all countries have not been sufficient to meet the many challenges posed. It is also necessary to implement user interfaces that are easy to use for people who are not accustomed to interacting with conventional information technologies. These technologies should therefore not appear as new constraints disrupting the daily activities but as real assistance tools which make life easier for the user. The aim of this research is to develop an intelligent alert system that warns caregivers about abnormal states of elderly people, patients, etc. The system consists of ontology and machine learning mechanisms capable of identifying behaviors that are potentially harmful to the well-being of the user.

Activity recognition is an innate act in humans. Indeed, every time as we observe a person, we cannot help thinking about what he is doing or what he intends to do. For example, a person, at 9:00 am, grab a cup and sit on the chair, we can associate this action, with the activity of having breakfast.

There are multiple activities that are carried out routinely by healthy people. However, people that lack autonomy, especially those people with dementia like Alzheimer's disease no longer realize these activities as naturally as before. They tend to forget what to do, or do their tasks partially, or on the contrary, do them several times. We must then find a way to effectively remind them of the activities they have to perform, without disturbing them by reminding them of an activity already done. It is at this level that the notion of ubiquitous intelligence comes into play: by collecting information from the smart home, it is possible to know what the person is doing. Once you know what the person is doing, it becomes possible to adapt the reminders of activities according to what it does, so remember several times an activity that would have been missed and conversely do not recall one that would have been done in advance or on time.

In this chapter, we propose a knowledge representation model for the smart home. The model to be developed must be able to express, easily and unambiguously, the complex relationships occurring in the perceptual environment. Knowledge needs to be shared by several inference modules; it is therefore necessary to adopt a model that offers a common vocabulary based on standard language as well as standard representations.

The rest of the chapter is organized as follows. Sections II and III introduce the background of our research work and description of some related work respectively. Section IV presents the proposed ontology and its components. Section V outlines the implementation of the proposed knowledge representation model. Finally, Section VI concludes the chapter and suggests further research work.

2. BACKGROUND

2.1. Activities of Daily Living (ADL)

The concept of ADL was described for the first time by Dr. Katz (1963) as the set of activities carried out by an individual in his routine to take care of himself. This includes activities such as meal preparation, clothing, sleeping, etc. Health professionals often assess a person's level of autonomy based on their ability or inability to exercise certain ADLs. In other words, the ADLs are a set of activities that a normal person is supposed to be able to perform to qualify as autonomous. Today, researchers distinguish two different types of ADL (Plantevin et al., 2017):

- **Basic ADL:** The basic activities of daily living are all of the activities that are fundamental and mandatory to meet the primary needs of a person. This includes the ability to move around, go to the bathroom, etc. These activities consist of only a few steps and do not require a real planning.
- **Instrumental ADL:** This kind of activity requires planning and involves manipulation of objects. These activities are necessary for living independently and autonomously. For a person, being able to perform all instrumental ADLs means to be relatively independent. This category includes activities such as preparing a meal, managing your money, and using a phone. Instrumental ADLs are more complex than basic ones. They are composed of a number of steps and they require better planning.

In the literature on technological assistance inside intelligent homes (Giroux et al., 2009), researchers generally generated ADLs without distinguishing them. However, most of the time research focuses on recognition and help of instrumental ADLs. The main reason is that the person who cannot successfully perform the basic ADL will have more comprehensive care needs than the help offered by technology inside a smart home.

2.2. Activity Recognition

The definition of activity recognition has evolved since the last decade. Many authors have attempted to adapt it to the very specific context of the activity recognition within a smart home (Bouchard et al., 2007; Patterson et al., 2005; Boger et al., 2006). The trend has been to refine the concept of the ambient environment so to formally link it to the challenge of the problem of activity recognition. For example, Goldman (Goldman et al., 1999) described the process as inferring an agent's behaviors according to observing its actions. The main difference from the previous definitions is that of the action of the observed entity and the observation perceived by the observer. This distinction reflects the fact that the actions are not directly observable in the smart home background. Patterson (Patterson et al., 2005) recently proposed to update the definition by specifying that observations are made from data collected from low level sensors. Moreover, this new definition adheres to the paradigm of ubiquitous computing (Weiser, 1991) and it is much closer to the reality of the problem. He encourages the creation of an improved environment where sensors will be integrated into objects so as not to be intrusive. This definition presents the problem as a whole observation made by different sensors in order to make decisions to help a resident with cognitive impairment. It presents the problem of activity recognition realistically, considering that we do not have access to basic actions executed by the observed entity.

2.3. Context

The context is defined as information about the environment of a computing system, or as conditions that determine an event. Describing all of these conditions and this information seems ambitious, what prompted the researchers to analyze, define and redefine the term context to use in their research works. Schilit et al. (1994) considered that context holds three important aspects that give answers to the following questions: where are you? With who? What resources do you have nearby?

Schilit and Theimer (1994) defined the context as the changes in the physical environment according to: location, description of people and objects in the environment and the changes of these objects. Later, Brown introduces the time, season, temperature, identity and location of the user to his definition of the context. Brown et al. (1997) presented a set of extensible elements to characterize the context that the basic elements are the location, the set of objects that the user needs, the time and spatial direction.

Simultaneously with Brown's work, definitions emerge with the explicit introduction of time and the notion of state. For example, Ryan et al. (1997) stated that "Context elements are user location, environment, identity and time". In another study, Pascoce (1998) describes the context as a subset of physical and conceptual states of interest to a particular entity. Chen and Kotz (2000) stated that context is a "set of environmental states and parameters that determine the behavior of an application or in which an event of the application takes place and of interest to the user".

Dey (2001) insists on the notion of relevance of information by providing a definition in which he tries to specify the nature of context-related entities: "any information that can be used to characterize the situation of an entity. Any entity is a person, or an object that is considered significant to the interaction between the user and the application, including the user and the application itself".

Winograd (2001) agreed with Dey's definition and defended the fact that it covers all existing work on the context. However, he considers that the expressions used by Dey as "all information" and "to characterize an entity" remain of a very general order and do not mark any limit to the notion of context (everything can be context). To further clarify Dey's definition, Winograd presents the context as a set of structured and shared information. He elaborates on this definition by saying: "the consideration of information as context is due to the way it is used and its inherent properties". He supports his idea by example: "the voltage of the electricity lines is part of the context if the system depends on it; otherwise, it can only be one parameter of the environment".

2.4. Ontology

The Artificial Intelligence and Knowledge Engineering community has proposed several definitions to define what computer ontology is and what ontology serves. In the early 1990s, Gruber proposed the following definition: An ontology is a "formal representation of knowledge" (Gruber et al., 1994). Then, he adds that an ontology is a "formal and explicit specification of a shared conceptualization" (Guarino, 1995). The knowledge must be not only formalized but also shared by several people. Guarino then reviews Gruber's definitions by defining an ontology as the "common and shared understanding of a domain that can be communicated between people and systems" (Cherlet et al., 2003).

The ontology must be understood by several people, but also understood by software. Charlet recalls that the representation of an ontology constitutes a set of classes of objects. It defines ontology as a "standard specification representing classes of objects recognized as existing in a domain." Building an ontology also means deciding how to be and exist objects in that domain" (Lassila & Swick, 1998). Bachimont emphasizes that an ontology refers to logic. He defines an ontology as a "rigorous and structured description of the vocabulary, in the logical sense, of a specialized field" (Bachimont, 2004).

Concretely, an ontology models the knowledge of a domain by a set of structured concepts in a network of dependencies of hierarchical relations (relation is-a inheritance of properties between the concepts) and of semantic relations (describing the properties of the concepts or roles between concepts).

2.5. Ontology Representation Standards

For an ontology to be manipulated by a program, a formal representation must be defined. The W3C (World Wide Web Consortium) then sets up working groups to propose several standards of knowledge representation for ontologies. We list some of these languages to show their diversity and their adequacy to represent an ontology.

2.5.1. Resource Description Framework (RDF)

RDF is a W3C recommendation developed to describe Web resources, that is, any entity referenced by a URI. This standard has been developed to manipulate and classify Web metadata, combine different resources to produce new information, and facilitate the processing of Web information by software agents. The RDF standard makes it possible to organize the resources of the Web in the form of triplets (subject, predicate and object) or again (resource, property, value). The elements of these triplets can be URIs (Universal Resource Identifiers), literals or variables. RDF data can be represented as a graph, whose nodes correspond to subjects (or resources) and objects (or values), and the arcs correspond to relationships between these elements. The RDF language is not very expressive because it only allows expressing relationships in the form of triplets. It does not allow to model the semantics of ontologies essentially at the level of classes.

2.5.2. Resource Description Framework Schema (RDFS)

The RDFS language (W3C recommendation) is an extension of RDF. It makes it possible to express semantic relations at the level of classes and properties (called RDF predicates). A schema (in the sense of XML) makes it possible to type the resources and the relations forming the triplets. RDFS relies on the XML syntax of RDF and extends it with new constructors.

- The element (rdfs: Class) allows to define classes;
- The element (rdfs: property) allows to name relations (or roles) between classes;
- The elements (rdfs: domain) and (rdfs: range) allow for a property to specify the source class and the target class of this property;
- The element (rdfs: subClassOf) makes it possible to define a hierarchy of classes;
- The element (rdfs: subPropertyOf) makes it possible to define a hierarchy of properties;
- The element (rdfs: type) binds an instance to a class.

2.5.3. Ontology Web Language (OWL)

The rapid development of ontology-based applications and the need to model more and more complex knowledge has led to some limitations in RDF / RDFS. Indeed, RDFS offers a simple vocabulary, limited to a hierarchy of classes, a hierarchy of relations and definitions of the domains of application of the latter ("domain" and "range") (Baget, 2005).

OWL (McGuinness, 2004) has been recommended by the W3C to enrich RDFS by defining a more complete vocabulary for the description of complex ontologies. The OWL language is dedicated to the definitions of classes and types of properties, and thus to the definition of ontologies. Inspired by the logics of descriptions and successor of two DAML+OIL6 languages, it provides a large number of constructors to express very finely the properties of defined classes. The OWL language provides mechanisms to create all the components of an ontology: classes, instances, properties, and axioms. OWL also relies on the syntax of RDF triples and reuses some RDFS elements. For example, it takes the subclassof relationship from RDFS, providing a mechanism for reasoning and property inheritance. The OWL language is divided into three sub-languages (OWL Lite, OWL DL and OWL Full) (Baget, 2005).

In this chapter we use the OWL language to formalize and implement ontologies. It is a choice imposed by the context in which the chapter is integrated. However, we would have made this choice because OWL is the most used language currently.

3. RELATED WORK

As already stated above, the activities recognition process is the most delicate in technological assistance, which explains the large number of researches devoted to this field (Ordóñez et al., 2013; Moutacalli et al., 2015; Suryadevara et al., 2013; Spriggs et al., 2009; Moutacalli et al., 2015). To better understand the activity recognition, Kautz (1991) defined this concept as being the set of activities that an individual performs as a routine for taking care of themselves, such as, preparing food, dressing, washing, etc. Authors in (Ye et all., 2012) define an activity as follows: "An activity is defined as an external semantic interpretation of the sensor data". Interpretation means that activities assign meanings to sensor data. In the above-mentioned definition, "external" means that interpretation is done at the application level, rather only at the level of the sensors. Also, "semantic" means that the interpretation assigns a sense to sensor data.

Activities recognition was defined by (Endsley, 1995) as "the perception of elements in the environment within an interval of time and space, the comprehension of their meaning, and the projection of their status in the near future". More recently, (Roy et al., 2013) add that it is possible to characterize the recognition of activities by the relationship between the observer and the observed. It allows to establish a classification according to whether the observed helps, prevents or remains neutral towards the process of recognition of the observing agent. In the first case, the observed will do everything to facilitate the work done by the observing agent.

The detection of the presence in various rooms of the home is essential to facilitate understanding of the activities in progress. In the work of (Valtonen et al., 2012), the position of the inhabitants is detected via ground and capacitive furniture (i.e. capable of detecting when touched). The floor is made up of tiles allowing a person to be located in the room. Although the results of their platform are satisfactory, this system requires heavy installation, in particular equipping the entire floor of an apartment. In addition, it is costly in time and money for such an installation. In addition, wearing shoes can disrupt the detection of the activity. Furthermore, this type of systems requires a prior knowledge of the furniture arrangement if one wishes to recognize activities via this system (the location of the person may suggest this or that activity depending on the surrounding equipment/furniture).

In the case of smart homes that are used by elderly or people with disabilities, the detection of posture can be useful to detect the fall of a person and thus be able to intervene if necessary. In the work presented by (Ricquebourg et al., 2007), several sensors are used to detect the person's posture: a camera with an omni-directional vision mirror, pressure sensors on the chairs and finally an accelerometer on the belt of the person. However, some equipment can be considered very problematic such as the camera. Cameras and microphones can be viewed as a violation of privacy and private life (Oulasvirta et al., 2012). Indeed, the users can have the impression of being listened to and/or observed by someone. This obviously poses significant problems of acceptability and comfort.

As an example of sensors installed in the environment and often used in smart homes, we can note the magnetic contact sensors, to detect if an opening, i.e. a door or a window, is closed or not. These sensors consist of a contact switch which closes a circuit when the magnet approaches the electronic part, or on the contrary open it by moving the magnet part away. This allows the system to know at all times whether the fitted door is open or closed.

In (Sezer et al., 2016), the authors proposed a multi-layered model that combines three paradigms; Big data, Internet of Things (IoT) and semantic web in order to build a new extended framework. They presented the various components of the system as well as the necessary connections which must be integrated. In addition, they provided a realistic use case that shows how the model can implement the desired functionality and reach the objectives of such a model. However, the proposed knowledge representation is not generic, as it does not model all the aspects of the domain context. Thus, the learning layer is not well developed. All those presented methods are efficient in their field of appliances. But the challenge to design and develop stable and general systems still persists, as most systems only solve specific problems in very particular environments.

To control its environment using sensors and actuators, researchers from CASAS project (Rashidi et al., 2011) modeled an apartment as an intelligent agent which perceives and monitors its environment. The goal of this project is to improve the comfort of users and decrease the energy costs. However, the authors worked on learning the habits of the user and not on the recognition of their activities (Vikramaditya et al., 2007; James et al., 1997; Chen & Cook, 2012).

Different methods exist in the literature to model the activity and behavior of the person (Sukor et al., 2019). However, people's activity and behavior change over time. If the problem of modeling the activities of the person has been frequently studied in the literature, the problem of detecting behavior changes has rarely been. The classification methods make it possible to characterize the person's activities so that they can then be identified. However, a restriction of this kind of methods is the use of labeled data, not allowing the detection of rare activities. Moreover, we involve strong constraints on the choice of sensors and on the supervision method. Indeed, the sensors used in our work must be acceptable to the resident and their privacy. It will therefore be difficult to envisage sensors such as cameras.

In this research work, we propose an ontology that aims to represent the knowledge in a semantic way through the valuation and interpretation of data from multiple heterogeneous sensors (IoT sensors, wearable gadgets, multimodal sensors, etc). The realization of this ontology will be extended based on the reuse and enrichment of resources. Indeed, the reuse of ontologies offers the advantage of using mature ontological resources, proven and validated by their applications, generally by the W3C. Our ontology is made up of a set of terms derived from several ontologies adapted to our needs, such as:

- The Semantic Sensor Network (SSN) ontology allows the description of sensors, observations, detection treatments, measurement capabilities and any other relative concept. This ontology provides the vocabulary required to describe the sensors used in our proposed model.
- The Geo-location ontology provides concepts to describe the spatial properties of resources.
- The Time ontology describes concepts related to time; the relation among instants and intervals, as well as, information about durations, and about temporal position including date-time information.

Once the ontology is formally defined, we can make inferences, i.e. reasoning. This reasoning will make it possible to generate a set of activity recognition; such as predicting if the user is eating, sleeping, etc. We use SWRL (Semantic Web Rule Language) to express the set of reasoning rules and SPARQL query language to retrieve and manipulate data from the proposed ontology.

4. CONTRIBUTION

4.1. Proposed Ontology

Our ontology aims to model and structure home assistance. It allows applications, humans or a community to collaborate to build and follow assistance scenarios to encourage home care. The use of the ontology is justified by the fact that it offers well-formed semantics and is adapted to provide answers to questions that require more information. It brings all the knowledge that the context needs to assist the well-being. Figure 1 represents the global ontological model for activity recognition in smart home. To model our ontology, we used several design elements that are paramount to describe human activity within context. The main elements are summarized below.

Figure 1. Proposed ontology

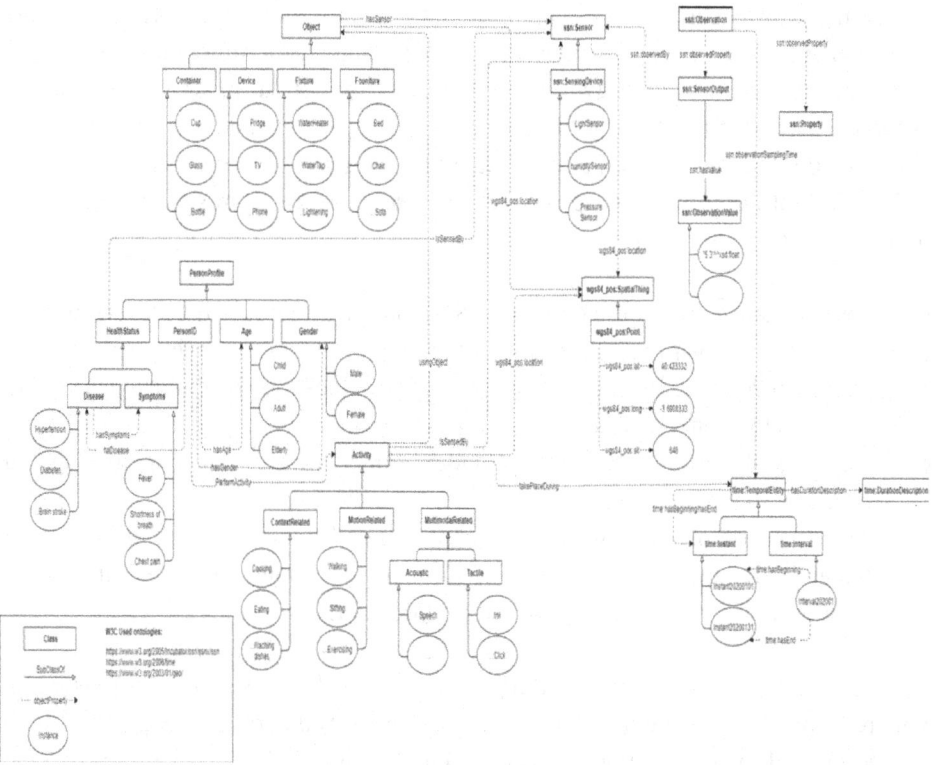

- **PersonProfile:** This class is devoted to the description of the person's profile, and consists of four subclasses. The first subclass describes the health status of the well-being, including the disease and its symptoms, while the *PersonID, Age, Gender* subclasses describe the other personal information of the person.

- **Object:** Four types of objects are defined in our ontology. The *Container, Device, Fixture* and *Furniture* subclasses, which describe the equipment making part of the person's environment in their smart home.

- **Activity:** It has three subclasses, namely *ContextRelated, MotionRelated* and *MultimodalRelated* subclasses. These concepts relate to the elements used in the inferences which have no relation to physical objects, such as the situation or activity concepts. In most cases, the instances of these concepts are the possible values resulting from an inference process. So, for example, cooking or sleeping are instances belonging to the Activity class.

- **SSN ontology:** In the proposed ontology, we reuse the SSN ontology for the description of the sensors. Sensors are the most obvious producers of an activity recognition solution dedicated to support well-being in smart home. In our work, sensors are divided into two categories: the sensors fixed inside the home and the wearable sensors, responsible for carrying out physiological measurements for instance. The SSN ontology is composed of different classes such as, ssn:*Sensor* class presenting the equipment responsible of sensing, detecting and measuring the ssn:*Propriety*.

- **Time ontology:** The ontology provides a vocabulary for expressing relations among instants and intervals. The main time classes are: *time:TemporalEntity* and its subclasses *time:Instant* and *time:Interval*. They are temporal entities with an extent or duration.
- **Wgs84_pos ontology:** This ontology describes the *lat*(latitude), *long*(longitude) and other information about spatially located things.

4.2. Inference Rules

Using rules increases expressiveness and possibilities of reasoning on the knowledge expressed with OWL ontologies (Horrocks, 2005). The language SWRL is certainly the most common among the existing inference languages (Horrocks et al., 2004). It is considered an extension of OWL-DL. It is based on logic programming, and more particularly on the Horn clauses. Also, it allows to manipulate the variables and the instances declared in the rules.

The data coming from various sensors is exploited in order to predict the different activities of the well-beings. The first task to be carried out is to set up a semantic reasoning to assist the person in carrying out ADLs. The reasoning mechanisms proposed in this study are based on SWRL language. Table 1 shows some examples of inference rules used in our work.

Table 1. SWRL rules

Activity	Rule
Sleeping	PersonID(?p) ^ Bedroom(?l) ^ Night(?t) ^ Bed(?o) ^ location(?p, ?l) ^ takePlaceDuring(?p, ?t) ^ usingObject(?p, ?o) → hasActivitySleeping
Preparing Breakfast	PersonID(?p) ^ Kitchen(?l) ^ Morning(?t) ^ Fridge(?o) ^ Cooker(?o) ^ Microwave(?o) ^ location(?p, ?l) ^ takePlaceDuring(?p, ?t) ^ usingObject(?p, ?o) → hasActivityPreparingBreakfast
Watching TV	PersonID(?p) ^ LivingRoom(?l) ^ Morning(?t) ^ Afternoon(?t) ^ Evening(?t) ^ Night(?t) ^ TV(?o) ^ Sofa(?o) ^ location(?p, ?l) ^ takePlaceDuring(?p, ?t) ^ usingObject(?p, ?o) → hasActivityWatchingTV
Toileting	PersonID(?p) ^ Bathroom(?l) ^ Morning(?t) ^ Afternoon(?t) ^ Evening(?t) ^ Night(?t) ^ ToiletFlush(?o) ^ ToiletDoor(?o) ^ location(?p, ?l) ^ takePlaceDuring(?p, ?t) ^ usingObject(?p, ?o) → hasActivityToileting

For instance, to predict the sleeping activity we use the following rule: if the person is located in the bedroom, using the bed during night time, then it inferred that the person is sleeping. In the next section we will introduce some SPARQL queries as well as some inference rules in order to validate the performance of our ontology.

4.3. SPARQL Queries

There are many query languages based on RDF, allowing the extraction of OWL ontological knowledge. The most used language is the SPARQL language (SPARQL Protocol and RDF Query Language), which has become an official recommendation of W3C. The SPARQL language allows to express a query in the form of an RDF graph (the variables of the query can be nodes or arcs of the graph). The answer of a query is to match that query to another graph representing the knowledge base made up of RDF descriptions.

Figure 2. An example of SPARQL query

```
Select   ?PersonID ?Object ?Activity
Where {
?PersonID   ARont:hasAge   ARont:Elderly
?Object   rdf:type   ARont:Bed
?Person ARont:performActivity ?Activity
?Activity ARont:usingObject   ?Object
}
```

Figure 2 shows the example of the SPARQL query for retrieving, from the ontology, activity that is performed by the elderly who is using the object bed.

5. IMPLEMENTATION AND EXPERIMENTAL RESULTS

To implement our proposed ontology, we used the open source software "Protégé". It is a knowledge acquisition tool that was created in the 80s, and dedicated from the start to medical applications, but became accessible for several types of tools and applications which are based on knowledge (Holger, 2004). Also, its modular and flexible architecture allowed to add several modules including the one based on OWL, allowing it to support the Semantic Web technology for the construction of ontologies.

The use of OWL is recommended by the W3C where processing the content of documents by machines is required. It aims to make the vocabulary as well as the relations between them more explicit (McGuiness & Harmelen, 2004; Latfi et al., 2007). OWL is a language for developing distributed ontologies, it is compatible with web standards, which will allow their internationalization, their extensibility and scalability across the web. Figure 3 shows the implementation of our ontological model.

Figure 3. Proposed ontology implementation

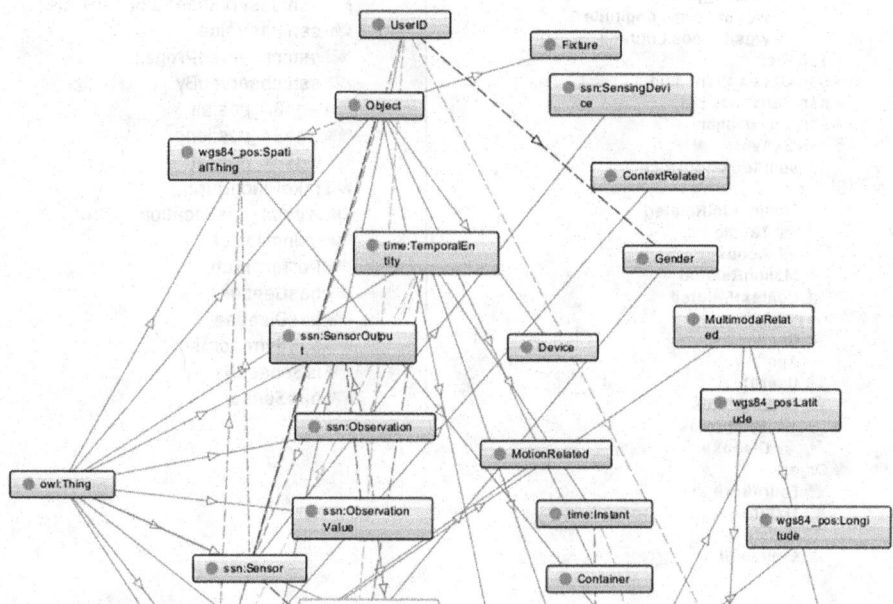

Figure 4 shows the snapshots of our ontology classes and its object properties as we had it designed using the Protégé-OWL editor.

Below, we introduce an experiment to evaluate the execution time of a simple query running on different environment. The aim to measure the response time of the SPARQL query given in section 4. For this experiment, we used two personal computers with different CPU speed as shown in Figure 5. Using JENA library, we performed the query simultaneously on 10, 20, 30, 50 and 100 triples.

As shown in Figure 5, the performance of execution time of the queries is based on the size of the ontology and the CPU speed. The figure shows an almost linear increase of the running time as the number of triples increases.

Figure 4. Ontology classes and object properties

Figure 5. SPARQL query execution time

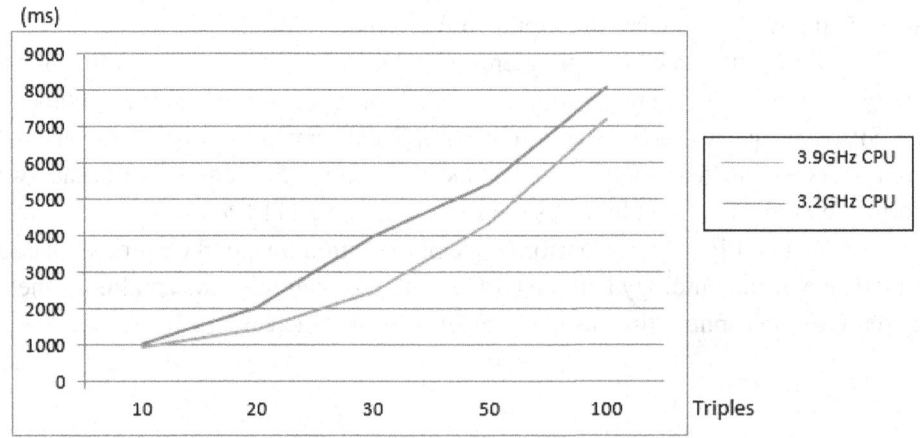

6. CONCLUSION

In this chapter, we introduced a novel ontological framework for modeling and predicting the activity of the well-being in smart home. The proposed approach is based on semantic web mechanisms to model the person's context inside the home. For the system to be reusable and deployable on many technologies, we adopted a model based on a number of standard W3C recommendations, such as the SSN, Time and Wgs84_pos ontologies. These are used to accurately describe the semantics of human activity within time and space context. We validated our model by performing a set of SWRL reasoning rules and SPARQL queries. Finally, we tested and measured the execution time of a simple query for performance evaluation. The proposed ontological framework looks robust in terms of accurately defining the main contextual components of an activity in smart home. However, in order to classify the activities and detect abnormal behavior, it is necessary to build some learning mechanisms on the top of this ontological structure. Our current research aims at achieving this goal.

REFERENCES

Allen, J. F., & Ferguson, G. (1994). Actions and events in interval temporal logic. *Journal of Logic and Computation, 4*(5), 531–579.

Bachimont, B. (2004, May). *Pourquoi n'y a-t-il pas d'expérience en ingénierie des connaissances?*

Baget, J. F. (2005, November). RDF entailment as a graph homomorphism. In *International Semantic Web Conference* (pp. 82-96). Springer, Berlin, Heidelberg.

Boger, J., Hoey, J., Poupart, P., Boutilier, C., Fernie, G., & Mihailidis, A. (2006). A planning system based on Markov decision processes to guide people with dementia through activities of daily living. *IEEE Transactions on Information Technology in Biomedicine, 10*(2), 323–333.

Bouchard, B., Giroux, S., & Bouzouane, A. (2007). A keyhole plan recognition model for Alzheimer's patients: First results. *Applied Artificial Intelligence, 21*(7), 623–658.

Brown, P. J., Bovey, J. D., & Chen, X. (1997). Context-aware applications: from the laboratory to the marketplace. *IEEE personal communications, 4*(5), 58-64.

Chen, C., & Cook, D. J. (2012, June). Behavior-based home energy prediction. In *2012 Eighth International Conference on Intelligent Environments* (pp. 57-63). IEEE.

Chen, G., & Kotz, D. (2000). *A survey of context-aware mobile computing research.* Dartmouth Computer Science Technical Report TR2000-381.

Cherlet, M., Schelkens, M., Croubels, S., & De Backer, P. (2003). Quantitative multi-residue analysis of tetracyclines and their 4-epimers in pig tissues by high-performance liquid chromatography combined with positive-ion electrospray ionization mass spectrometry. *Analytica Chimica Acta, 492*(1-2), 199–213.

Dey, A. K. (2001). Understanding and using context. *Personal and Ubiquitous Computing, 5*(1), 4–7.

Endsley, M. R. (2017). Toward a theory of situation awareness in dynamic systems. In *Situational awareness* (pp. 9–42). Routledge.

Goldman, R. P., Geib, C. W., & Miller, C. A. (2013). *A new model of plan recognition.* arXiv preprint arXiv:1301.6700.

Gottfried, I. B. B., & Aghajan, H. (2009). The praxis of cognitive assistance in smart homes. Behaviour Monitoring and Interpretation-BMI. *Smart Environments, 3*, 183.

Gruber, T. R., & Olsen, G. R. (1994, January). An ontology for engineering mathematics. In *Principles of Knowledge Representation and Reasoning* (pp. 258–269). Morgan Kaufmann.

Guarino, N. (1995). Formal ontology, conceptual analysis and knowledge representation. *International Journal of Human-Computer Studies, 43*(5-6), 625–640.

Horrocks, I. (2005). OWL rules, ok?. *Rule Languages for Interoperability*, 34.

Horrocks, I., Patel-Schneider, P. F., Boley, H., Tabet, S., Grosof, B., & Dean, M. (2004). SWRL: A semantic web rule language combining OWL and RuleML. *W3C Member submission, 21*(79), 1-31.

Jakkula, V. R., Crandall, A. S., & Cook, D. J. (2007, October). Knowledge discovery in entity based smart environment resident data using temporal relation based data mining. In *Seventh IEEE International Conference on Data Mining Workshops (ICDMW 2007)* (pp. 625-630). IEEE.

Katz, S., Ford, A. B., Moskowitz, R. W., Jackson, B. A., & Jaffe, M. W. (1963). Studies of Illness in the Aged. *Journal of the American Medical Association, 185*(12), 914. doi:10.1001/jama.1963.03060120024016

Kautz, H. A. (1991). A formal theory of plan recognition and its implementation. Reasoning about plans, 69-125.

Knublauch, H., & Musen, M. A. (2004, June). Weaving the biomedical semantic web with the Protégé OWL plugin. In *Proceedings of the First International Conference on Formal Biomedical Knowledge Representation-Volume 102* (pp. 39-47).

Lassila, O., & Swick, R. R. (1998). *Resource description framework (RDF) model and syntax specification.*

Latfi, F., Descheneaux, C., & Lefebvre, B. (2007). *Habitat intelligent en télé–Santé: ontologie de l'équipement.* Canada: FICCDAT Toronto.

McGuinness, D. L., & Van Harmelen, F. (2004). OWL web ontology language overview. *W3C recommendation, 10*(10), 2004.

Moutacalli, M. T., Bouchard, B., & Bouzouane, A. (2015), Sensors Activation Times Prediction in Smart Home. In *8th ACM International Conference on Pervasive Technologies Related to Assistive Environments.*

Moutacalli, M. T., Bouzouane, A., & Bouchard, B. (2015). The behavioral profiling based on times series forecasting for smart homes assistance. *Journal of Ambient Intelligence and Humanized Computing, 6*(5), 647–659.

Ordóñez, F. J., Iglesias, J. A., De Toledo, P., Ledezma, A., & Sanchis, A. (2013). Online activity recognition using evolving classifiers. *Expert Systems with Applications, 40*(4), 1248–1255.

Oulasvirta, A., Pihlajamaa, A., Perkiö, J., Ray, D., Vähäkangas, T., Hasu, T., ... Myllymäki, P. (2012, September). Long-term effects of ubiquitous surveillance in the home. In *Proceedings of the 2012 ACM Conference on Ubiquitous Computing* (pp. 41-50).

Pascoe, J. (1998, October). Adding generic contextual capabilities to wearable computers. In Digest of papers. second international symposium on wearable computers (cat. no. 98ex215) (pp. 92-99). IEEE.

Patterson, D. J., Fox, D., Kautz, H., & Philipose, M. (2005, October). Fine-grained activity recognition by aggregating abstract object usage. In *Ninth IEEE International Symposium on Wearable Computers (ISWC'05)* (pp. 44-51). IEEE.

Plantevin, V., Bouzouane, A., & Gaboury, S. (2017). The light node communication framework: A new way to communicate inside smart homes. *Sensors (Basel), 17*(10), 2397.

Ramos, C., Augusto, J. C., & Shapiro, D. (2008). Ambient intelligence—the next step for artificial intelligence. *IEEE Intelligent Systems, 23*(2), 15–18.

Rashidi, P., Cook, D. J., Holder, L. B., & Schmitter-Edgecombe, M. (2010). Discovering activities to recognize and track in a smart environment. *IEEE Transactions on Knowledge and Data Engineering, 23*(4), 527–539.

Ricquebourg, V., Delafosse, M., Delahoche, L., Marhic, B., Jolly-Desodt, A., & Menga, D. (2007). *Fault detection by combining redundant sensors: a conflict approach within the tbm framework. Cognitive Systems with Interactive Sensors.* COGIS.

Ricquebourg, V., Menga, D., Durand, D., Marhic, B., Delahoche, L., & Loge, C. (2006, December). The smart home concept: our immediate future. In *2006 1st IEEE international conference on e-learning in industrial electronics* (pp. 23-28). IEEE.

Roy, P. C., Bouchard, B., Bouzouane, A., & Giroux, S. (2013). Ambient Activity Recognition in Smart Environments for Cognitive Assistance. *International Journal of Robotics Applications and Technologies, 1*(1), 29–56. doi:10.4018/ijrat.2013010103

Ryan, N. S., Pascoe, J., & Morse, D. R. (1998). Enhanced reality fieldwork: the context-aware archaeological assistant. In *Computer applications in archaeology.* Tempus Reparatum.

Schilit, B., Adams, N., & Want, R. (1994, December). Context-aware computing applications. In *1994 First Workshop on Mobile Computing Systems and Applications* (pp. 85-90). IEEE.

Schilit, B. N., & Theimer, M. M. (1994). Disseminating active map information to mobile hosts. *IEEE Network, 8*(5), 22–32.

Sezer, O. B., Dogdu, E., Ozbayoglu, M., & Onal, A. (2016, December). An extended iot framework with semantics, big data, and analytics. In *2016 IEEE International Conference on Big Data (Big Data)* (pp. 1849-1856). IEEE.

Spriggs, E. H., De La Torre, F., & Hebert, M. (2009, June). Temporal segmentation and activity classification from first-person sensing. In *2009 IEEE Computer Society Conference on Computer Vision and Pattern Recognition Workshops* (pp. 17-24). IEEE.

Sukor, A. S. A., Zakaria, A., Rahim, N. A., Kamarudin, L. M., Setchi, R., & Nishizaki, H. (2019). A hybrid approach of knowledge-driven and data-driven reasoning for activity recognition in smart homes. *Journal of Intelligent & Fuzzy Systems, 36*(5), 4177–4188.

Suryadevara, N. K., Mukhopadhyay, S. C., Wang, R., & Rayudu, R. K. (2013). Forecasting the behavior of an elderly using wireless sensors data in a smart home. *Engineering Applications of Artificial Intelligence, 26*(10), 2641–2652.

Valtonen, M., Vuorela, T., Kaila, L., & Vanhala, J. (2012). Capacitive indoor positioning and contact sensing for activity recognition in smart homes. *Journal of Ambient Intelligence and Smart Environments*, *4*(4), 305–334.

Weiser, M. (1991). The Computer for the 21st Century. *Scientific American*, *265*(3), 94–105.

Winograd, T. (2001). Architectures for context. *Human-Computer Interaction*, *16*(2-4), 401–419.

Ye, J., Dobson, S., & McKeever, S. (2012). Situation identification techniques in pervasive computing: A review. *Pervasive and Mobile Computing*, *8*(1), 36–66.

Chapter 8
Sliding Mode Control for PV Grid–Connected System With Energy Storage

Saloua Marhraoui
 https://orcid.org/0000-0003-2077-2334
Mohammed V University in Rabat, Morocco

Ahmed Abbou
Mohammed V University in Rabat, Morocco

Zineb Cabrane
Mohammed V University in Rabat, Morocco

Salahddine Krit
 https://orcid.org/0000-0003-3868-472X
Ibn Zohr University, Morocco

ABSTRACT

We need to solve the problem due to the nonlinearity and power fluctuation in the photovoltaic (PV) connected storage system and grid; for that, the authors develop an algorithm to obtain the maximum power point tracking (MPPT) via control of the duty cycle of DC/DC boost converter. Consequently, they design an MPPT based on the second-order sliding mode control. Next, generating the law control founded on the Lyapunov theory can augment the robustness and stability of the PV connected grid. Then, they add a battery energy storage system (BESS) with a control management algorithm in the DC/DC side to eliminate any fluctuation of the output power of the PV system because of the temperature and irradiation variation. On the grid side, they control the DC/AC inverter side by the three-phase voltage source inverter control (VSIC) as a charge controller for the grid parameters.

DOI: 10.4018/978-1-7998-4381-8.ch008

INTRODUCTION

Nowadays, the grid-connected photovoltaic (PV) with batteries energy storage system (BESS) show in Figure 1 is growth because of the rapid of using of a PV as source energy otherwise the system controller confronted with a lot challenges of keeping a grid reliability and stability (R. Teodorescu, M. Liserre, P. Rodrlguez, 2011). In Figure.1, two global factors key of PV grid-connected with the energy storage system should be obtained (Mehmet Yesilbudak, Ramazan Bayindir, 2018). The first factor is the effects of climatic conditions present in the irradiation, temperature and other meteorological conditions. The second factor is the influence of the DC/AC inverter (J. M. Carrasco, L. G. Franquelo, J. T. Bialasiewicz, et al, 2006; Raghvendraprasad Deshpande, 2019). Surely, the climatic conditions are uncontrollable, for that reason, a Batteries-Energy-Storage-System is added in the DC side of the PV system to compensate the fluctuation of PV output power when we have a changing in the level of the irradiation and temperature (J. Liu, W. Luo, X. Yang, et al, 2016; Y. Yang, Q. Ye, L. Tung, et al, 2018). We design a control of the DC/DC boost converter via a Maximum Power Point Tracking (MPPT) effectively to keep the output power at the demanded quality of the PV system. The grid transfer the maximum power of a PV power system into the high value, which is the global goal of a PV grid-connected and BESS, the system of power generation, i.e., despite the changes in climatic conditions, the factor of a power of the PV grid-connected system is independent (A. Kouchaki, H. Iman-Eini, B. Asaei, 2013; M. H. Moradi, A. R. Reisi, 2011).

Figure 1. Topology of BESS connected PV and grid

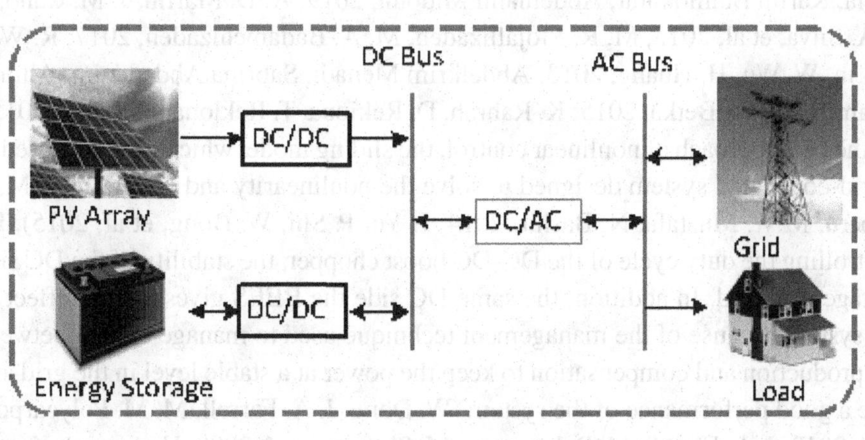

Actually, many studies oriented to PV-grid connected BEES. In reference (Dezhi Xua, Gang Wanga, Wenxu Yana, Xinggang Yanb, 2019), a second-order sliding mode controller applied to a DC/DC Boost Converter and we realized the control. Tracking of the output voltage. When the PV panel connected to the DC side, the maximum power point tracking (MPPT) established via the use of sliding mode control based on the Lyapunov function-Theory we can be designed a good controller to determine the stability of the system. The work in (Kok Soon Tey ; Saad Mekhilef ; Mehdi Seyedmahmoudian; Ben Horan; Amanullah Than Oo; Alex Stojcevski, 2018) the control strategy proposed of the voltage and frequency for the energy resource system, were simulated the control impact of different types of loads, also, of multiple resource units. However, we notice in the anthers simulations as (S. Kolesnik and A. Kuperman, 2016) the author does not use the compensation of energy resources in the simulation, and there is no energy storage unit in the system. However, in our work, the control algorithm designed without the last limitations, we improve that with a batteries energy storage system unit was considered in the global PV system under study.

In the classical linear control methods, there is a lot of investigation into power converters (J. Ahmed and Z. Salam, 2016; Mohammed Ali Elgendy, David John Atkinson, Khalifa University, 2016; Heng Po-Chen, Peng Bo-Rei, Liu Yi-Hua, Cheng Yu-Shan, Huang Jia-We, 2015). We know the nonlinear of the dynamic of the power converters, and accurately we cannot measure some parameters. Hence, the development of a lot of the number of nonlinear control strategies to find a solution problem of the nonlinearity of power converters (M. Guisser, A. EL-Jouni, EL. H. Abdelmounim, 2014; Mohamed Keddar, Mamadou Lamine Doumbia, Mohamed Della, Karim Belmokhtar, Abdelhami Midoun, 2019; A. D. Martin, J. M. Cano, J. F. A. Silva, et al, 2015; M. R. Mojallizadeh, M. A. Badamchizadeh, 2017; R. Wai, C. Lin, W. Wu, H. Huang, 2013; Abdelkrim Menadi, Sabrina Abdeddaim, Ahmed Ghamri, Achour Betka, 2015; K. Rahrah, D. Rekioua, T. Rekioua, S. Bacha, 2015). As the best approach of nonlinear control, the sliding mode, which is widely used in a grid-connected system designed to solve the nonlinearity and uncertainty (M.Y. Suberu, M.W. Mustafa, N. Bashir, 2014; J. Yu, P. Shi, W. Dong, et al, 2015). By controlling the duty cycle of the DC-DC boost chopper, the stability of the DC side voltage achieved. In addition, the same DC side the BEES gives a good effect of the system because of the management technique used to manage energy between the production and compensation to keep the power at a stable level in the grid and give a good performance of the system (W. Dong, J. A. Farrell, M. M. Polycarpou, et al, 2012; J. A. Farrell, M. Polycarpou, M. Sharma, et al, 2009; H. Amresh Kumar Singh, Ikhlaq Hussain, Bhim Singh, 2018).

In the DC/AC side, we ensure that the harmonic misalignment rate of the DC/AC inverter by VSIC. Consequently, we achieved the effect of the VSI controller by simulation. In order to estimate the system parameters and this technique gives satisfactory control and good performance (Alok Kumar Pani, Niranjan Nayak, 2019). We can show that the VSIC control impact is better than proportion-integral PI (C. Aouadi, A. Abouloifa,A. Hamdoun, Y. Boussairi, 2014; S.Sivakumar,M. JagabarSathik, P.S.Manoj, G.Sundararajan, 2016).

In this work, we designed and improved an MPPT based on the second sliding mode controller for the BESS-PV-grid-connected system, which we stabilized DC bus side voltage and control output power of the PV system via MPPT controller also, with management technique for BESS. In this proposed controller, an adaptive law based on Lyapunov theory used to estimate the uncertain parameters (DC-link capacitor, output voltage, and inductance) of the boost converter. In the grid-connected inverter side, we consider the demand of loads with grid and we estimated the value demanded, via the implementation of the VSIC algorithm to control the different parameters in the grid as powers, voltage, current, and frequency. In the VSIC design procedure, we solve the problems of harmonics by using the RL filter and compensation. The system and the controllers demonstrated in MATLAB/Simulink.

CONVECTIONAL POWER GRID

In general, the PV-Grid connects with BEES under study as shown in Figure.2 characterized by global operations: power production, energy transmission, energy distribution, and control (K.K. Zame, C.A. Brehm, A.T. Nitica, C.L. Richard, G.D. Schweitzer, 2017; S.O. Geurin, A.K. Barnes, B.J. Carlos, 2012). Power production is centralized by the now stability of PV power out due to climatic change (Irradiation and Temperature). Also, due to the load and Grid monitoring. Actually, the functionalities integrated with many algorithms to manage the energy smartly and automatically in PV-Grid connected with BEES. The information flows are from the system to design a good MPPT algorithm. Furthermore, integration of the BEES at the distribution system helps to keep the grid by the power compensating in demand of the system, as well as the PV panel source know by the fluctuating behavior. The integration of the BEES with the management of the power grid. The last one is the key role in PV-grid management power stability. In addition, we need to add one of the principal algorithms of the stability of energy, therefore, it can eliminate the effect of the harmonic is given in (F. Mohamad, J. Teh, C.M. Lai, L.R. Chen, 2018).

Figure 2. Block diagram of the system under study

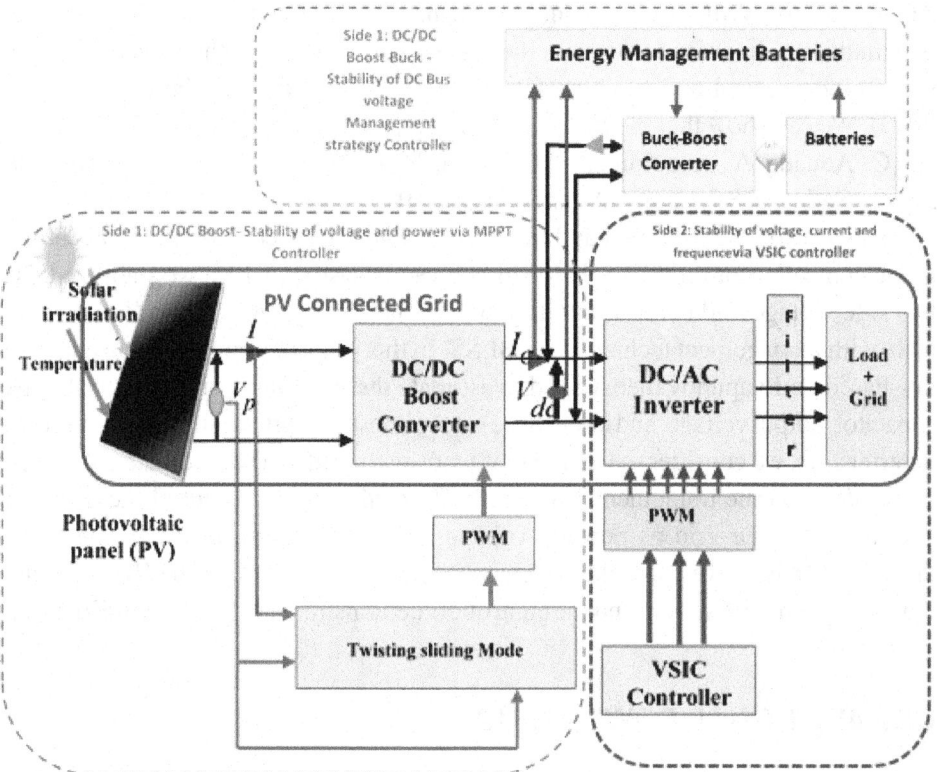

APPLICATION

Figure 3 shows the proposed PV-grid connected BEES configuration dedicated to smart house applications in smart grids. We can be connected to this configuration with wind turbine and diesel engine generators the battery chargers for electric vehicles can be connected to the DC bus. Therefore, we can use the three-phase inverter with the designed control techniques. The proposed system operates in different areas modes. Because the solar photovoltaic energy system presents a clean and sustainable source of power to the connected load. It can charge the BESS and at the same time injects power into the grid during peak hour demand. The solar photovoltaic panel feeds the load. In two modes, the BESS balance the power in the system and compensate the power when we have decrease energy and store the energy when we have excess energy.

Figure 3. Block diagram of smart grid

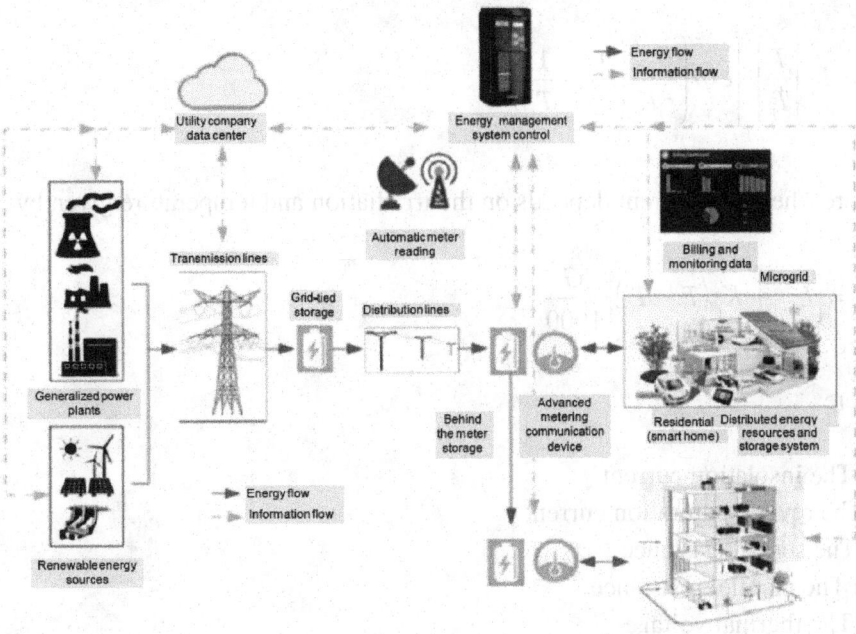

CONTROL SYSTEM

We reserved this section to develop the control techniques for DC-DC boost and three-phase inverter.

PV Mathematical Model

The PV generator (PVG) designed in electrical circuit as show in Figure.4. The PVG panel formed of photovoltaic cell series to product electricity under solar irradiation and temperature. The mathematical model of PVG modelled in the equation linking current-voltage with weather conditions (Asma Mlayah, Adel Khedher, 2018; Bharath K R, Harsha Choutapalli, Kanakasabapathy P, 2018):

$$I = I_{ph} - I_s \left[exp\left(\frac{Vp + IR_s}{N_\gamma V_T}\right) \right] - \frac{Vp + R_s I}{R_{sh}} \tag{1}$$

Where

$$I_s = I_{rr} \left[\frac{T}{T_r}\right]^3 \left[exp\left(\frac{E_{G0}}{\gamma K}\left(\frac{1}{T_r} - \frac{1}{T}\right)\right)\right] \tag{2}$$

Where The photocurrent depends on the irradiation and temperature given by:

$$I_{ph} = \left[I_{scr} + K_i\left(T - T_r\right)\right]\frac{G}{1000} \tag{3}$$

Where

I_{ph}: The insolation current
I_s: The reverse saturation current
R_s: The series resistance
R_{sh}: The parallel resistance
V_T: The thermal voltage
K: The Boltzman constat
T: The temperature in Kelvin
q: The charge of an electron
γ γ: The factor of ideality
T_r: The cell reference temperature
I_n: The cell reverse saturation at temperature T_r
E_{G0}: The band gap of the semiconductor used in the cell.
I_{scr}: The cell short-circuit current at reference temperature and irradiation
G: The solar irradiation in w/m²

We analyzed the performance of a PVG via using the curves. Figure 5 gives characteristic current-voltage and Figure.6 present power-voltage characteristic of the PVG with conditions: Solar irradiation G = 1000 w/m². Temperature T= 0°C, Temperature T= 25°C, Temperature T= 50°C

The condition of maximum power point following by the equation 4:

$$\frac{\partial Pp}{\partial Vp} = \frac{\partial Vp}{\partial Vp} = i + Vp\frac{\partial Vp}{\partial Vp} = 0 \tag{4}$$

Where $P_p = v_p i$ the PVG output power.

Figure 4. Model of the generator photovoltaic

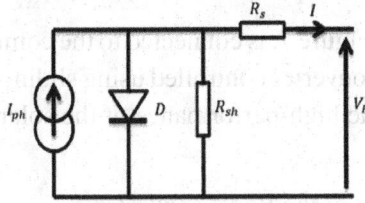

Figure 5. PV Current-Voltage characteristic

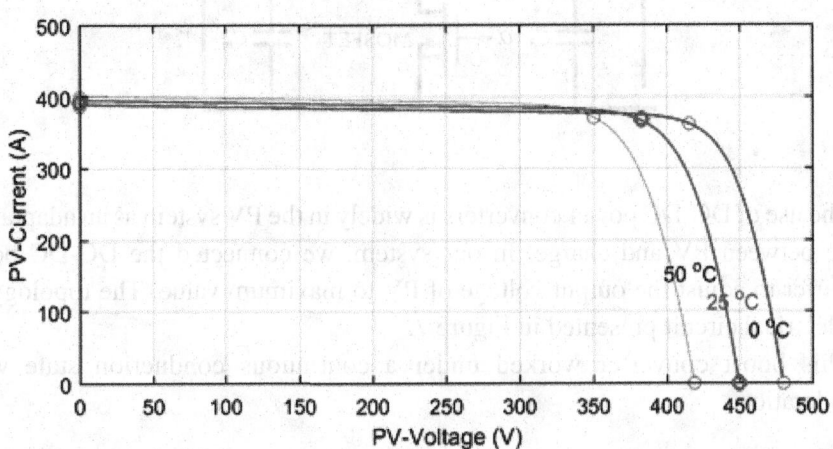

Figure 6. PV Power-Voltage characteristic

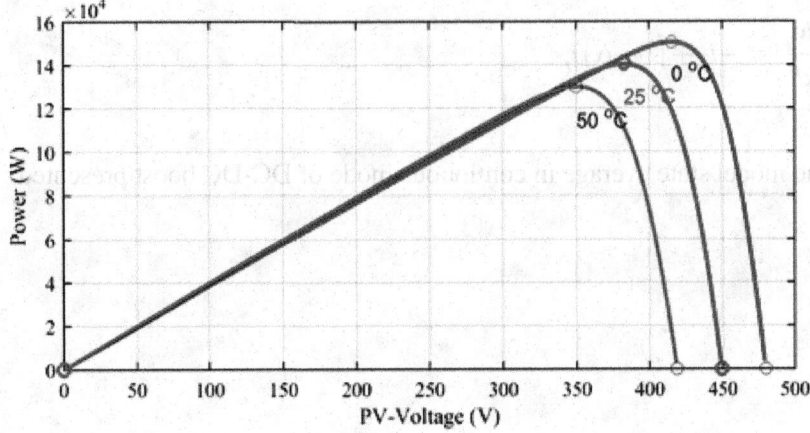

Mathematical Model of the DC-DC Boost Converter

The PV panel is shown in Figure 7, is connected to the common DC bus Via DC-DC Boost. the DC-DC boost converter controlled using sliding mode control to achieve the MPPT and obtained the high-performance of the solar photovoltaic.

Figure 7. Electrical circuit of the Boost converter

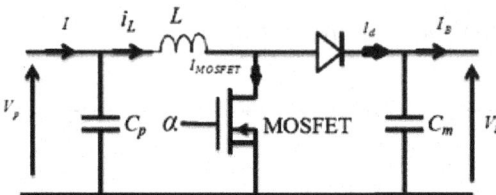

The use of DC-DC power converters is widely in the PV system as an adaptation phase between PV and charge. In our system, we connected the DC-DC boost converter to adjust the output voltage of PV to maximum value. The topology of the electrical circuit presented in Figure 7.

This boost converter worked under a continuous conduction state with consideration.

$$
\begin{cases}
c_p \dfrac{dv_p}{dt} = i - i_L \\[2mm]
L \dfrac{di_L}{dt} = v_p - (1-\alpha) v_{dc} \\[2mm]
c_m \dfrac{dv_{dc}}{dt} = -i_c + (1-\alpha) i_L
\end{cases} \tag{5}
$$

The model state average in continuous mode of DC-DC boost presented:

$$\begin{cases} \dfrac{dv_p}{dt} = \dfrac{1}{C_p} i - \dfrac{1}{C_p} i_L \\[3mm] \dfrac{di_L}{dt} = \dfrac{1}{L} v_p - \dfrac{1}{L}(1-\alpha) v_{dc} \\[3mm] \dfrac{dv_{dc}}{dt} = -\dfrac{1}{C_m} i_C + \dfrac{1}{C_m}(1-\alpha) i_L \end{cases} \tag{6}$$

Where:

v_p: Input voltage
v_{dc}: Output voltage
i_L: Inductor current
L: The inductor
C_p: Input capacitor
C_m: Output capacitor
α: Signal switch state

The nonlinear time-invariant system is written as:

$$\dot{X} = f(X) + g(X)\alpha \tag{7}$$

Where: $X \in \begin{bmatrix} I & V_{dc} \end{bmatrix}^T$ and $\alpha \in \begin{bmatrix} 0 & 1 \end{bmatrix}$

MPPT Based on Second Order Siding Mode Control for DC-DC Boost Converter

To obtain the maximum of the PV power output, we use the second order-sliding mode based on the super twisting controller. We developed the super twisting with degree one in order to eliminate the chatter. Its advantage is not needed any information on the time derivative of the sliding mode variable and it keeps all the controller characterizes.

The characteristics of the sliding plan trajectories presented by twisting around the origin. We developed the sliding mode control on two phases. The first phase is the determination of a sliding surface; the second phase is to detect a control law in order to push the system on the sliding surface.

Sliding Surface

The first phase is for choosing stable hyper planes state.

The sliding mode S(t):

$$S(t) = \left\{ X \setminus S(X,t) = \dot{S}(X,t) = 0 \right\} \tag{8}$$

To push the system states to the sliding surface and get the maximum power output, we choose the sliding surface as given in:

$$\frac{\partial P_p}{\partial v_p} = 0 \tag{9}$$

By developing (9), we obtain

$$\frac{\partial P_p}{\partial v_p} = v_p \left(\frac{\partial i}{\partial v_p} + \frac{i}{v_p} \right) \tag{10}$$

Then, the sliding surface defined by

$$S(t, X) = \frac{\partial i}{\partial v_p} + \frac{i}{v_p} \tag{11}$$

Controller Design

To limit the trajectories of the PV system (6), a second-order sliding mode controller designed to obtain and stay on the sliding surface (11). The total control law is the sum of two control terms: the equivalent control μ_{eq} and the super twisting control μ_{st}. The super twisting strategy was chosen to stabilize the system, eliminate chattering effects and make the system to converge to the desired target infinite time.

$$\mu = \mu_{eq} + \mu_{st} \tag{12}$$

The control ueq obtained by considering the conditions mentioned:

$$\begin{cases} S(t) = 0 \\ \dot{S}(t) = 0 \end{cases} \tag{13}$$

The control strategy obtained by solving the equation:

$$\dot{S} = \left[\frac{dS}{dX}\right]^T \dot{X} = \left[\frac{dS}{dX}\right]^T \left(f(X) + g(X)\mu_{eq}\right) = 0 \tag{14}$$

Hence

$$\mu_{eq} = \frac{\left[\dfrac{dS}{dX}\right]^T f(X)}{\left[\dfrac{dS}{dX}\right]^T g(X)} = 1 - \frac{v_p}{V_{dc}} \tag{15}$$

The super twisting control law expression is introduced:

$$\mu_{st} = »|S|^j \, sign(S) - W \int sign(S) \tag{16}$$

S: Sliding surface.
To save the finite-time convergence, the conditions to the sliding:

$$\begin{cases} W > \dfrac{\varnothing}{\Gamma_m} \\ \Gamma^2 \geq \dfrac{\varnothing}{\Gamma^2{}_m} \dfrac{\Gamma_M\left(W + \varnothing\right)}{\Gamma_m\left(W + \varnothing\right)} \\ 0 \leq j \leq \dfrac{1}{2} \end{cases} \tag{17}$$

Where W, β_m, β_M, Γ, and ρ are positive constants.

Theorem. The control of the PV system (6) with the second order Sliding Mode controller (12) where Γ and W are positive constants and the sliding surface given by (11). The closed-loop system saved with finite-time convergence for the MPPT.

In order to get the stability condition and obtain the stable convergence of our proposed controller, the Lyapunov function is chosen:

$$V(t) = \frac{1}{2} S^2(t) \tag{18}$$

With

$$\begin{cases} V(0) = 0 \\ V(t) > 0 \ for \ S(t) \neq 0 \\ \dot{S}(t) \neq 0 \end{cases} \tag{19}$$

We obtained the stability via the derivative of the Lyapunov function is negative and meets the condition:

$$\begin{cases} \dot{V} < 0 \\ S(t) \neq 0 \\ \overset{\text{Ù}}{S}(t) \neq 0 \end{cases} \tag{20}$$

The first derivative of the sliding surface giving by:

$$\dot{S} = \left[\frac{\partial S}{\partial X} \right]^T \dot{X} \tag{21}$$

$$\dot{S} = \left[\frac{\partial S}{\partial i} \right]^T \left(\frac{V_{dc}}{L}(1 - \alpha) + \frac{v_p}{L} \right) \tag{22}$$

The first term of (22) expressed by

$$\left[\frac{\partial S}{\partial i} \right] = \frac{\partial}{\partial i} \left[\frac{\partial i}{\partial v_p} + \frac{i}{v_p} \right] \tag{23}$$

With

$$\frac{\partial}{\partial i}\left[\frac{\partial i}{\partial v_p} + \frac{i}{v_p}\right] = \frac{i}{v_p} - \frac{i}{v_p^2}\frac{\partial v_p}{\partial i} \tag{24}$$

And

$$\frac{\partial}{\partial i}\left[\frac{\partial i}{\partial v_p} + \frac{i}{v_p}\right] = \frac{q}{K_b AT}\frac{\partial i}{\partial v_p}\frac{\partial v_p}{\partial i} \tag{25}$$

Then:

$$\frac{\partial S}{\partial i} = \frac{i}{v_p} - \frac{i}{v_p^2}\frac{\partial v_p}{\partial i} + \frac{q}{K_b AT}\frac{\partial i}{\partial v_p}\frac{\partial v_p}{\partial i} \tag{26}$$

The first derivative of the PV current defined:

$$\frac{\partial i}{\partial v_p} = \frac{q}{K_b AT}I_d \exp\left(\frac{q}{K_b AT}\right) < 0 \tag{27}$$

The PV voltage can be written:

$$v_p = \frac{K_b AT}{q}\ln\left(\frac{I_{ph} + I_d + i}{I_d}\right) \tag{28}$$

Its first derivative:

$$\frac{\partial v_p}{\partial i} = \frac{K_b AT}{q}\frac{I_{ph} + I_d + i}{I_d} < 0 \tag{29}$$

Finally, we determine the sign of the first term

$$\frac{\partial S}{\partial i} > 0 \tag{30}$$

It remains only to verify the sign of the second term. We have:

$$\dot{X} = \frac{v_{dc}(1-\alpha)}{L} + \frac{v_p}{L} \tag{31}$$

$$= \frac{v_{dc}(1-\alpha)}{L} + \frac{v_p}{L}$$

$$= \frac{v_{dc}\left[1 - \left(1 - \dfrac{v_p}{v_{dc}}\right) - \mu_{st}\right]}{L} + \frac{v_p}{L}$$

$$= \frac{v_{dc}}{L}\left(-»|S|^{1/2} sign(S) - W \int sign(S)\right)$$

Finally

$$S\dot{S} = S\left[\frac{\partial S}{\partial i}\right]\frac{v_{dc}}{L}\left(-»|S|^{1/2} sign(S) - W \int sign(S)\right) \tag{32}$$

From (32), we can conclude that the stability is ensured since S and $\dot{S}$ always have an opposite sign.

MODEL OF THE DC/AC CONVERTER, FILTER AND GRID

The main function of the inverter is to transform the current produced by the solar generator, into three-phase alternating current via DC/DC Boost converter. The three-phase voltage source inverter control diagram is given in Figure 8.

Figure 8. DC/AC inverter connected microgrid system

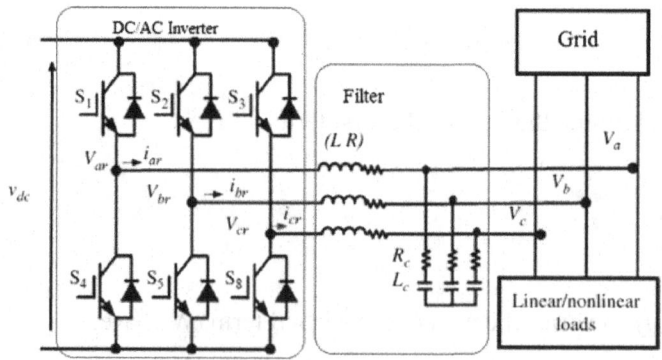

Modeling of the RI Filter Associated With the DC/AC Converter

The connection to the grid is carried out via a first-order RL input filter, aiming to respect the alternation of sources and to prevent the components due to the switches to spread over the grid.

The application of the mesh law for each phase, at the connection point of the filter. We give the equations, which link the voltages modulated, by the converter and the currents passing the filter:

$$
\begin{cases}
v_{ar} = R + L\dfrac{di_{ar}}{dt} + V_a \\[2mm]
v_{br} = R + L\dfrac{di_{br}}{dt} + V_b \\[2mm]
v_{cr} = R + L\dfrac{di_{br}}{dt} + V_c
\end{cases}
\tag{33}
$$

Such:

$$
\begin{cases}
i_{ar} = \dfrac{1}{R + LP}\left(v_{ar} + v_a\right) \\[2mm]
i_{br} = \dfrac{1}{R + LP}\left(v_{br} + v_b\right) \\[2mm]
i_{cr} = \dfrac{1}{R + LP}\left(v_{cr} + v_c\right)
\end{cases}
\tag{34}
$$

Electrical Grid Modeling

Loads are elements that consume electrical energy in a system. The consumption of this electrical power depends on the characteristics of the load. A correct modelling of these characteristics is essential to represent the load behavior connected to the voltage inverter. Which can be a charge R or a low voltage network through a filter.

Our electrical grid modelled by the following equations:

$$\begin{cases} v_{ar} = \sqrt{2}V_{dc}\sin\omega_s t \\ v_{br} = \sqrt{2}V_{dc}\sin\left(\omega_s t - \dfrac{2\pi}{3}\right) \\ v_{cr} = \sqrt{2}V_{dc}\sin\left(\omega_s t + \dfrac{2\pi}{3}\right) \end{cases} \qquad (35)$$

Active and Reactive Power Control (P And Q)

The purpose of the command is to impose the active and reactive power values injected into the grid. The control circuit by acting on the hysteresis control of the currents must impose the instantaneous value of the current supplied by the inverter so that the current supplied by the grid is sinusoidal and in phase with the corresponding voltage. This control is represented on the following block diagram and its principle is presented on the following Figure 9.

Figure 9. Block diagram of the inverter connected grid

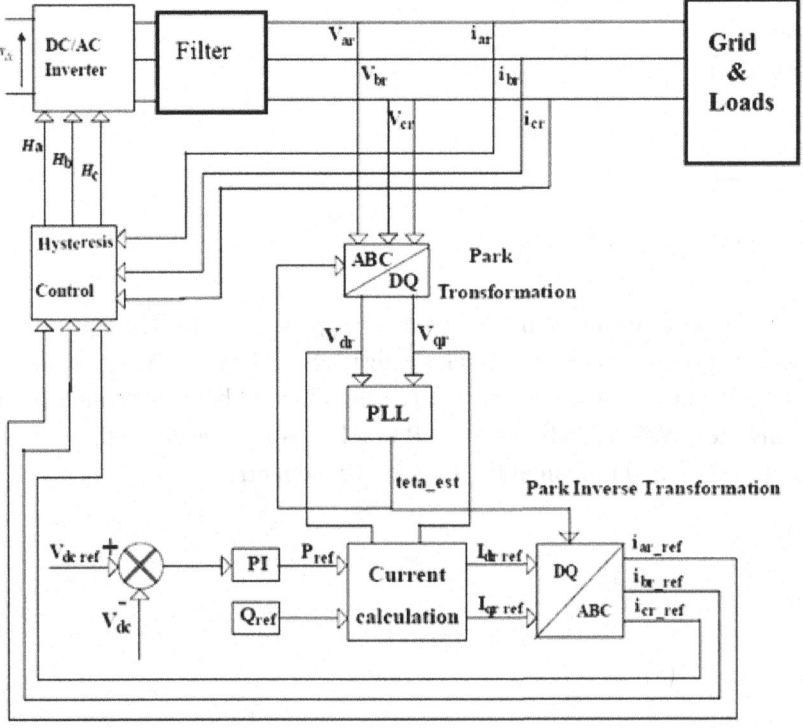

The various simplifications carried out after analysis of the system allowed us to conclude that the set point currents at the output of the upstream control will be injected at the connection point of the PV production. These currents are calculated using power references and measuring the voltage at the connection point, these will be calculated in the Park reference system according to the following equations:

$$\begin{cases} P_{ref} = V_{dr}I_{dr\_ref} . V_{qr}I_{qr\_ref} \\ Q_{ref} = V_{qr}I_{dr\_ref} . V_{qr}I_{qr\_ref} \end{cases} \tag{36}$$

Then:

$$\begin{cases} I_{qr_{ref}} = \dfrac{P_{ref}V_{dr}+Q_{ref}V_{qr}}{V_{dr}^2+V_{dr}^2} \\ I_{dr_{ref}} = \dfrac{P_{ref}V_{dr}+Q_{ref}V_{qr}}{V_{dr}^2+V_{dr}^2} \end{cases} \tag{37}$$

Where: P_{ref} and Q_{ref} : are the reference powers of PV production. V_{dr} and V_{qr} : are the direct and quadratic components of the voltage, measured at the point of connection of PV production, in the Park repository. I_{dr_ref} and I_{qr_ref} are the direct and quadratic components of the current produced by reference by the PV production on the network to which it is connected. These currents, therefore, depend on the reference powers as well as on the voltage measured at the production connection point. This measured voltage is transformed into the Park reference system before the currents are calculated. A Phase-Locked Loop (PLL) is used to synchronize Park's transformation on the pulsation of the voltage measured on the network.

Study of Converter Synchronization on the Grid

In order to connect sources to the electrical grid, the voltage of the generator of production with that of the grid, this is why the phase and network voltage frequency is required.

PLL Three-Phase in the Park Domain

The basic principle of three-phase PLL is to apply a transformation Park inverse on the three-phase grid voltages. The component of axis q generated by this transformation is slaved to zero by action on the angle of Park's coordinate system $\left(\theta_{est}\right)$ In regime established the angle $\left(\theta_{est}\right)$ is equal to the angle $\left(\theta_r\right)$ of the grid.

The principle of PLL in the Park domain is given in Figure 10.

Figure 10. Principle of the PLL

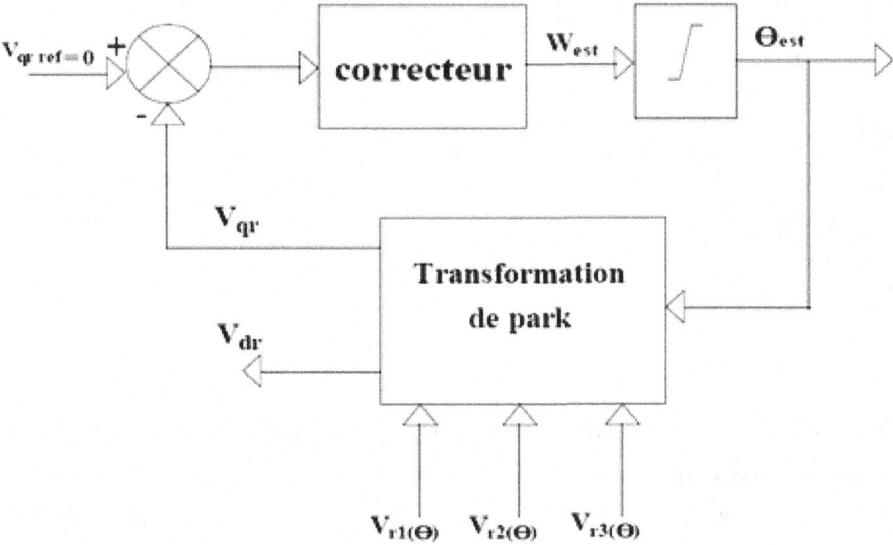

BATTERY ENERGY STORAGE SYSTEM BEES WITH MANAGEMENT

Our Battery Energy Storage System connected to PV in the same time with grid. The BEES model present in Figure 11. This PV-Grid-BEES system has batteries connected to DC bus via Buck-Boost converter with control management (Borekci Selim, Kandemir Ekrem, Kircay Ali, 2015).

The Battery Energy Storage System with a control to save the DC bus voltage at the optimum value determined via MPTT. The principle of BEES strategy present in Figure 12. We keep the DC reference voltage V_{ref} at 500V by the PI corrector calculates the reference current of DC bus $I_{dc_{ref}}$.

The reference current $I_{dc_{ref}}$ also reference current of batteries $I_{bat_{ref}}$ of the DC bus which are determined by the DC bus controller.

Figure 11. Battery energy storage system

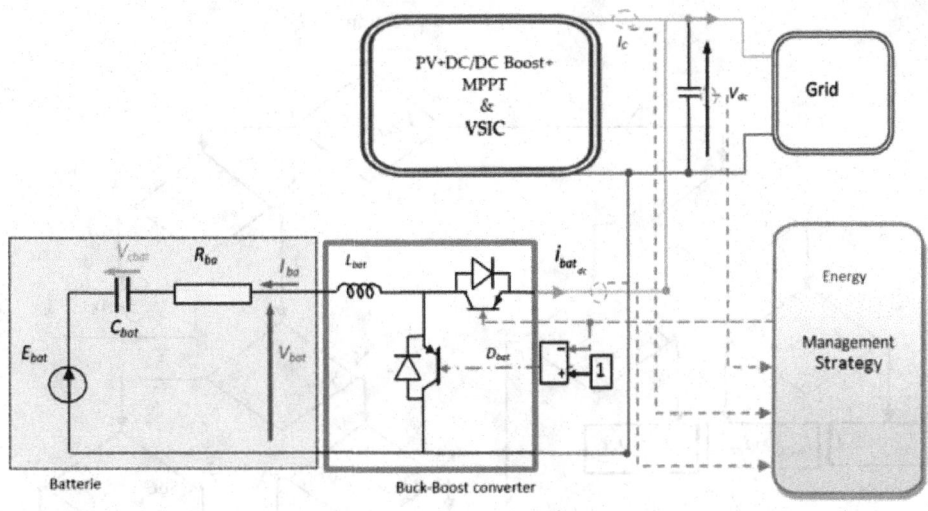

Figure 12. Control the DC bus voltage

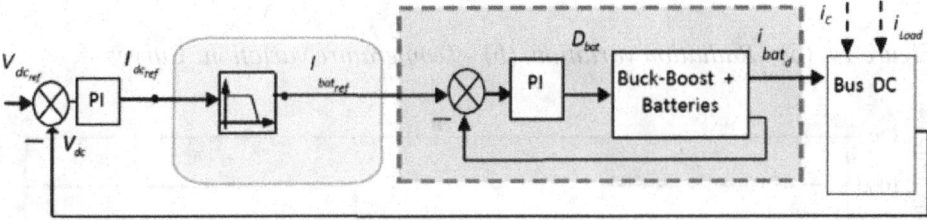

The control algorithm of the batteries shown in Figure 13. The charge of batteries state in PV system SoC_{bat} is between [25% and 95%]. The $I_{bat_{ref}}$ based on the passage of the DC bus reference current I_{dc} ref from a low-pass filter.

RESULTS AND DISCUSSION

Conditions and Parameters of System

The system simulated with variable values of temperature and solar irradiation in the MATLAB/Simulink environment. Figure 14 shows the proposed variation of temperature between the value in the interval chosen [25°C - 50°C] and solar irradiation interval [600W/m² -1000W/m²] for all system under study (See Tables 1, 2, and 3).

Figure 13. Control algorithm of storage energy in the batteries

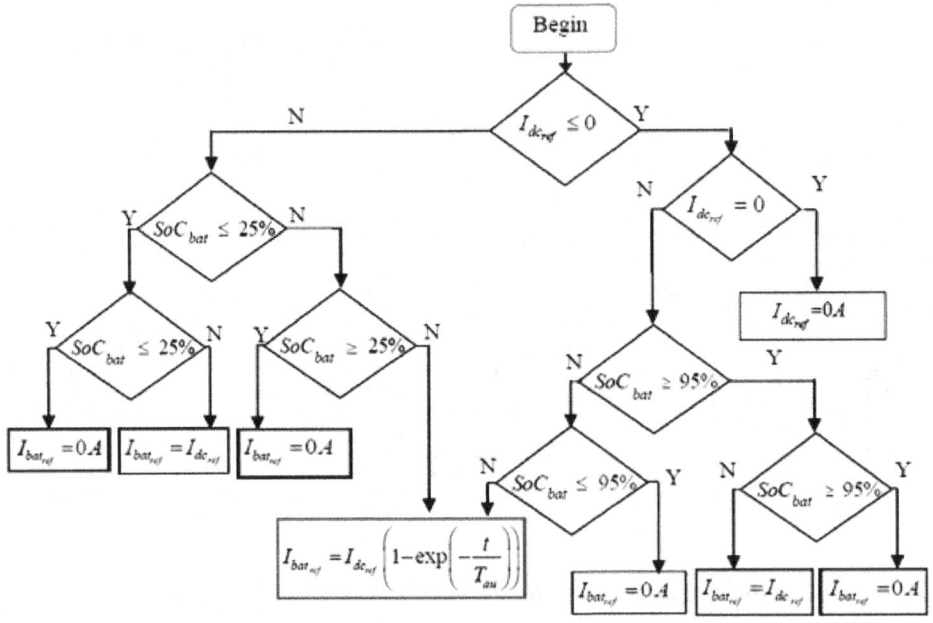

Figure 14. (a) - Radiation variation, (b) - Temperature variation: Curves

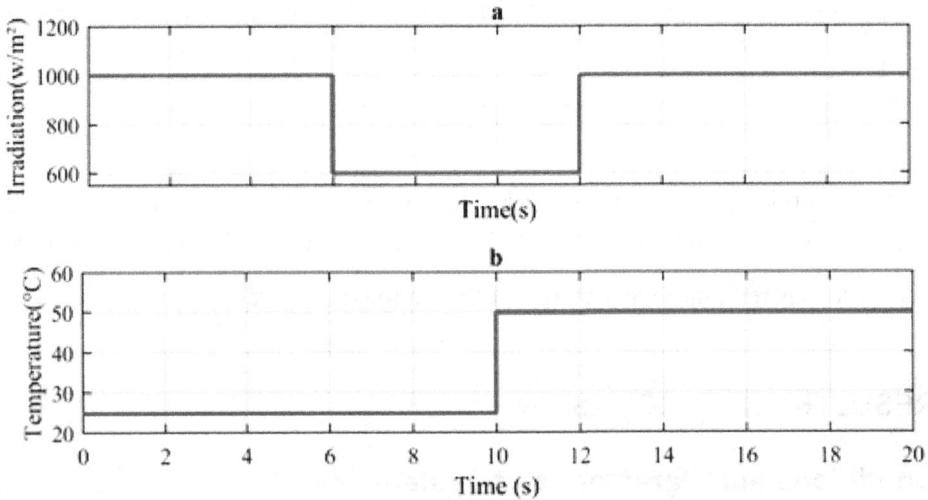

Table 1. PV 330 * SunPower SPR-305E-WHT-D values

Component	Input data value
N_s	66
E_{G0}	1.12
K	$1.3805.10^{-23}$ J/K°
q	$1.6201.10^{-19}$°C
R_s	153.2529 Ohms
R_{sh}	0.44959 Ohms
γ	1.4

Table 2. DC/DC Boost values

Component	Input data value
C_p	$47e^{-3}$F
C_m	$4.7e^{-3}$F
L	$3.5e^{-3}$H

Table 3. Global characteristics of the energy storage system

Component	Input data value
Maximal Power of batteries charge	48 kW
Switching frequency	5000 kHz
Bus DC	V_{dc} = 500V k_{idc} = 0.36 k_{Pdk} = 0.0025
Batteries, Buck-Boost and PI controller	$C_{buck\_boost\_out}$ = 5.4mF L_{buck_boost} = 15mH k_{ibat} = 8.3 k_{Pbat} = 0.037 L_{bat} = $15e^{-3}$H

- We filtrate the harmonics produced by VSC via10-kvar capacitor bank.
- The transformer three-phase coupling 100-kVA 260V/25kV.
- The grid utility: Distribution 25-kV + Equivalent transmission system 120 kV.

Conditions and Parameters of System

Figure 15, Figure 16 and Figure 17 show successively the PV output voltage, PV current and PV power curves. We notice that the power generate provided by the PV varies with the intensity of irradiation and the temperature of the climatic conditions. To solve the fluctuation of the PV power we add battery-energy-storage system on the DC side to compensate the power fluctuation. Figure 18 presents the battery state of charge and discharge (SOC in %) with climatic conditions and Figure 19 and Figure 20 shows successively the battery power response between compensation and energy-storage in the PV system and current variation of battery pending the charge and discharge with PV system.

Figure 15. PV voltage variation with climatic conditions

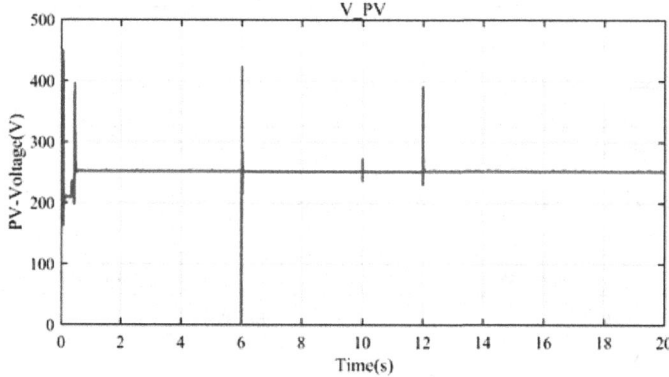

Figure 16. PV Current variation with climatic conditions

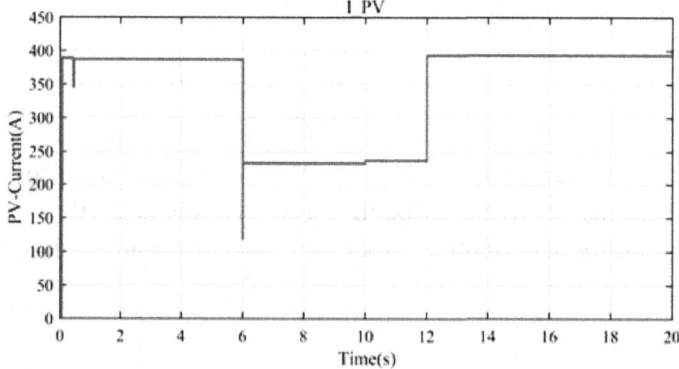

Figure 17. PV Power variation with climatic conditions

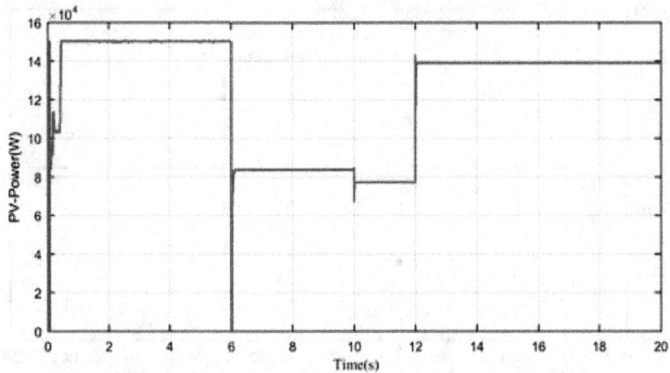

Figure 18. Battery's SOC response withe climatic condition

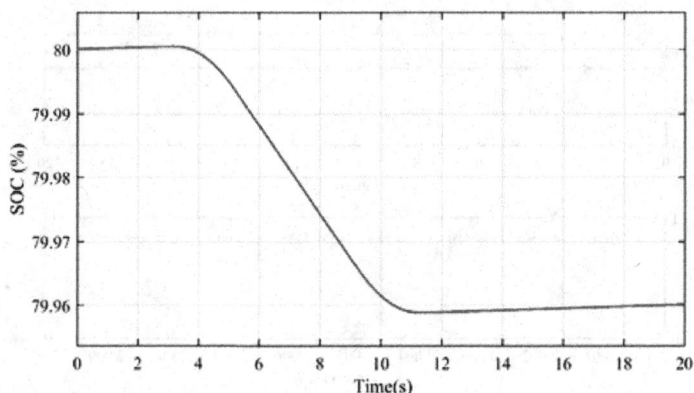

Figure 19. The battery's currentresponse withe climatic condition

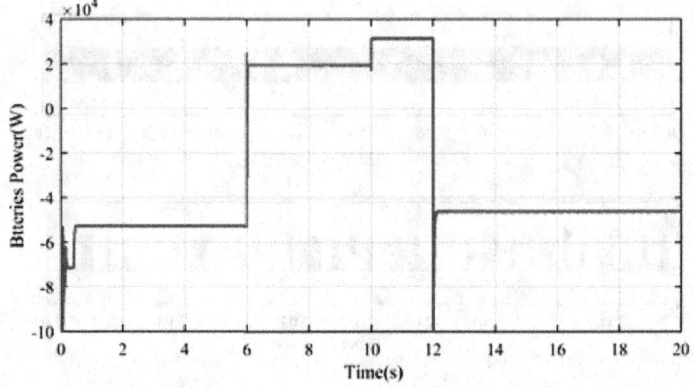

Figure 20. The battery's current response withe climatic condition

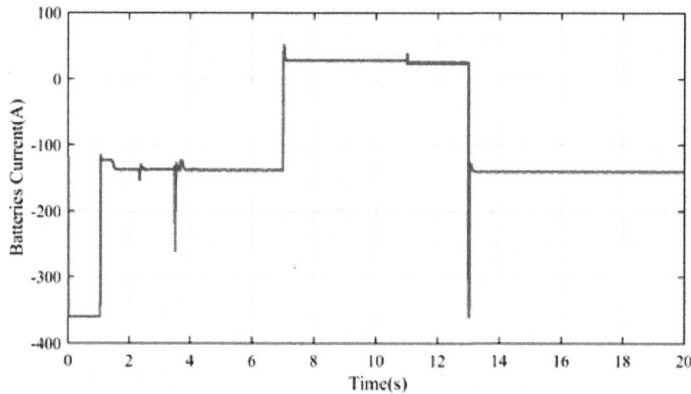

Figure 21. (a)- Duty cycle signal, (b)- Zoom of duty cycle signal

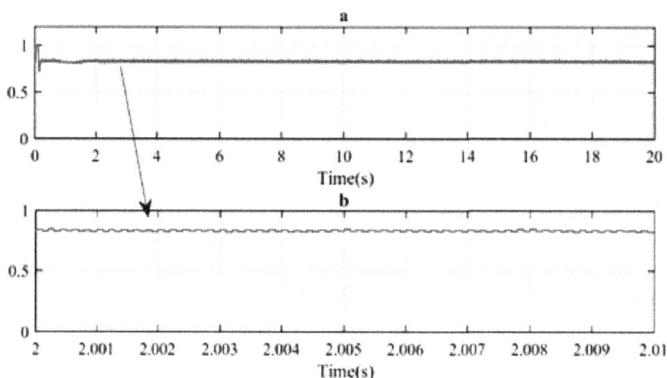

Figure 22. (a)- Sliding mode surface signal, (b)- Zoom of Sliding mode surface signal

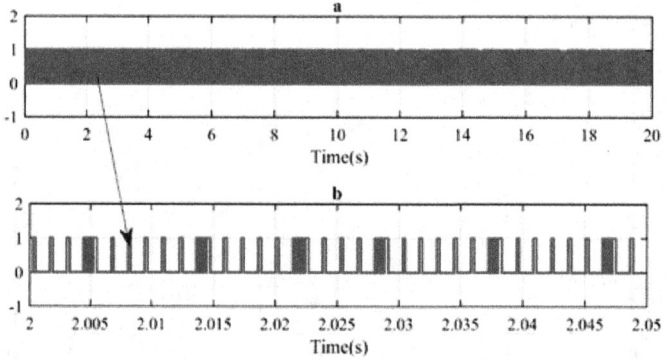

Figure 21a presents the duty cycle signal generated via sliding mode controller and its zoom in Figure 21b. Therefore, Figure 22a shows the sliding mode surface and its zoom in Figure 22b. Figure 23 shows the output voltage obtained from DC/DC boost. As a result, we attained the maximum power point of the photovoltaic generator, after a transition/response relying on sunlight and temperature changes for the PV. We observe that the MPPT algorithm by using sliding mode which give excellent results. Hence, we evaluated the efficiency of the control system with temperature changes and solar irradiation. In addition, we can see that the controller used in this work has a good performance and robustness and has an exact following effect at the desired values in each part of PV system.

Figure 23. Output boost voltage and reference voltage

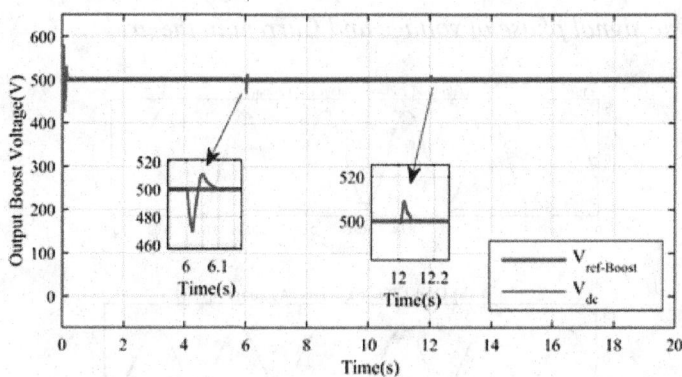

Figure 24. 3-Phase voltage of grid and its Zoom

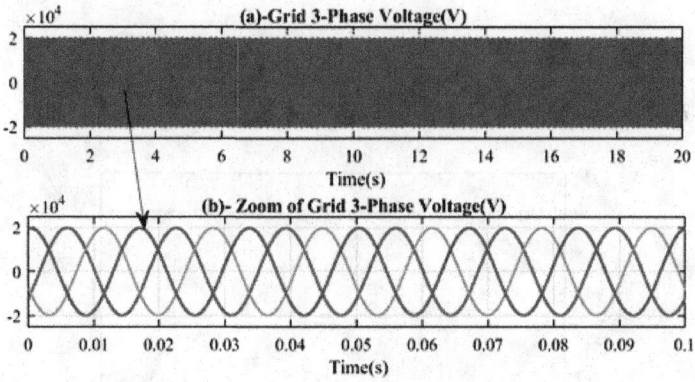

Figure 25. 3-Phase current of grid and its Zoom

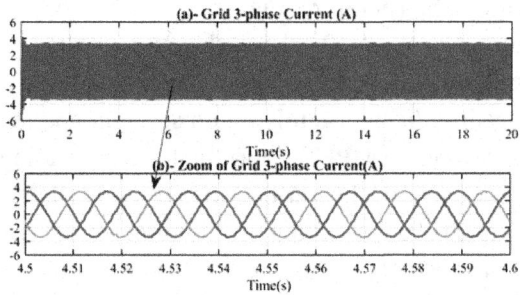

Figure 26. One signal phase of voltage and Current in the grid

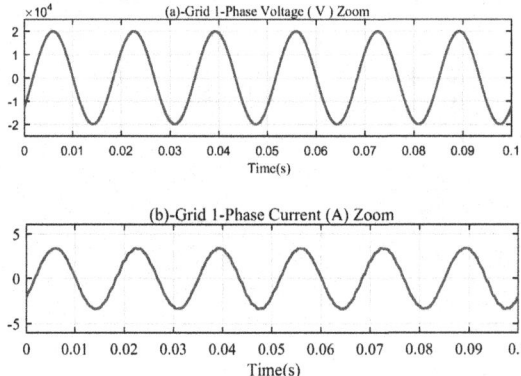

Figure 27. Power in the grid

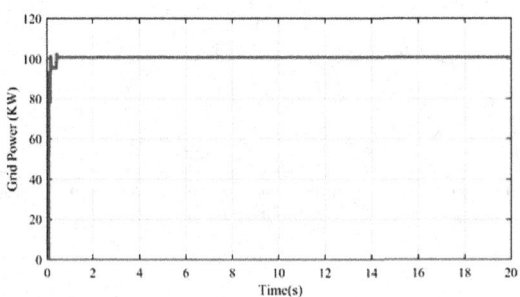

Figure 24a indicated 3-phase of voltage and its zoom in Figure 24b, Figure 25 indicated 3-phase current and its zoom in Figure 25b, Figure 26 shows the single-phase the grid voltage with grid current and Figure 27 power of grid controlled by VISC. It is worth noting that the performance kept at the wished stable value, no matter what the values of solar irradiance and temperature conditions. By doing so we make sure, we have a stable operation of the PV connected grid system with sliding mode and Battery Energy Storage System. Furthermore, the VSIC gives satisfactory results on the grid side.

In general, result, we realize the target of our work, as discussed how we can design a PV system connected grid with stable power via three major factors: Obtain maximum power from the PV array with all climatic conditions. We add Battery Energy Storage System management to manage the energy obtained from PV. We injected in the grid after, the VSIC granted a good value in the grid side. The aim of work presents a success PV system with high performance.

CONCLUSION

In our work, we connected the PV to the grid and BEES simulated under different values of irradiation and temperature. When identifying an MPPT controller based on sliding mode, high performance and good efficiency during operation that controls a duty cycle of the DC/DC boost converter. The MPPT used a synthesized strategy focus on the sliding mode controller. The global target-present to obtain a rapid response, stability, and, robustness determined by the Lyapunov theory. In addition, we add a BESS with management as a regulator of energy generated via PV because it saves the power in the constant value all the time via compensation for lack of power or absorption of excess power. The use of the VSIC controller makes the grid current and voltage obtained after a DC/AC, in good values and balanced in 3-phases make the currents and the voltages in phases in the grid.

REFERENCES

Ahmed, J., & Salam, Z. (2016, May). A Modified P&O Maximum Power PointTracking Method With Reduced Steady-State Oscillation and Improved Tracking Efficiency. *IEEE Transactions on Sustainable Energy*, 7(4), 1506–1515. doi:10.1109/TSTE.2016.2568043

Aouadi, Abouloifa, Hamdoun, & Boussairi. (2014). Backstepping Based Control of PV system Connected to the Grid. *International Journal of Computer and Information Technology, 3*, 1021-1026.

Bharath, K. R. (2018, November). Control of Bidirectional DC-DC Converter in Renewable based DC Microgrid with Improved Voltage Stability. *International Journal of Renewable Energy Research, 8*, 872–1630.

Carrasco, J. M., Franquelo, L. G., Bialasiewicz, J. T., Galvan, E., PortilloGuisado, R. C., Prats, M. A. M., Leon, J. I., & Moreno-Alfonso, N. (2006). Power-electronic systems for the grid integration of renewable energy sources: A survey. *IEEE Transactions on Industrial Electronics, 53*(4), 1002–1016. doi:10.1109/TIE.2006.878356

Deshpande, R. (2019, March). Analysis of Power Quality Variations in Electrical Distribution System with Renewable Energy Sources. *International Journal of Renewable Energy Research, 9*, 282–289.

Dong, W., Farrell, J. A., & Polycarpou, M. M. (2012). Command filtered adaptive backstepping. *IEEE Transactions on Control Systems Technology, 20*(3), 566–580. doi:10.1109/TCST.2011.2121907

Elgendy, M. A., Atkinson, D. J., & Zahawi, B. (2016, February). Khalifa University, "Experimental investigation of the incremental conductance maximum power point tracking algorithm at high perturbation rates. *IET Renewable Power Generation, 10*(2), 133–139. doi:10.1049/iet-rpg.2015.0132

Farrell, J. A., Polycarpou, M., Sharma, M., & Wenjie Dong. (2009). Command filtered backstepping. *IEEE Transactions on Automatic Control, 54*(6), 1391–1395. doi:10.1109/TAC.2009.2015562

Geurin, S. O., Barnes, A. K., & Carlos, B. J. (2012). Smart grid applications of selected energy. In Innovative Smart Grid Technologies (ISGT). IEEE PES.

Guisser, El-Jouni, & Abdelmounim. (2014). Robust Sliding Mode MPPT Controller Based on High Gain Observer of a Photovoltaic Water Pumping System. *International Review of Automatic Control, 7*, 225-232.

Heng, P.-C., Bo-Rei, P., Liu, Y.-H., Cheng, Y.-S., & Jia-We, H. (2015, June). Optimization of a fuzzy-logic-control-based MPPT algorithm using the particle swarm optimization technique. *Energies, 8*(6), 5338–5360. doi:10.3390/en8065338

Keddar, M., Doumbia, M. L., Della, M., Belmokhtar, K., & Midoun, A. (2019, September). Interconnection Performance Analysis of Single Phase Neural Network-Based NPC and CHB Multilevel Inverters for Grid-connected PV Systems. *International Journal of Renewable Energy Research, 9*, 1451–1461.

Kolesnik, S., & Kuperman, A. (2016, November). On the Equivalence of Major Variable-Step-Size MPPT Algorithms. *IEEE Journal of Photovoltaics, 6*(2), 590–594. doi:10.1109/JPHOTOV.2016.2520212

Kouchaki, A., Iman-Eini, H., & Asaei, B. (2013). A new maximum power point tracking strategy for PV arrays under uniform and non-uniform insolation conditions. *Solar Energy, 91*, 221–232. doi:10.1016/j.solener.2013.01.009

Kumar Singh, H. A., Hussain, I., & Singh, B. (2018, May). Double-Stage Three-Phase Grid-Integrated Solar PV System With Fast Zero Attracting Normalized Least Mean Fourth Based Adaptive Control. *IEEE Transactions on Industrial Electronics, 65*(5), 3921–3931. doi:10.1109/TIE.2017.2758750

Liu, J., Luo, W., Yang, X., & Wu, L. (2016). Robust model-based fault diagnosis for PEM fuel cell air-feed system. *IEEE Transactions on Industrial Electronics, 63*(5), 3261–3270. doi:10.1109/TIE.2016.2535118

Martin, A. D., Cano, J. M., Silva, J. F. A., & Vazquez, J. R. (2015). Backstepping control of smart grid-connected distributed photovoltaic power supplies for telecom equipment. *IEEE Transactions on Energy Conversion, 30*(4), 1496–1504. doi:10.1109/TEC.2015.2431613

Menadi, A., Abdeddaim, S., Ghamri, A., & Betka, A. (2015, September). Implementation of fuzzy-sliding mode based control of a grid-connected photovoltaic system. *ISA Transactions, 58*, 586–594. doi:10.1016/j.isatra.2015.06.009 PMID:26243440

Mlayah, A., & Khedher, A. (2018, September). Sliding Mode Control Strategy for Solar Charging of High Energy Lithium Batteries. *International Journal of Renewable Energy Research, 8*(3), 1621–1623.

Mohamad, Teh, Lai, & Chen. (2018). Development of energy storage systems for power network reliability: a review. *Energies, 11*(9).

Mojallizadeh, M. R., & Badamchizadeh, M. A. (2017). Second-order fuzzy sliding-mode control of photovoltaic power generation systems. *Solar Energy, 149*, 332–340. doi:10.1016/j.solener.2017.04.014

Moradi, M. H., & Reisi, A. R. (2011). A hybrid maximum power point tracking method for photovoltaic systems. *Solar Energy, 85*(11), 2965–2976. doi:10.1016/j.solener.2011.08.036

Pani, A. K., & Nayak, N. (2019, November). A Short Term Forecasting of PhotoVoltaic Power Generation Using Coupled Based Particle Swarm Optimization Pruned Extreme Learning Machine. *International Journal of Renewable Energy Research, 9*, 1190–1202.

Rahrah, K., Rekioua, D., Rekioua, T., & Bacha, S. (2015, October). Photovoltaic pumping system in Bejaia climate with battery storage. *International Journal of Hydrogen Energy, 40*(39), 13665–13675. doi:10.1016/j.ijhydene.2015.04.048

Selim, B., Ekrem, K., & Ali, K. (2015, November). A simpler single-phase single-stage grid connected PV system with maximum power point tracking controller. *Elektronika ir Elektrotechnika, 21*, 44–47.

Sivakumar, JagabarSathik, Manoj, & Sundararajan. (2016). An assessment on performance of DC–DC converters for renewable energy applications. *Elsevier Renewable and Sustainable Energy Reviews, 58*, 1475-1485.

Suberu, M. Y., Mustafa, M. W., & Bashir, N. (2014). Energy storage systems for renewable energy power sector integration and mitigation of intermittency. *Energy Rev., 35*, 499–514.

Teodorescu, R., Liserre, M., & Rodriguez, P. (2011, March 15). *Grid Converters for Photovoltaic and Wind Power Systems*. Retrieved from https://www.bookdepository.com/Grid-Converters-for-Photovoltaic-Wind-Power-Systems-Remus-Teodorescu/9780470057513

Tey, Mekhilef, Seyedmahmoudian, Horan, Oo, & Stojcevski. (2018). Improved Differential evolution-based MPPT Algorithm Using SEPIC for PV Systems Under Partial Shading Conditions and Load Variation. *IEEE Transactions on Industrial Informatics, 14*, 22 – 43.

Wai, R., Lin, C., Wu, W., & Huang, H. (2013). Design of backstepping control for high-performance inverter with stand-alone and grid-connected power-supply modes. *IET Power Electronics, 6*(4), 752–762. doi:10.1049/iet-pel.2012.0579

Xua, D., Gang, W., Yana, W., & Yanb, X. (2019, January). A novel adaptive command-filtered backstepping sliding mode control for PV grid-connected system with energy storage. *Solar Energy, 178*, 1–17. doi:10.1016/j.solener.2018.12.033

Yang, Y., Ye, Q., Tung, L., Greenleaf, M., & Li, H. (2018). Integrated size and energy management design of battery storage to enhance grid integration of large-scale PV power plants. *IEEE Transactions on Industrial Electronics*, *65*(1), 394–402. doi:10.1109/TIE.2017.2721878

Yesilbudak, Colak, & Bayindir. (2018). What are the Current Status and Future Prospects in Solar Irradiance and Solar Power Forecasting? *International Journal of Renewable Energy Research, 8*, 636-646.

Yu, J., Shi, P., Dong, W., & Yu, H. (2015). Observer and command-flter-based adaptive fuzzy output feedback control of uncertain nonlinear systems. *IEEE Transactions on Industrial Electronics*, *62*(9), 5962–5970. doi:10.1109/TIE.2015.2418317

Zame, Brehm, Nitica, Richard, & Schweitzer III. (2017). Smart grid and energy storage: policy recommendations. *Renew. Sustain. Energy Rev., 82*(1), 1646–1654.

Chapter 9

Comparison Analysis of MAC Protocols for Wireless Sensor Networks:
A Comprehensive Survey

Tapaswini Samant
Kalinga Institute of Industrial Technology, India

Yelithoti Sravana Kumar
iD https://orcid.org/0000-0001-6397-5524
Kalinga Institute of Industrial Technology, India

Swati Swayamsiddha
Kalinga Institute of Industrial Technology, India

ABSTRACT

Wireless sensor networks (WSN) are rapidly emerging as an interesting and challenging area of research in the field of communication engineering. This review work is different from other state-of-the-art literature as the MAC protocols discussed here are applicable both for homogeneous and heterogeneous networks. Performances like energy efficiency, cost optimization, throughput, bandwidth utilization, and scalability of the sensor network depend on MAC protocols, which are application-based. In the study, the authors have surveyed different MAC protocols with different merits and demerits. Based on the study, it is very hard to recommend any particular protocol as a standard for implementation as these are exclusively application dependent. The work can be further extended in terms of hybrid protocols, which may carry the advantages of the respective protocols along with energy-efficient criteria for practical implementation. Further cooperative WSN communication can be used for internet of things (IoT)-based systems, where the node placements and multi-operations concepts are of main concern.

DOI: 10.4018/978-1-7998-4381-8.ch009

INTRODUCTION

Wireless Sensor Network (WSN) consists of distributed autonomous sensor nodes which are used to monitor different environmental conditions like temperature, pressure, humidity, noise level, light intensity, etc. These nodes cooperatively send their information to the central unit or base station. WSN finds applications in remote monitoring, target tracking, automation and traffic control (Pegatoquet et.al 2018, Salman 2014, Heidemann & Estrin 2004, Qian & Waltenegus 2012). A WSN consists of wireless units called nodes. Each sensor node incorporates a processing unit, a power unit and a communication unit as illustrated in Figure 1.

Figure 1. Structure of a WSN node

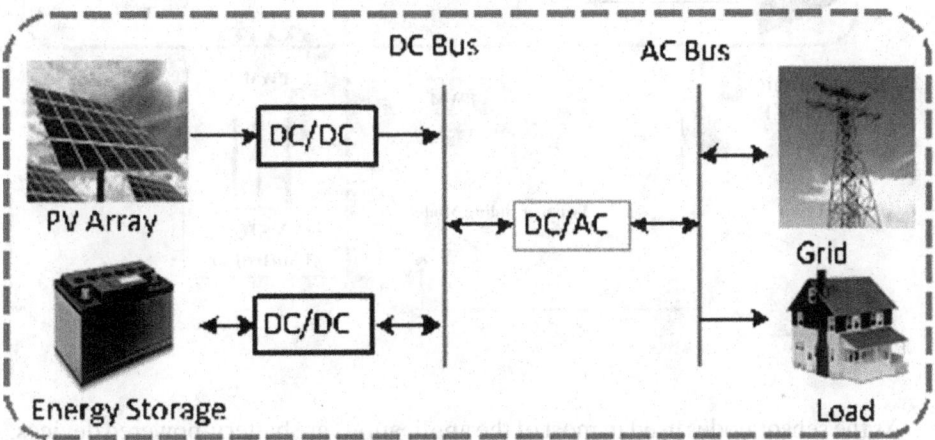

Sensor nodes located in monitored areas coordinate to form a network through the access mechanism of self-organization. These nodes transmit the monitored data through other nodes one after another (Demirkol et.al. 2006). Figure 2 shows the architecture of WSN, which consists of a source (as of end-user) and a sink connected through the internet and the sensing operation is performed using the wireless sensors.

Figure 2. Architecture of WSN

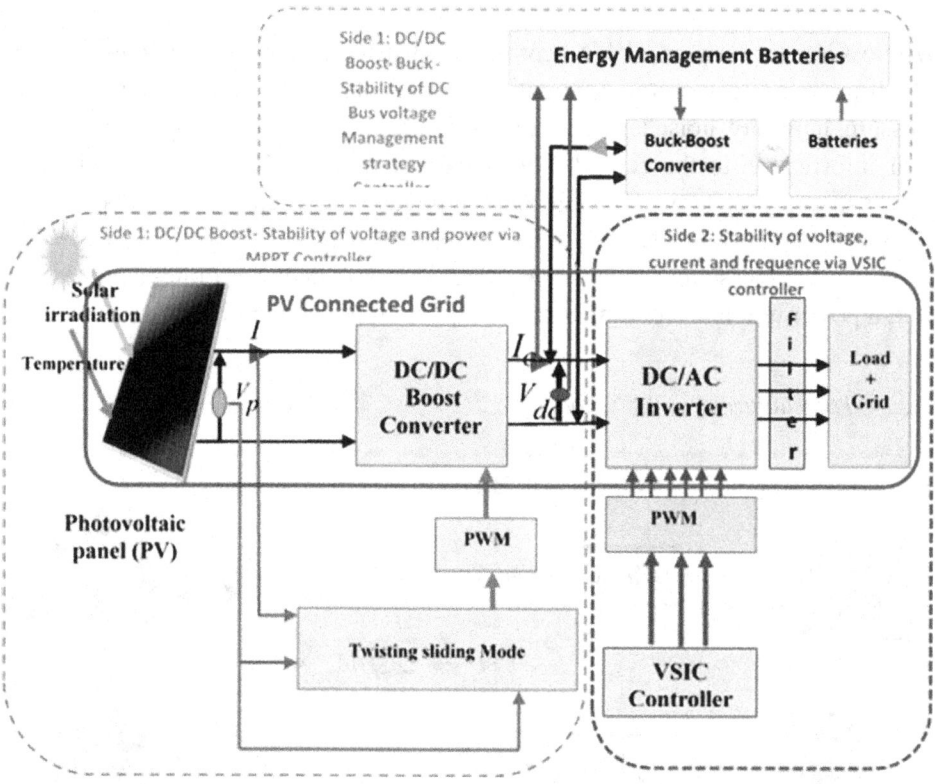

As the sensor nodes used in most of the applications are battery-powered devices, most of the challenges are focused on how to increase the lifetime of each sensor node and how to reduce energy consumption in the nodes, thus implicating greater network energy efficiency. Network lifetime is represented as the time passed from the path operation starts till the first node (or last node) in the path expends its energy (dies). Generally, the activity of the nodes depends upon its transmission and reception of information. On the other way, it performs other activities like redundant control packets broadcast, re-transmission due to collision, overhearing, and idle listening. These unnecessary node operations increase energy consumption and hence reduce the network lifetime. So, to avoid these unnecessary activities, many energy-efficient MAC protocols are proposed. Here, WSN multiple nodes may transmit information on the same shared channel at the same time. In such a situation, the transmitted information would get distorted unless a suitable medium access attribution scheme is deployed. Usually to perform this task MAC protocols are implemented in the data link layer. The traditional wireless MAC protocols (IEEE 802.11) are having many constraints. The main problem is that the nodes used are battery-driven, so sensor network applications are limited (Yang et al. 2018, Liu et al. 2018, Halkes et al. 2005).

This review work is different from other state-of-the-art review papers as the MAC protocols discussed here are applicable both for homogeneous and heterogeneous networks. In this paper, we discussed the different MAC protocols and their classifications. The rest of the paper is organized as follows. In section 2, we discuss the different challenges of MAC protocols in WSNs. Different merits and demerits of MAC are described in section 3. Then a comparison of all protocols is discussed in section 4. Finally, the behavioral analysis of MAC protocols is explained and concluded in section 5.

MEDIUM ACCESS PROTOCOLS

MAC protocols play an important role in accessing the same channel in the wireless communication system. For the design of MAC Protocols, there are many factors like collisions, overhead, overhearing, complexity, idle listening, Quality of Service (QoS), etc. where maximum energy consumption is required. Since energy consumption is very important for sensor nodes as they are battery powered, the design of MAC protocols throughput, fairness, and the end to end delay are relevant to the Quality of Service(Enz et al. 2004, Ray & Turuk 2009, Li et al. 2018).

Factors Influencing WSN MAC Design

According to (Huang et al. 2012), to design an efficient MAC protocol for WSN, attributes such as Energy-efficiency, scalability and adaptivity, Latency, Channel utilization, throughput, fairness, etc. are to be considered. In this section, a brief description of each of the attributes is discussed.

Energy-Efficiency

The sensor nodes run on battery power. And it is almost impossible to change or recharge the batteries of the nodes. The radio is the major consumer of energy in many hardware platforms. MAC protocols are classified into different categories: Contention free protocols, Hybrid protocols, Contention-based protocols, and Cross-Layer MAC. So, the MAC layer needs to consider this issue as it directly controls radio activities and can enhance the network's lifetime.

Scalability and Adaptivity

The number of sensor nodes deployed in studying a phenomenon may vary. Depending on the application, it can be thousands or any extreme number. Some nodes in the network may die, also at times, new nodes may be added to the network. Also because of mobility, it finds a scenario where the existing nodes may move to a new region. All such characteristics must be accommodated by a good MAC protocol.

Latency

Latency refers to the time-delay between the time when a packet is sent by the sender and the time when the receiver receives the packet successfully. Based on the sensor network application, the latency requirement varies and accordingly reported to the sink node. These actions are carried out in real-time to implement the required measures.

Channel Utilization

This reflects how well the entire bandwidth of the channel is utilized in communication. Bandwidth is a valuable resource in wireless communication. So, the MAC protocols designed for WSN should maximize the utilization of this scarce resource.

Throughput

It is the total bulk of successful data that is sent from a sender node to the receiver node at any given instance of time. The throughput is recorded as bits or bytes per second. Like latency, the throughput also varies for different applications.

Fairness

Fairness is referred to as how effectively the sensor nodes equally access the common transmission channel. As the nodes in WSN cooperate to perform a common and single task, it is vital to achieving per-node fairness along with a focus on the quality of service for the overall task.

Causes of Energy Depletion

1. **Collision**: We can say that the collision occurs at times when the sensor node receives multiple packets at the same time. This leads to data loss as the packets

get corrupted and they are discarded from the network. So, one of the important resources i.e. energy is wasted.

2. **Overhearing**: This means some nodes think the data they are listening to belongs to them as a destination but they are destined for some other nodes. This also leads to one of the causes of energy loss.

3. **Control Packet Overhead**: Some packets are by default used to control the data transmission like Request-to-Send (RTS) and Clear-to-Send (CTS) signals. These also use energy. So we can focus to use minimum required control packets which will carry out the data transmission.

4. **Idle Listening**: Normally a node remains active throughout. That means to trace whether the channel is available or not, the node continues its active state. So the energy is consumed continuously. Integrating such factor in a broader concept the amount of energy may be even greater than that is required actually for effective communication.

5. **Over Emitting**: Sometimes during the transmission the receiver is either not ready or the receiving signal strength is lower than the transmitter. This causes the over emitting of data packets.

Classification of MAC Protocol

The WSN MAC protocols are classified based on energy efficiency, latency, bandwidth utilization, fairness, throughput, and other parameters and discussed as follows (Ye et al. 2006, Zhou et al. 2006):

Classification Based on Energy Efficiency

As shown in Figure 3, based on the energy efficiency, MAC protocols are being classified into different categories. Initially, these are divided based on the operational techniques that are contention-based, contention-free TDMA, Hybrid or Cross layer-based. In the later sections, these techniques are discussed elaborately.

Figure 3. Taxonomy of MAC protocols based on energy efficiency in WSNs

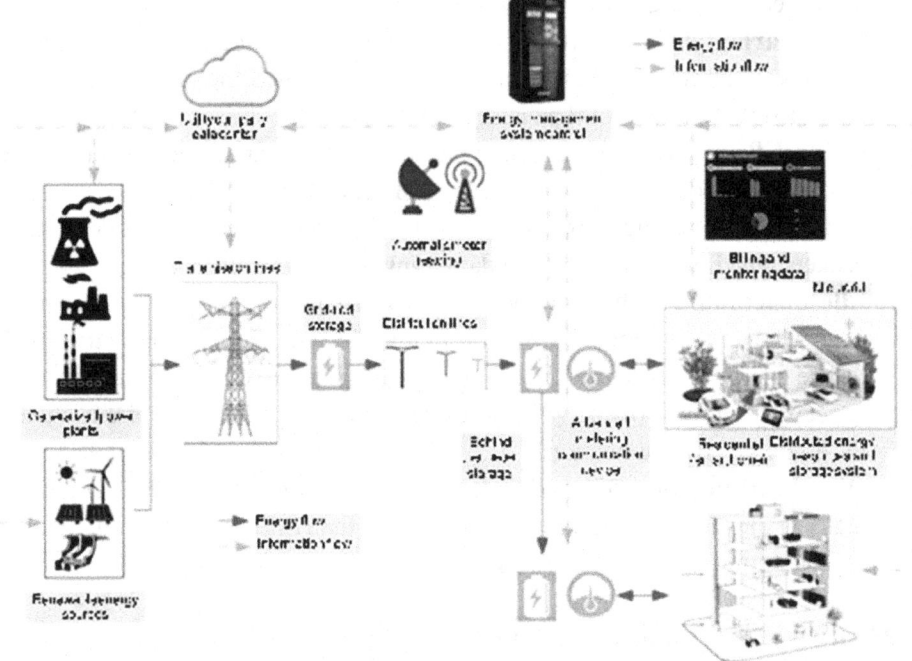

Contention-Based MAC Protocols

In this case, the MAC protocols are operated on wireless communication that allows multiple users to use the same channel without pre-co-ordination. So, there is a chance of collision. Carrier Sense Multiple Access (CSMA) is used when a sensor transmits the signal to the network, otherwise, a collision occurs which results in loss of data packet or frame. Here the channel is sensed before any transmission is initiated. If the channel is found to be busy then it has to wait till the communication is over. There may be a scenario when multiple devices sense the channel and repeat the same approach and the probability of collision is still not avoided. So they undergo a time slot as a waiting state and then re-initiate the process. Carrier Sense Multiple Access with Collision Avoidance (CSMA/CA) works to avoid a collision before the occurrence but CSMA with Collision Detection (CSMA/CD) handles transmission only after a collision has taken place (Incel et al. 2011).

"Sense before transmit" is the basic principle of CSMA which is the advantage of MAC protocol. Under this MAC protocol, there are several other protocols like S-MAC, PAMAS, and T-MAC, U-MAC, B-MAC.

Power-Aware Multi-Access with Signaling (PAMAS) Protocol

Power-Aware Multi-Access with Signaling (PAMAS) protocol works under the principle of MAC protocol having individual signaling channels avoiding the over-hearing problem. In this protocol, the energy node uses signaling channels that transfer the frames of RTS-CTS for accessing data. Every node which has gained the data turns themselves off while allowing others to transmit collision-free data frames.

Though the problem of over-hearing is avoided the use of an additional channel leads to an increase in the cost of the network (Hossein & Fard 2011).

S-MAC Protocol

Based on the MAC protocols, the Sensor MAC (S-MAC) protocol solves the energy consumption problem such as idle listening, collision, and over-hearing. In this protocol, every node does not need to be awake all the time. S-MAC protocol solves unwanted energy consumption due to collision, over-hearing, idle listening and based on fixed duty cycles, the overhead turns the signal OFF-ON in WSN.

The demerit of S-MAC lies on the fixed duty cycles. So the communication subsystem remains active throughout. This principle uses lots of energy which is wasted unnecessarily(Hamid et al. 2010, Ali et al. 2019, Kalaivaani & Rajeswari 2013).

Time-out MAC (T-MAC) Protocol

As the continuous listening overhead consumed more energy, the Timeout T-MAC protocol avoids the issue. It is based on S-MAC concepts. Here the nodes go to a sleep mode when it finds no activations either or any event is triggered for the duration of the set period. It follows an active/inactive dynamic adaptive duty cycle which incurs the benefit in energy saving over classic CSMA and S-MAC protocols (Doudou et al. 2014).

But the early sleep issues also exist in T-MAC which can cause reduced throughput.

B-MAC Protocol

The sampling mechanism is used in the Berkeley Media Access Control (B-MAC) protocol. Here the protocol samples the channel at the set periods. The samples help to let the node know the status of the channel by listening to it. When a node is ready to transmit, initially it senses the channel and then if it finds it free, it takes small back-off and sends the preamble and then the data packet. This preamble is a signal, which is a bit greater than the sampling period length. When the receiver detects channel activity after it wakes up, it checks the channel for any preamble. Then it waits for the preamble to end and receives the respective packet. If not destined for the receiver, it goes to idle mode (Sherazi et al. 2018).

U-MAC

Ultra-Wide Band MAC U-MAC is used for the effective use of energy. The SMAC approaches are used here with many enhancements like selective sleeping, changeable duty cycles as per requirement and tuning of duty-cycle which has a greater impact on the performance.

Contention Free MAC Protocols

The contention-free MAC protocol when implemented leads to no collision. Here contention-free TDMA protocol is discussed. In TDMA based protocols, each sensor node is allotted with a specific time slot that makes it contention-free. The sensor nodes act actively in their respective slots and listen to the channel. Then they become inactive or we can say that the sensor nodes go to sleep mode. Though this mechanism possesses the advantage of collision-free the throughput is higher only when the traffic is greater. But in lower-traffic, the throughput is significantly reduced. Traffic-Adaptive Medium Access (TRAMA) is a TDMA based protocol that uses traffic-based scheduling to avoid wasting slots (Djiroun & Djenouri 2017). Other TDMA based MAC protocols are discussed below.

TRAMA Protocol

TRAMA protocol works in the distributed methodology which creates the environment for flexible time slots. It works on a distributed TDMA method and thus implements dynamic scheduling of time slots. Also, the contention-based issues are avoided as of time slot allocations to the sensor nodes. The transmission schedules are managed by the Neighbour Protocol (NP) and the Schedule Exchange Protocol (SEP) (Huang et al. 2012).

µ-MAC Protocol

The µ-MAC protocol significantly records the high sleep ratios and maintains the reliability and latency of the message. It works by scheduling the task and accesses the shared medium. This is observed and predicted as per the traffic. the µ-MAC protocol uses the single time-slotted channel (Yigitel et al. 2011).

DEE-MAC Protocol

In this approach, the synchronization of the various nodes occurs. This is done at the cluster head and the nodes which are retaining the idle listen mode are forcibly made as sleep mode. This significantly reduces energy consumption.

SPARE MAC Protocol

This protocol considers the overhearing and idle listening and thus, SPARE MAC works as an energy-efficient protocol. A distributed scheduling solution is implemented to allocate specific different time slots. Thus, collisions issues are also avoided (Ali et al. 2005).

Hybrid Protocols

The hybrid protocol is a combination of TDMA and CSMA. Here the devices access the channels in a contention oriented approach or maybe in the collision-free period. Z-MAC, W-MAC, and A-MAC are the hybrid type protocols, robust to synchronization error.

Zebra MAC (Z-MAC) Protocol

Integrating the features of CSMA and TDMA protocol the Z-MAC is a hybrid new approach that is helpful in a dynamic scenario. It is also adaptive to various topology changes and different time failure issues of synchronization establishment. A level of contention is followed in both approaches. That means, in low contention, it follows the approach of CSMA, otherwise that of TDMA (Halkes et al. 2005).

Wise MAC (W-MAC) Protocol

This protocol is proposed with a modified preamble sampling approach. Considering the methodology of spatial TDMA and CSMA, the sampling is decided dynamically in changing different scenarios. This also helps as the traffic changes and accordingly the preamble. The nodes are allotted with two different channels, where one is for data and another regulates the control by following the TDMA and CSMA approaches respectively. One of the disadvantages here is the redundant transmission of data packets. This happens as the sleep-listen states are decentralized. So, each node transmits to its entire neighbor till individual awakening. In addition to this, the issues of having a collision also exist as it uses the persistent CSMA approach.

A-MAC Protocol

A-MAC is proposed in a certain scenario where the receiver is having prior notice of receiving the data. So when the receiver/sender is having data to receive/send the channel is busy. This reduces idle-listening and over-hearing problems as well. Such protocols are useful in surveillance and monitoring that occurs for a longer amount of time.

Cross-Layer MAC Protocol

We have discussed the above MAC protocols based on a single MAC layer design. In all the previous literature, the MAC layer information is taken as a parameter but the co-relation with the other is not considered. Integrating the physical layer and MAC layer characteristics a new cross-layer and accordingly cross-layer protocols are proposed. Combining both, a better sleep/wake period and respective energy efficiency are determined. The following are two cross-layer protocols that follow the above considerations (Ye et al. 2002).

MAC-CROSS Protocol

MAC-CROSS Protocol tries to improve the energy by considering the properties of the MAC layer and routing layer. The routing table information is taken into consideration and accordingly the sleep times of the nodes are maximized for idle nodes.

CLMAC Protocol

Here the network layer and transport layer are removed for simplification of the CLMAC protocol stack. The features are merged to MAC/physical layers and the Application layer. We have observed that in the preamble field, B-MAC works on the routing distance and the CLMAC protocol follows the B-MAC approach. The nodes work to reduce the control traffic routing overhead.

The goal is achieved by integrating the cross layers features where only the important information is communicated. In the MAC layer, the routing assessment transforms the contention level. Thus, it helps to identify the better wireless links and the data packets are sent accordingly (Jian et al. 2008).

Classification Based on Other Parameters

So far we have discussed the MAC protocols by considering energy constraints, adaptivity, and scalability. Also based on the latency, bandwidth utilization, fairness, throughput, and other parameters, the MAC protocols are categorized as Synchronous, Asynchronous, Multichannel, and Frame-Slotted protocols(Saxena et al. 2008). The taxonomy is shown in Figure 4.

Figure 4. Taxonomy of MAC protocols based on other parameters in WSNs

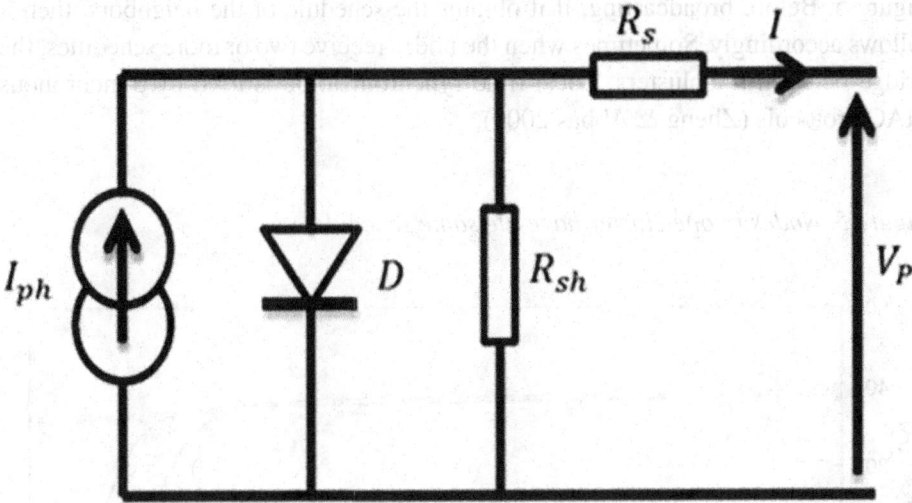

Asynchronous Protocols

The sensor node here emphasizes to consider and maintain its schedule rather than synchronizing all its neighbor so that the information is processed in a faster way. Not only the synchronizing cost is reduced but also the low duty cycle is achieved. As keeping a node active throughout takes more energy, the node goes to sleep mode and wakes up periodically. The preamble sampling method is used and sometimes the long preambles also lead to limited throughput. Thus, the new dynamic models are discussed that follow dynamic preamble and keeps an aim to have higher throughput (Dong & Dargie 2012).

Synchronous Protocols

The clustering technique is used in Synchronous MAC protocols . Combining the nodes in a cluster, a common active/sleep schedule is derived. But there may be a scenario where a node can be a part of more than one cluster. So it may face conflicts due to different schedules. In Fig. we can have such a scenario where the node is in two clusters. This must also be taken into consideration and the extra coordination overhead is to be made negligible. The node in the active state tries to listen to the channel periodically to get a schedule. Accordingly, they set the next wake up time and advertise on their own when they are unable to find a schedule as shown in Figure 5. Before broadcasting, if it obtains the schedule of the neighbors, then it follows accordingly. Sometimes when the nodes receive two or more schedules, the bridge between the clusters. Local time synchronization is used by Synchronous MAC protocols (Zheng & Abbas 2009).

Figure 5. Nodes in one cluster have the same schedule

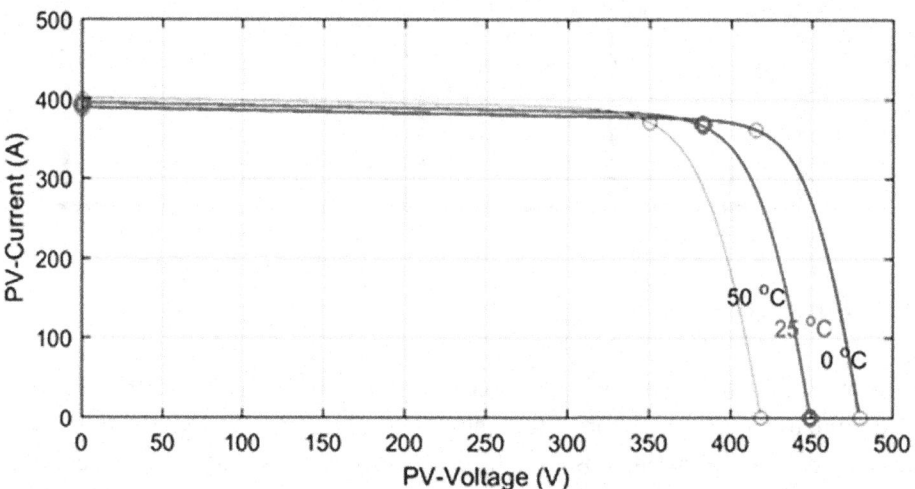

Table 1. Different asynchronous MAC protocols

Protocols	Technique	Advantage	Disadvantages
B-MAC (6 joules)	Low power listening	Delay tolerant and no frames required (RTS, CTS)	Large overhead created by a long preamble
X-MAC (70% of duty cycle)	Preamble length reduction	Latency operation is low and energy efficient	After seeing packet gaps, the neighbor mistakenly transmits data
Speck MAC	Preamble sampling continuous	Receiver energy-reduced	Redundancy causes waste in transmission power
RC-MAC	In receiver-initiated transmission, collision is reduced	In heavy traffic load, high throughput	Increase in delay
PW-MAC (10% of duty cycle)	The timer at the receiver (estimated wakeup)	Low delay and avoid collisions due to pseudo-random schedules	Overhead created using idle listening and beacons
DPS-MAC	Short strobed and LPL preamble	Energy-efficient in low traffic applications which decreases the idle listening.	Switching time of radio affects the short size of preamble

Table 2. Different Synchronous MAC Protocols

Protocols	Technique	Advantages	Disadvantages
S-MAC	Adaptive listening	adaptive with the topological change, energy-efficient	Need to maintain loose sync., RTS, CTS increase energy
T-MAC	Predicting the next instant request to be sent	active optimal period is achieved	overhearing problem
RMAC	Sleep period from the data transmission shift	overhearing problem solved	hidden terminals collisions
D-MAC	Staggered schedule	active slots can be increased or decreased, low delay	at sink contention may occur, long idle listening occurs,
SCP-MAC	Adaptive duty cycle	low schedule maintenance	long listen-interval, Synchronization overhead

Table 3. Different Frame-Slotted MAC Protocols

Protocols	Technique	Advantage	Disadvantage
Tree MAC	Maximize throughput at sink	Fairness in the flow instead of an individual node	lower channel priority and time requirement to join the tree during the adoption of CSMA
TRAMA	Adaptive assignment	Channel utilization is more than Z-MAC	Due to sequential node priority, low spatial reuse of time
Crankshaft	by switching sending slots to receiving slots, the duty cycle is reduced	Data nodes can only wake up	Time slots assigned to receiver hence collision may occur
Z-MAC	Slot Stealing	under low contention possess high throughput	Introduces additional overhead to detect abandoned slot

Multichannel MAC Protocols

The higher data requirements need the nodes to transmit data in a parallel approach. But using one channel it becomes difficult to implement. So, the multichannel MAC protocols attract the researchers for such requirements. There is need for a multichannel approach, as using the single channel, the parallel transmission cannot be implemented. This leads to interference and at times collision may also occur. If the nodes can be assigned with different channels, the data transfer can be significantly higher. But, in a parallel channel, the nodes participate effectively for sending the information simultaneously which increases the throughput and data rates.

Table 4. Multichannel MAC protocols

Protocol	Technique	Advantage	Disadvantage
TMCP	Metric optimization	Minimize intra-tree Interference value among all trees	Metric does not reflect the actual interference intensity.
TMMAC	Cross channel communication	Dynamic ATIM window to increase flexibility	loss of energy in nodes by toggle transmission and toggle snooping
Y-MAC	TDMA/FD MA for receiving	Dynamic Channel Selection Scheme introduced.	contentions may occur when two receivers hop the same channel
MC-LMAC	TDMA/FD MA for sending	Semi Dynamic Channel switching i.e. no frequent channel Switching	channel/slot utilization is low and high overhead
MMSN	Cross channel communication	among neighboring nodes, possible parallel transmission	powerful radio size is not required

Frame-Slotted MAC Protocols

TDMA protocols use the time slots which are also applied in Frame-Slotted MAC protocols. When cluster overlapping occurs during the active periods, a collision takes place. The TDMA works on global time synchronization which results in higher throughput. Here within the two-hop communication, no two neighborhood nodes are assigned the same slot. By implementing this, occurrences of the collisions are avoided but the hidden terminal problem still exists. We can analyze that in a certain scenario when some of the nodes have to send data, the channel utilization is low. Also, the distinct time slots assigned to neighboring nodes are not utilized properly and thus, wasted.

Merits and Limitations of Different MAC Protocols

As shown in tables below, the comparison is shown for different asynchronous, synchronous, frame-slotted and multichannel MAC protocols, based on their operation techniques along with the merits and demerits (Malik & Sharma 2019, Kumar & Balakrishna 2019, Radha et al. 2019).

CONCLUSION

In this chapter, we have discussed various types of MAC protocols of WSN. Energy efficiency is the main concern for WSN as the sensor nodes' power is controlled by the batteries. Thus, in case of a particular application, where the sensor node setup is to be moved, because of the power source for the nodes, the whole system performance hampers. The different parameters regarding these protocols are studied and compared. Many different protocols have been suggested but to date, none is accepted as standard one as these are application dependent. Due to lower delays in CSMA techniques, fewer traffic loads and good throughput are obtained in the case of WSN. Further, the technique is improved by implementing collision detection and collision avoidance. The synchronization problem occurs in the case of TDMA, and it changes in some nodes, and thus needs further attention. Although it is collision-free access, in a decentralized environment, there exist the hardships of changing the time slot assignments. Thus, from the study, for an efficient WSN system, the collective decision of all the layers can be more impulsive. Creating more overhead certainly increases the consumption of energy and thus the routing information based on MAC is very much essential before starting the process of communication. Hybrid protocols and application-centric protocol designs may be an interesting research area for an extension of this survey work.

REFERENCES

Ali, A., Shah, G. A., & Shoaib, M. (2019). Energy-Efficient Uplink MAC Protocol for M2M Devices. *IEEE Access: Practical Innovations, Open Solutions*, 7, 35952–35962. doi:10.1109/ACCESS.2019.2903647

Ali, M., Tashfeen, S., & Zartash, A. U. (2005). MMAC: A mobility-adaptive, collision-free mac protocol for wireless sensor networks. In *Proceedings of 24th International Performance, Computing, and Communications Conference*. IEEE. 10.1109/PCCC.2005.1460597

Demirkol, I., Cem, E., & Fatih, A. (2006). MAC protocols for wireless sensor networks: A survey. *IEEE Communications Magazine*, 44(4), 115–121. doi:10.1109/MCOM.2006.1632658

Djiroun, F. Z., & Djenouri, D. (2017). MAC Protocols With Wake-Up Radio for Wireless Sensor Networks: A Review. *IEEE Communications Surveys and Tutorials*, 19(1), 587–618. doi:10.1109/COMST.2016.2612644

Dong, Q., & Dargie, W. (2012). A survey on mobility and mobility-aware MAC protocols in wireless sensor networks. *IEEE Communications Surveys and Tutorials*, 15(1), 88–100. doi:10.1109/SURV.2012.013012.00051

Doudou, M., Djenouri, D., Badache, N., & Bouabdallah, A. (2014). Synchronous contention-based MAC protocols for delay-sensitive wireless sensor networks: A review and taxonomy. *Journal of Network and Computer Applications*, 38, 172–184. doi:10.1016/j.jnca.2013.03.012

Enz, C., El-Hoiydi, A., Decotignie, J.-D., & Peiris, V. (2004). WiseNET: An Ultralow-Power Wireless Sensor Network Solution. *IEEE Computer*, 37(8), 62–70. doi:10.1109/MC.2004.109

Halkes, G. P., Tijs, V. D., & Langendoen, K. G. (2005). Comparing energy-saving MAC protocols for wireless sensor networks. *Mobile Networks and Applications*, 10(5), 783–791. doi:10.100711036-005-3371-x

Hamid, M., Abdullah-Al-Wadud, M., & Chong, I. (2010). A schedule-based Multi-channel MAC Protocol for Wireless Sensor Networks. *Sensors (Basel)*, 10(10), 9466–9480. doi:10.3390101009466 PMID:22163420

Heidemann, W. Ye. J., & Estrin, D. (2004). Medium Access Control With Coordinated Adaptive Sleeping for Wireless Sensor Networks. *IEEE/ACM Transactions on Networking*, 12(3), 493–506. doi:10.1109/TNET.2004.828953

Hossein, G., & Fard, E. (2011). Multi-channel Medium Access Control Protocols for Wireless Sensor Networks: A Survey. Department of Computer Engineering, Lahijan Branch, Islamic Azad University, Guilan, Iran.

Huang, P., Xiao, L., Soltani, S., Mutka, M. W., & Xi, N. (2012). The evolution of MAC protocols in wireless sensor networks: A survey. *IEEE Communications Surveys and Tutorials*, *15*(1), 101–120. doi:10.1109/SURV.2012.040412.00105

Incel, O. D., Hoesel, L. V., Jansen, P., & Havinga, P. (2011). MC-LMAC: A multi-channel MAC protocol for wireless sensor networks. *Ad Hoc Networks*, *9*(1), 73–94. doi:10.1016/j.adhoc.2010.05.003

Jian, Q. (2008). Overview of MAC protocols in wireless sensor networks. *Journal of Software*, *19*(2), 389–403. doi:10.3724/SP.J.1001.2008.00389

Kalaivaani, P. T., & Rajeswari, A. (2013). An energy-efficient analysis of S-MAC and H-MAC protocols for wireless sensor network. *International Journal of Computer Network & Communication*, *5*(2), 83–94. doi:10.5121/ijcnc.2013.5207

Kumar, K. S. A., & Balakrishna, R. (2019). *Performance Analysis of Reliability and Throughput Using an Improved Receiver—Centric MAC Protocol and Itree-MAC Protocol in Wireless Sensor Networks. In Cognitive Informatics and Soft Computing* (pp. 83–90). Singapore: Springer.

Li, C., Zhang, B., Yuan, X., Ullah, S., & Vasilakos, A. V. (2018). MC-MAC: A multi-channel based MAC scheme for interference mitigation in WBANs. *Wireless Networks*, *24*(3), 719–733. doi:10.100711276-016-1366-0

Liu, Y., Ota, K., Zhang, K., Ma, M., Xiong, N., Liu, A., & Long, J. (2018). QTSAC: An energy-efficient MAC protocol for delay minimization in wireless sensor networks. *IEEE Access: Practical Innovations, Open Solutions*, *6*, 8273–8291. doi:10.1109/ACCESS.2018.2809501

Malik, M., & Sharma, M. (2019). A Novel Approach for Comparative Analysis on Energy Effectiveness of H-MAC and S-MAC Protocols for Wireless Sensor Networks. *Advanced Science, Engineering and Medicine*, *11*(1-2), 29–35. doi:10.1166/asem.2019.2326

Pegatoquet, A., Trong, N. L., & Michele, M. (2018). A Wake-Up Radio-Based MAC Protocol for Autonomous Wireless Sensor Networks. *IEEE/ACM Transactions on Networking*, *27*(1), 56–70. doi:10.1109/TNET.2018.2880797

Qian, D., & Waltenegus, D. (2012). A survey on mobility and mobility-aware MAC protocols in wireless sensor networks. *IEEE Communications Surveys and Tutorials*, *15*(1), 88–100.

Radha, S. (2019). ScP Protocol for Wireless Sensor Networks. In *Proceedings of 5th International Conference on Advanced Computing & Communication Systems (ICACCS)*. IEEE.

Ray, N. K., & Turuk, A. K. (2009). A review of energy-efficient MAC protocols for wireless LANs. In *Proceedings of International Conference on Industrial and Information Systems (ICIIS)*, IEEE. 10.1109/ICIINFS.2009.5429875

Salman, H. M. (2014). Survey of routing protocols in wireless sensor networks. *International Journal of Sensors and Sensor Networks*, *2*(1), 1–6.

Saxena, N., Roy, A., & Shin, J. (2008). Dynamic duty cycle and adaptive contention window based QoS-MAC protocol for wireless multimedia sensor networks. *Computer Networks*, *52*(13), 2532–2542. doi:10.1016/j.comnet.2008.05.009

Sherazi, H. H., Raza, L., Alfredo, G., & Gennaro, B. (2018). A comprehensive review of energy harvesting MAC protocols in WSNs: Challenges and tradeoffs. *Ad Hoc Networks*, *71*, 117–134. doi:10.1016/j.adhoc.2018.01.004

Yang, X., Wang, L., Su, J., & Gong, Y. (2018). Hybrid MAC protocol design for mobile wireless sensors networks. *IEEE Sensors Letters*, *2*(2), 1–4. doi:10.1109/LSENS.2018.2828339

Ye, W., Heidemann, J., & Estrin, D. (2002). An energy-efficient MAC protocol for wireless sensor networks. In *Proceedings of Twenty-First Annual Joint Conference of the IEEE Computer and Communications Societies* (vol. 3). IEEE.

Ye, W., Silva, F., & Heidemann, J. (2006). Ultra-Low Duty Cycle MAC with Scheduled Channel Polling, In *Proceedings of the 4th International Conference on Embedded Networked Sensor Systems* (pp. 321-334). ACM. 10.1145/1182807.1182839

Yigitel, M. A., Ozlem, D. I., & Cem, E. (2011). QoS-aware MAC protocols for wireless sensor networks: A survey. *Computer Networks*, *55*(8), 1982–2004. doi:10.1016/j.comnet.2011.02.007

Zheng, J., & Abbas, J. (2009). *Wireless sensor networks: a networking perspective*. John Wiley & Sons. doi:10.1002/9780470443521

Zhou, G., Huang, C., Yan, T., He, T., Stankovic, J. A., & Abdelzaher, T. F. (2006). Mmsn: Multi-frequency media access control for wireless sensor networks. In *Proceedings of Infocom*. IEEE. doi:10.1109/INFOCOM.2006.250

Compilation of References

Abbott, L. F. (1997). Synaptic depression and cortical gain control. *Science, 275*(5297), 220–224. doi:10.1126cience.275.5297.221 PMID:8985017

Abbott, L. F., & Kepler, T. B. (1990). Model neurons: From hodgkin-huxley to Hopfield. *Statistical Mechanics of Neural Networks, 18*, 5–18.

Abdelbar, A. M., & Hedetniemi, S. M. (1998). The complexity of approximating MAP explanation. *Artificial Intelligence, 102*, 21–38. doi:10.1016/S0004-3702(98)00043-5

Abideen. (2017). An IOT based robust healthcare model for continuous health monitoring. In *23rd International conference*. IEEE. doi:10.23919/IConAC.2017.8082012

Abraham, J., & Ramanatha, K. S. (2006). An Efficient Protocol for Authentication and Initial Shared Key Establishment in Clustered Wireless Sensor Networks. In *Proc. Third International conference on Wireless and optical Communication networks* (pp. 5-10). Bangalore, India: IEEE. 10.1109/WOCN.2006.1666568

Ahmed, E., Yaqoob, I., & Gani, A. (2016). Internet-of-thingsbased smart environments: state of the art, taxonomy, and open research challenges. *IEEE Wireless Commun, 23*(5), 10–16.

Ahmed, J., & Salam, Z. (2016, May). A Modified P&O Maximum Power PointTracking Method With Reduced Steady-State Oscillation and Improved Tracking Efficiency. *IEEE Transactions on Sustainable Energy, 7*(4), 1506–1515. doi:10.1109/TSTE.2016.2568043

Alam, M. M., Malik, H., Khan, M. I., Pardy, T., Kuusik, A., & Le Moullec, Y. (2018). A survey on the roles of communication technologies in IoT-based personalized healthcare applications. *IEEE Access, 6*, 36611–36631.

Al-Fuqaha, A., Guizani, M., Mohammadi, M., Aledhari, M., & Ayyash, M. (2015). Internet ofthings: A survey on enabling technologies, protocols, and applications. *IEEE Communications Surveys & Tutorials, 17*(4), 2347–2376, https://iot-analytics.com

Ali, A., Shah, G. A., & Shoaib, M. (2019). Energy-Efficient Uplink MAC Protocol for M2M Devices. *IEEE Access: Practical Innovations, Open Solutions, 7*, 35952–35962. doi:10.1109/ACCESS.2019.2903647

Ali, M., Tashfeen, S., & Zartash, A. U. (2005). MMAC: A mobility-adaptive, collision-free mac protocol for wireless sensor networks. In *Proceedings of 24th International Performance, Computing, and Communications Conference*. IEEE. 10.1109/PCCC.2005.1460597

Allen, J. F., & Ferguson, G. (1994). Actions and events in interval temporal logic. *Journal of Logic and Computation*, 4(5), 531–579.

Ambika, N., & Raju, G. T. (2014). ECAWSN: Eliminating compromised node with the help of auxiliary nodes in wireless sensor network. *International Journal of Security and Networks*, 9(2), 78–84. doi:10.1504/IJSN.2014.060743

Amendola, S., Lodato, R., Manzari, S., Occhiuzzi, C., & Marrocco, G. (2014). RFID technology for IoT-based personal healthcare in smart spaces. *IEEE Internet of Things Journal*, 1(2), 144–152.

Anitha, C. L., & Sumathi, R. (2014). Comparative Analysis of Data Aggregation Algorithms Under Various Architectural Models in Wireless Sensor Networks. *International Journal of Information Technology*, 6(2), 757–763.

Antipolis. (2014). *New ETSI specification for Internet of Things and Machine to Machine Low Throughput Networks*. https://www.etsi.org/news-events/news/827-2014-09-news-etsi-new-specification-for-internet-of-things-and-machine-to-machine-low-throughput-networks

Aouadi, Abouloifa, Hamdoun, & Boussairi. (2014). Backstepping Based Control of PV system Connected to the Grid. *International Journal of Computer and Information Technology*, 3, 1021-1026.

Aravinthan, Namboodiri, Sunku, & Jewell. (2011). Wireless AMI application and security for controlled home area networks. *2011 IEEE Power and Energy Society General Meeting*, 1–8.

Arrue, B.C., Ollero, A., & De Dios, J.M. (2000). An intelligent system for false alarm reduction in infrared forest-fire detection. *IEEE Intell. Syst. Appl.*, 15, 64–73.

Aslan, Y.E., Korpeoglu, I., & Ulusoy, Ö. (2012). A framework for use of wireless sensor networks in forest fire detection and monitoring. *Computers, Environment and Urban Systems*, 36, 614–625.

Atibalentja, N., & Eastburn, D. M. (1997). Evaluation of inoculation methods for screening horseradish cultivars for resistance to *Verticillium dahliae*. *American Phythopathology Society*, 81(4), 356–362. doi:10.1094/PDIS.1997.81.4.356 PMID:30861815

Atzori, Iera, & Morabito. (2010). The Internet of Things: A survey. *Computer Networks*, 54(15), 2787–2805.

Bachimont, B. (2004, May). *Pourquoi n'y a-t-il pas d'expérience en ingénierie des connaissances?*

Baek, J., Foo, E., Tan, H., & Zhou, J. (2009). Securing wireless sensor networks—threats and Counter measures. In *Security and Privacy in Wireless and Mobile*. Troubador Publishing.

Baget, J. F. (2005, November). RDF entailment as a graph homomorphism. In *International Semantic Web Conference* (pp. 82-96). Springer, Berlin, Heidelberg.

Bajaj. (2016). *IOT is the game changer fro retailer, aspire systems.* https://www.aspiresys.com/WhitePapers/IoT-is-the-game-changer-for-Retailers.pdf

Balsamo, D., Merrett, G. V., & Zaghari, B. (2017). Wearable and autonomous computing for future smart cities: open challenges. In *Proceedings of the 25th international conference on software, telecommunications and computer networks (SoftCOM).* New York: IEEE.

Bandoyopadhayam. (2011). *Internet of Things - Applications and Challenges in Technology and Standardization.* arvix

Bansal, D., Khan, M., & Salhan, A. K. (2009). A Computer Based Wireless System For Online Acquisition, Monitoring and Digital Processing of ECG Waveforms. *Computers in Biology and Medicine, 39*(4), 361-367.

Bansal, D. (2013). Design Of 50 Hz Notch Filter Circuits For Better Detection of Online ECG. *International Journal of Biomedical Engineering and Technology, 13*(1), 30–48. doi:10.1504/IJBET.2013.057712

Barick, S., & Bagha, S. (2013). Removal of 50Hz Power Line Interference For Quality Diagnosis of ECG Signal. *International Journal of Engineering Science and Technology, 5*(05), 1149–1155.

Bekara. (2016). *Security issues and challenges for the IoT based smart grid.* https://www.sciencedirect.com/science/article/pii/S1877050914009193

Benson, D. M., & Baker, R. (1974). Epidemiology of *Rhizoctonia solani* pre-emergence damping-off of radish: Influence of pentachloro-nitrobenzene. *Phytopathology, 64*(1), 38–40. doi:10.1094/Phyto-64-38

Bharath, K. R. (2018, November). Control of Bidirectional DC-DC Converter in Renewable based DC Microgrid with Improved Voltage Stability. *International Journal of Renewable Energy Research, 8*, 872–1630.

Blanca, B. L., Juan, A., Navas-Cortés, R., & Jiménez-Díaz, M. (2004). Integrated management of Fusarium wilt of chickpea with sowing date, host resistance, and biological control. The American Phytopathological Society, 94, 946-960.

Blestos, F., Thanassoulopoulous, C., & Roupakias, D. (2003). Effect of grafting on growth yield and *Verticillium* wilt of eggplant. *Horticultural Science, 38*, 183–186.

Boger, J., Hoey, J., Poupart, P., Boutilier, C., Fernie, G., & Mihailidis, A. (2006). A planning system based on Markov decision processes to guide people with dementia through activities of daily living. *IEEE Transactions on Information Technology in Biomedicine, 10*(2), 323–333.

Bokde, P. R., & Choudhari, N. K. (2015). Implementation of Digital Filter on FPGA For ECG Signal Processing. *International Journal of Emerging Technology and Innovative Engineering, 1*(2), 175–181.

Bortolan, G., & Christov, I. (2014). Dynamic Filtration of High-Frequency Noise in ECG Signal. *Computing in Cardiology, 41*, 1089-1092.

Botta. (2016). Integration of cloud computing and internet of things: A survey. *Future Generation Computer Systems, 56,* 684–700.

Bouchard, B., Giroux, S., & Bouzouane, A. (2007). A keyhole plan recognition model for Alzheimer's patients: First results. *Applied Artificial Intelligence, 21*(7), 623–658.

Boyen, X., & Koller, D. (1998). Tractable inference for complex stochastic processes. *UAI98 – Proceedings of the Fourteenth Conference on Uncertainty in Uncertainty in Artificial Intelligence,* 33–42.

Breivold. (2015). IOT for industrial automation. *IEEE International conference on data science and data intensive systems,* 532-539. doi:10.1109/DSDIS.2015.11

Brindha. (2017). Involuntary Nutrients Dispense System for Soil Deficiency using IOT. *International Journal of ChemTech Research.*

Brown, P. J., Bovey, J. D., & Chen, X. (1997). Context-aware applications: from the laboratory to the marketplace. *IEEE personal communications, 4*(5), 58-64.

Buratti, C., Conti, A., Dardari, D., & Verdone, R. (2009). An overview on wireless sensor networks technology and evolution. *Sensors (Basel), 9,* 6869–6896.

Burke, D. W., & Hall, R. (1991). Fusarium root rot. In R. Hall (Ed.), Compendium of Bean Diseases (pp. 9-10), St. Paul, MN: The American Phytopathological Society.

Butun, I., & Sankar, R. (2011). Advanced Two Tier User Authentication Scheme for Heterogeneous Wireless Sensor Networks. In *Proc. IEEE Consumer communications and networking conference* (pp. 169-171). Las Vegas, NV: IEEE. 10.1109/CCNC.2011.5766446

Campbell, C. L., & Neher, D. A. (1994). Estimating disease severity and incidence. In C. L. Campbell & D. M. Benson (Eds.), Epidemiology and management of root diseases (pp. 117-147). Springer. doi:10.1007/978-3-642-85063-9_5

Carrasco, J. M., Franquelo, L. G., Bialasiewicz, J. T., Galvan, E., PortilloGuisado, R. C., Prats, M. A. M., Leon, J. I., & Moreno-Alfonso, N. (2006). Power-electronic systems for the grid integration of renewable energy sources: A survey. *IEEE Transactions on Industrial Electronics, 53*(4), 1002–1016. doi:10.1109/TIE.2006.878356

Chaitin, G. J., Auslander, M. A., Chandra, A. K., Cocke, J., Hopkins, M. E., & Markstein, P. W. (1981). Register allocation via coloring. *Computer Languages, 6*(1), 47–57. doi:10.1016/0096-0551(81)90048-5

Chandrakar, B., Yadav, O. P., & Chandra, V. K. (2013). A Survey of Noise Removal Techniques For ECG Signals. *International Journal of Advanced Research in Computer and Communication Engineering, 2*(3), 1354–1357.

Chatzigiannakis, I., Nikoletseas, S., & Strikos, A. (2006). Experimental evaluation of the performance of multi-hop wireless sensor networks. In Proc. 5th Communication Systems, Networks and Digital Signal Processing, (pp. 19-21). Patras, Greece: Academic Press.

Chauhan, S., Sharma, L., & Mehra, R. (2013). Cost Analysis of Digital FIR Filter Using Different Window Techniques. *International Journal of Electrical, Electronics and Data Communication, 1*(6), 25-29.

Cheikhrouhou, O., Koubaa, A., Boujelbenl, M., & Abid, M. (2010). A lightweight user Authentication Scheme for Wireless Sensor Networks. In *International Conference on Computer Systems and Applications - AICCSA 2010* (pp. 1-7). Hammamet, Tunisia: IEEE. 10.1109/AICCSA.2010.5586995

Chen, G., & Kotz, D. (2000). *A survey of context-aware mobile computing research.* Dartmouth Computer Science Technical Report TR2000-381.

Chen, C., & Cook, D. J. (2012, June). Behavior-based home energy prediction. In *2012 Eighth International Conference on Intelligent Environments* (pp. 57-63). IEEE.

Cheng, J., & Druzdzel, M. (2000). AIS-BN: An adaptive importance sampling algorithm for evidential reasoning in large Bayesian networks. *Journal of Artificial Intelligence Research, 13,* 155–188. doi:10.1613/jair.764

Chentoufi, J. A., El Imrani, A., & Bouroumi, A. (2007). A multipopulation cultural algorithm using fuzzy clustering. *Applied Soft Computing, 7*(2), 506–519. doi:10.1016/j.asoc.2006.10.010

Cherlet, M., Schelkens, M., Croubels, S., & De Backer, P. (2003). Quantitative multi-residue analysis of tetracyclines and their 4-epimers in pig tissues by high-performance liquid chromatography combined with positive-ion electrospray ionization mass spectrometry. *Analytica Chimica Acta, 492*(1-2), 199–213.

Ching & Singh. (2016). Wearable technology devices security and privacy vulnerability analysis. *International Journal of Network Security & Its Applications, 8*(3), 19. doi:10.5121/ijnsa.2016.8302

Chiwewe, T.M., Mbuya, C.F., & Hancke, G.P. (2015). Using cognitive radio for interference-resistant industrial wireless sensor networks: An overview. *IEEE Transactions on Industrial Informatics, 11,* 1466–1481.

Chliyeh, M., Rhimini, Y., Selmaoui, K., Ouazzani Touhami, A., Filali-Maltouf, A., El Modafar, C., ... Douira, A. (2014). Survey of the fungal species associated to olive-tree (Olea europaea L.) in Morocco. *International Journal of Recent Biotechnology, 2,* 15–32.

Choi, Y. B., Kadakkuzha, B. M., Liu, X. A., Akhmedov, K., Kandel, E. R., & Puthanveettil, S. V. (2014). Huntingtin is critical both presynaptically and postsynaptically for long-term learning-related synaptic plasticity in Aplysia. *PLoS One, 9*(7), e103004. doi:10.1371/journal.pone.0103004 PMID:25054562

Cho, K., Jo, M., Kwon, T., Chen, H.-H., & Lee, D. H. (2013). Classification and Experimental Analysis for Clone Detection Approaches in Wireless Sensor Networks. *IEEE Systems Journal, 7*(1), 26–35. doi:10.1109/JSYST.2012.2188689

Chuchaisri, P., & Newman, R. (2012). Fast response PKC-based broadcast authentication in wireless sensor network. *Mobile Networks and Applications, 17*(4), 508–525. doi:10.100711036-011-0349-8

Cinar. (2018). Optimization algorithm used in smart grids and comparison of results. *IOSR Journal of Electrical and Electronics Engineering, 13*(6), 14-20.

Cintuglu, Mohammed, Akkaya, & Uluagac. (2017). A Survey on Smart Grid Cyber-Physical System Testbeds. *IEEE Communications Surveys and Tutorials, 19*(1), 446–464.

Cleveland. (2008). Cyber security issues for Advanced Metering Infrasttructure (AMI). *2008 IEEE Power and Energy Society General Meeting - Conversion and Delivery of Electrical Energy in the 21st Century*, 1–5.

Cottenie, A. (1980). Soil and Plant testing as a basis of fertilizer recommendations. *Food and Agriculture Organization Bulletin, 38*(2), 119.

Dasgupta, K. K., & Namjoshi, P. (2003). An efficient clustering-based heuristic for data gathering and aggregation in sensor networks. In *IEEE Wireless Communications and Networking Conference* (pp. 1948-1953). New Orleans, LA: IEEE. 10.1109/WCNC.2003.1200685

Dayan, P., & Abbott, L. F. (2001). Theoretical Neuroscience: Computational and Mathematical Modeling of Neural Systems. *Neuroscience, 39*(3), 460.

Demirkol, I., Cem, E., & Fatih, A. (2006). MAC protocols for wireless sensor networks: A survey. *IEEE Communications Magazine, 44*(4), 115–121. doi:10.1109/MCOM.2006.1632658

Deshpande, Pitale, & Sanp. (2016). Industrial automation using IoT. *International Journal of Advanced Research in Computer Engineering and Technology, 5*(2).

Deshpande, R. (2019, March). Analysis of Power Quality Variations in Electrical Distribution System with Renewable Energy Sources. *International Journal of Renewable Energy Research, 9*, 282–289.

Dey, A. K. (2001). Understanding and using context. *Personal and Ubiquitous Computing, 5*(1), 4–7.

Divya & Rao. (2016, August). Measurement and Monitoring of Soil Moisture using Cloud IoT and Android System. *Indian Journal of Science and Technology*.

Dixit, H. V., & Gupta, V. (2012). IIR Filters Using System Generator For FPGA Implementation. *International Journal of Engineering Research and Applications, 2*(5), 303–306.

Djiroun, F. Z., & Djenouri, D. (2017). MAC Protocols With Wake-Up Radio for Wireless Sensor Networks: A Review. *IEEE Communications Surveys and Tutorials, 19*(1), 587–618. doi:10.1109/COMST.2016.2612644

Dominguez, F., & Ochoa, X. (2017). Smart objects in education: an early survey to assess opportunities and challenges. In *2017 Fourth International Conference on eDemocracy & eGovernment (ICEDEG)*, (pp. 216–220). IEEE. doi:10.1109/ICEDEG.2017.7962537

Dong, Q., & Dargie, W. (2012). A survey on mobility and mobility-aware MAC protocols in wireless sensor networks. *IEEE Communications Surveys and Tutorials, 15*(1), 88–100. doi:10.1109/SURV.2012.013012.00051

Dong, W., Farrell, J. A., & Polycarpou, M. M. (2012). Command filtered adaptive backstepping. *IEEE Transactions on Control Systems Technology, 20*(3), 566–580. doi:10.1109/TCST.2011.2121907

Dorobantu, B. (2016, December 5). *Video ZF Live. Dana Războiu, preşedinta Asociaţiei Naţie prin Educaţie: 50 de echipe de elevi pasionaţi de robotică au primit kituri şi echipamente în valoare de 280.000 de euro pentru a construi roboţi.* Retrieved from https://www.zf.ro/business-hi-tech/video-zf-live-dana-razboiu-presedinta-asociatiei-natie-educatie-50-echipe-elevi-pasionati-robotica-au-primit-kituri-echipamente-valoare-280-000-euro-construi-roboti-16009941

Doudou, M., Djenouri, D., Badache, N., & Bouabdallah, A. (2014). Synchronous contention-based MAC protocols for delay-sensitive wireless sensor networks: A review and taxonomy. *Journal of Network and Computer Applications, 38*, 172–184. doi:10.1016/j.jnca.2013.03.012

Dyk, Chmielewski, & Najgebauer. (2017). Combat triage support using the Internet of Military Things. In Proc. of the 2017 Federated Conference on Computer Science and Information Systems Murugesan, S.: Harnessing green IT: principles and practices. *IT Professional, 10*(1), 24–33. doi:10.1109/MITP.2008.10

Eastburn, D. M., Degennaro, M. M., Delucia, E. H., Dermody, O., Elrone, M., & Elevated, A. J. (2010). Atmospheric carbon dioxide and ozone alter soybean diseases at Soybean. *Global Change Biology, 16*(1), 320–330. doi:10.1111/j.1365-2486.2009.01978.x

ECDL. (2017, October). *Lansare proiect imprimante 3D pentru şcolile din România. 3DUTECH - Modelează viitorul. Printează-l 3D!* Retrieved from https://www.ecdl.ro/en/news-article/lansare-proiect-imprimante-3d-pentru-scolile-din-romania.-3dutech-modeleaza-viitorul.-printeaza-l-3d_677.html

Elgendy, M. A., Atkinson, D. J., & Zahawi, B. (2016, February). Khalifa University, "Experimental investigation of the incremental conductance maximum power point tracking algorithm at high perturbation rates. *IET Renewable Power Generation, 10*(2), 133–139. doi:10.1049/iet-rpg.2015.0132

Elijah, O., Rahman, T. A., Orikumhi, I., Leow, C. Y., & Hindia, M. N. (2018). An overview of Internet of Things (IoT) and data analytics in agriculture: Benefits and challenges. *IEEE Internet of Things Journal, 5*, 3758–3773.

Elmaghraby & Losavio. (2014). Cyber security challenges in Smart Cities: Safety, security and privacy. *Journal of Advanced Research, 5*(4), 491–497. doi:10.23919/IConAC.2017.8082057 PMID:25685517

Endsley, M. R. (2017). Toward a theory of situation awareness in dynamic systems. In *Situational awareness* (pp. 9–42). Routledge.

Enz, C., El-Hoiydi, A., Decotignie, J.-D., & Peiris, V. (2004). WiseNET: An Ultralow-Power Wireless Sensor Network Solution. *IEEE Computer, 37*(8), 62–70. doi:10.1109/MC.2004.109

Erciyes, K. (2013). *Distributed graph algorithms for computer networks.* Springer Science & Business Media. doi:10.1007/978-1-4471-5173-9

Escrivá, L., Font, G., Manyes, L., & Berrada, H. (2017). Studies on the Presence of Mycotoxins in Biological Samples. *An Overview Toxins, 9*(8), 251. doi:10.3390/toxins9080251 PMID:28820481

Eslahchi, C., & Onagh, B. N. (2006). Vertex-strength of fuzzy graphs. *International Journal of Mathematics and Mathematical Sciences, 2006,* 43614–1. doi:10.1155/IJMMS/2006/43614

Fabri, F., Buratti, C., & Verdone, R. (2008). A multi-sink multi-hop wireless sensor network over a square region: connectivity and energy consumption issues. In *GLOBECOM Workshop* (pp. 1-6). New Orleans, LA: IEEE. 10.1109/GLOCOMW.2008.ECP.38

Farooq, U., Marrakchi, Z., & Mehrez, Z. (2012). *Tree Based Heterogeneous FPGA Architecture* (Vol. 16). Springer. doi:10.1007/978-1-4614-3594-5

Farrell, J. A., Polycarpou, M., Sharma, M., & Wenjie Dong. (2009). Command filtered backstepping. *IEEE Transactions on Automatic Control, 54*(6), 1391–1395. doi:10.1109/TAC.2009.2015562

Fatima, S., Ahmad, S., & Khan, P. M. (2014). Certificate Based Security Services in Adhoc Sensor Network. *International Journal of Information Technology, 6*(2), 783–790.

Fernandes, Jung, Prakash. (2016). Security Analysis of Emerging Smart Home Applications. *2016 IEEE Symposium on Security and Privacy (SP),* 636–654.

Francesco Ricci, F., Adornettoa, G., & Palleschi, G. (2012). A review of experimental aspects of electrochemical immunosensors. ElectroChemica acta.

Gaikwad, Gabhane, & Golait. (2015). A survey based on Smart Homes system using Internet-of-Things. *2015 International Conference on Computation of Power, Energy, Information and Communication (ICCPEIC),* 330–335.

García, E.M., Serna, M.Á., Bermúdez, A., & Casado, R. (2008). Simulating a WSN-based wildfire fighting support system. *Proceedings of the International Symposium on Parallel and Distributed Processing with Applications.*

Garey & Johnson. (1990). Computers and Intractability; a Guide to the Theory of Np-Completeness. W. H. Freeman & Co.

Geurin, S. O., Barnes, A. K., & Carlos, B. J. (2012). Smart grid applications of selected energy. In Innovative Smart Grid Technologies (ISGT). IEEE PES.

Gharaibeh, A., Salahuddin, M. A., Hussini, S. J., Khreishah, A., Khalil, I., & Guizani, M. (2017). Smart cities: A survey on data management, security, and enabling technologies. *IEEE Communications Surveys & Tutorials, 19*(4), 2456–2501.

Goldman, R. P., Geib, C. W., & Miller, C. A. (2013). *A new model of plan recognition.* arXiv preprint arXiv:1301.6700.

Gómez, D., Montero, J., & Yáñez, J. (2006). A coloring fuzzy graph approach for image classification. *Information Sciences, 176*(24), 3645–3657. doi:10.1016/j.ins.2006.01.006

Gong, P., & Van Leeuwen, C. (2007). Dynamically maintained spike timing sequences in networks of pulse-coupled oscillators with delays. *Physical Review Letters. APS., 98*(4), 048104. doi:10.1103/PhysRevLett.98.048104 PMID:17358818

Gottfried, I. B. B., & Aghajan, H. (2009). The praxis of cognitive assistance in smart homes. Behaviour Monitoring and Interpretation-BMI. *Smart Environments, 3*, 183.

Goudos, S. K., Dallas, P. I., Chatziefthymiou, S., & Kyriazakos, S. (2017). A survey of IoT key enabling and future technologies: 5G, mobile IoT, sematic web and applications. *Wireless Personal Communications, 97*(2), 1645–1675.

Grandinetti, L. (2013). *Pervasive Cloud Computing Technologies: Future Outlooks and Interdisciplinary Perspectives.* IGI Global.

Gray, L. E., Achenbach, L. A., Duff, R. J., & Lightfoot. D. (1999). Pathogenicity of *Fusarium solani f. sp. glycines* isolates on soybean and green bean plants. *Journal of Phytopathology, 147*, 281-284. doi:10.1016/j.bios.2016.09.091

Grover, N., & Soni, M.K. (2014). Design of FPGA Based 32-bit Floating Point Arithmetic Unit and Verification of Its VHDL Code Using MATLAB. *International Journal of Information Engineering and Electronic Business, 1*, 1-14. Doi:10.5815/ijieeb.2014.01.01

Gruber, T. R., & Olsen, G. R. (1994, January). An ontology for engineering mathematics. In *Principles of Knowledge Representation and Reasoning* (pp. 258–269). Morgan Kaufmann.

Guarino, N. (1995). Formal ontology, conceptual analysis and knowledge representation. *International Journal of Human-Computer Studies, 43*(5-6), 625–640.

Gubbi, Buyya, Marusic, & Palaniswami. (2013). Internet of Things (IoT): A vision, architectural elements, and future directions. *Future Generation Computer Systems, 29*, 1645–1660.

Gubbi, J., Buyya, R., Marusic, S., & Palaniswami, M. (2013). Internet of Things (IoT): A vision, architectural elements, and future directions. *Future Generation Computer Systems, 29*(7),1645–1660.

Guillemin. (2015). *Internet of Things Position Paper on Standardization for IoT technologies.* European research cluster on the internet of things.

Guisser, El-Jouni, & Abdelmounim. (2014). Robust Sliding Mode MPPT Controller Based on High Gain Observer of a Photovoltaic Water Pumping System. *International Review of Automatic Control, 7*, 225-232.

Hadded, Zagrouba, Laouiti, Muhlethaler, & Saidane. (2015). A multi-objective genetic algorithm-based adaptive weighted clustering protocol in vanet. *IEEE Congress on Evolutionary Computation (CEC),* 994-1002.

Halkes, G. P., Tijs, V. D., & Langendoen, K. G. (2005). Comparing energy-saving MAC protocols for wireless sensor networks. *Mobile Networks and Applications, 10*(5), 783–791. doi:10.100711036-005-3371-x

Hamid, M., Abdullah-Al-Wadud, M., & Chong, I. (2010). A schedule-based Multi-channel MAC Protocol for Wireless Sensor Networks. *Sensors (Basel), 10*(10), 9466–9480. doi:10.3390101009466 PMID:22163420

Han, K., Shon, T., & Kim, K. (2010). Efficient Mobile Sensor Authentication in Smarth Home and WPAN. *IEEE Transactions on Consumer Electronics, 56*(2), 591–596. doi:10.1109/TCE.2010.5505975

Hassan, M. M., Lin, K., Yue, X., & Wan, J. (2015). A multimedia healthcare data sharing approach through cloud-based body area network. *Future Generation Computer Systems, 66,* 48–58. doi:10.1016/j.future.2015.12.016

Heidemann, W. Ye. J., & Estrin, D. (2004). Medium Access Control With Coordinated Adaptive Sleeping for Wireless Sensor Networks. *IEEE/ACM Transactions on Networking, 12*(3), 493–506. doi:10.1109/TNET.2004.828953

Heng, P.-C., Bo-Rei, P., Liu, Y.-H., Cheng, Y.-S., & Jia-We, H. (2015, June). Optimization of a fuzzy-logic-control-based MPPT algorithm using the particle swarm optimization technique. *Energies, 8*(6), 5338–5360. doi:10.3390/en8065338

Horrocks, I. (2005). OWL rules, ok?. *Rule Languages for Interoperability*, 34.

Horrocks, I., Patel-Schneider, P. F., Boley, H., Tabet, S., Grosof, B., & Dean, M. (2004). SWRL: A semantic web rule language combining OWL and RuleML. *W3C Member submission, 21*(79), 1-31.

Hossein, G., & Fard, E. (2011). Multi-channel Medium Access Control Protocols for Wireless Sensor Networks: A Survey. Department of Computer Engineering, Lahijan Branch, Islamic Azad University, Guilan, Iran.

Huang, P., Xiao, L., Soltani, S., Mutka, M. W., & Xi, N. (2012). The evolution of MAC protocols in wireless sensor networks: A survey. *IEEE Communications Surveys and Tutorials, 15*(1), 101–120. doi:10.1109/SURV.2012.040412.00105

Huisman, O. C., & Gerik, J. S. (1989). *Dynamics o colonization of plant roots by Verticillium dahlia and other fungi in Vascular wilt disease of plants.* Berlin: Springer-Verlag.

I.Akyildiz, Su, W., & Sankarasubramaniam, Y. (2002). A Survey On Sensor Networks. *IEEE Communications,* 102-114.

Ibriq, J., & Mahgoub, I. (2007). A Hierarchical Key Establishment Scheme forWireless Sensor Networks. In *21st International Conference on Advanced Networking and Applications* (pp. 210-219). Niagara Falls, Ontario, Canada: IEEE. 10.1109/AINA.2007.14

Incel, O. D., Hoesel, L. V., Jansen, P., & Havinga, P. (2011). MC-LMAC: A multi-channel MAC protocol for wireless sensor networks. *Ad Hoc Networks, 9*(1), 73–94. doi:10.1016/j.adhoc.2010.05.003

Isam, O., & Hussain, S. (239-242). An intelligent multi-hop routing for wireless sensor networks. In *IEEE/WIC/ACM International Conference on Web intelligence and Intelligent agent technology workshops* (pp. 239-242). Hong Kong, China: IEEE.

Islam, R., Sarker, R., Saha, S., & Nokib Uddin, A. F. M. (2012). *Design of A Programmable Digital IIR Filter Based on FPGA*. IEEE. doi:10.1109/ICIEV.2012.6317409

Izhikevich, E. M., Gally, J. A., & Edelman, G. M. (2004). Spike-timing dynamics of neuronal groups. Cerebral Cortex, 14(8), 933–944. doi:10.1093/cercor/bhh053

Jaballah, W. B., Mosbah, M., Youssef, H., & Zemmari, A. (2013). Lightweight Source Authentication Mechanisms for Group Communications in Wireless Sensor Networks. In *IEEE 27th International Conference on Advanced Information Networking and Applications (AINA)* (pp. 598-605). Barcelona, Spain: IEEE.

Jacobsson, Boldt, & Carlsson. (2016). A risk analysis of a smart home automation system. *Future Generation Computer Systems, 56*, 719–733.

Jakkula, V. R., Crandall, A. S., & Cook, D. J. (2007, October). Knowledge discovery in entity based smart environment resident data using temporal relation based data mining. In *Seventh IEEE International Conference on Data Mining Workshops (ICDMW 2007)* (pp. 625-630). IEEE.

Jan, M., Nanda, P., Usman, M., & He, X. (2016). PAWN: a payload-based mutual authentication scheme for wireless sensor networks. *Concurrency and Computation Practice and Experience, 29*(17), 1-32.

Jensen, T. R., & Toft, B. (2011). *Graph Coloring Problems*. John Wiley & Sons.

Jian, Q. (2008). Overview of MAC protocols in wireless sensor networks. *Journal of Software, 19*(2), 389–403. doi:10.3724/SP.J.1001.2008.00389

Jin, I., Udo, H., Rayman, J. B., Puthanveettil, S., Kandel, E. R., & Hawkins, R. D. (2012). Spontaneous transmitter release recruits postsynaptic mechanisms of long-term and intermediate-term facilitation in Aplysia. *Proceedings of the National Academy of Sciences of the United States of America, 109*(23), 9137–9142. doi:10.1073/pnas.1206846109 PMID:22619333

Joshi. (2017). IoT based Air and Sound Pollution Monitoring System. *International Journal of Computer Applications, 178*(7).

Joshi, P. J., Patkar, V. P., Pawar, A. B., Patil, P. B., Bagal, U. R., & Mokal, B. D. (2013). ECG Denoising Using MATLAB. *International Journal of Scientific and Engineering Research*, *4*(5), 1401–1405.

Kalaivaani, P. T., & Rajeswari, A. (2013). An energy-efficient analysis of S-MAC and H-MAC protocols for wireless sensor network. *International Journal of Computer Network & Communication*, *5*(2), 83–94. doi:10.5121/ijcnc.2013.5207

Kamal, N., Saad, M.H.M., Kok, C.S., & Hussain, A. (2018). Towards revolutionizing stem education via IoT and blockchain technology. *Int. J. Eng. Technol.* doi:10.14419/ijet.v7i4.11.20800

Karlof, C., & Wagner, D. (2003). Secure routing in wireless sensor networks: Attacks and countermeasures. *Ad Hoc Networks*, *1*(2-3), 293–315. doi:10.1016/S1570-8705(03)00008-8

Kasetwar, A. R., & Gulhane, S. M. (2013). Adaptive Power Line Interference Canceller: A Survey. *International Journal of Advances in Engineering and Technology*, *5*(2), 319–326.

Kassab, M., DeFranco, J., & Voas, J. (2018). Smarter education. *IT Professional*, *20*(5), 20–24. doi:10.1109/MITP.2018.053891333

Katz, S., Ford, A. B., Moskowitz, R. W., Jackson, B. A., & Jaffe, M. W. (1963). Studies of Illness in the Aged. *Journal of the American Medical Association*, *185*(12), 914. doi:10.1001/jama.1963.03060120024016

Kaufmann, A. (1975). *Introduction à la théorie des sous-ensembles flous à l'usage des ingénieurs (fuzzy sets theory)*. Academic Press.

Kaur & Kalra. (2016). A review on IoT based smart grid. *International Journal of Energy, Information and Communication, 7*(3), 11-22.

Kaur, M., & Singh, B. (2011). Comparisons of Different Approaches For Removal of Baseline Wander From ECG Signal. *International Journal of Computer Applications*, 30-36.

Kautz, H. A. (1991). A formal theory of plan recognition and its implementation. Reasoning about plans, 69-125.

Keddar, M., Doumbia, M. L., Della, M., Belmokhtar, K., & Midoun, A. (2019, September). Interconnection Performance Analysis of Single Phase Neural Network-Based NPC and CHB Multilevel Inverters for Grid-connected PV Systems. *International Journal of Renewable Energy Research*, *9*, 1451–1461.

Kempter, R., Gerstner, W., & Van Hemmen, J. (1999). Hebbian learning and spiking neurons. *Physical Review E. APS*, *59*(4), 4498–4514. doi:10.1103/PhysRevE.59.4498

Keshavarz, E. (2016). Vertex-coloring of fuzzy graphs: A new approach. *Journal of Intelligent & Fuzzy Systems*, *30*(2), 883–893. doi:10.3233/IFS-151810

Khan, Khan, Zaheer, & Khan. (2012). Future Internet: The Internet of Things architecture, possible applications and key challenges. *Proc. 10th Int. Conf. FIT*, 257–260.

Khatoun & Zeadally. (2017). Cybersecurity and Privacy Solutions in Smart Cities. *IEEE Communications Magazine, 55*(3), 51–59.

Knublauch, H., & Musen, M. A. (2004, June). Weaving the biomedical semantic web with the Protégé OWL plugin. In *Proceedings of the First International Conference on Formal Biomedical Knowledge Representation-Volume 102* (pp. 39-47).

Kolawole, E. S., Ali, W. H., Cofie, P., Fuller, J., Tolliver, C., & Obiomon, P. (2015). Design and Implementation of Low- Pass, High- Pass And Band- Pass Finite Impulse Response (FIR) Filters Using FPGA. *Circuit and System, 6*(02), 30–48. doi:10.4236/cs.2015.62004

Kolesnik, S., & Kuperman, A. (2016, November). On the Equivalence of Major Variable-Step-Size MPPT Algorithms. *IEEE Journal of Photovoltaics, 6*(2), 590–594. doi:10.1109/JPHOTOV.2016.2520212

Kopetz. (2011). Internet of things. In *Real-time systems*, (pp. 307-323). Springer US.

Kouchaki, A., Iman-Eini, H., & Asaei, B. (2013). A new maximum power point tracking strategy for PV arrays under uniform and non-uniform insolation conditions. *Solar Energy, 91*, 221–232. doi:10.1016/j.solener.2013.01.009

Kumar & Shoghli. (2018). A review of IOT applications in supply chain optimization of construction materials. *34th International symposium on automation and robotics in construction.*

Kumar Singh, H. A., Hussain, I., & Singh, B. (2018, May). Double-Stage Three-Phase Grid-Integrated Solar PV System With Fast Zero Attracting Normalized Least Mean Fourth Based Adaptive Control. *IEEE Transactions on Industrial Electronics, 65*(5), 3921–3931. doi:10.1109/TIE.2017.2758750

Kumar, K. S. A., & Balakrishna, R. (2019). *Performance Analysis of Reliability and Throughput Using an Improved Receiver—Centric MAC Protocol and Itree-MAC Protocol in Wireless Sensor Networks. In Cognitive Informatics and Soft Computing* (pp. 83–90). Singapore: Springer.

Kuon, I., & Rose, J. (2007). Measuring The Gap Between FPGAs and ASICs. *IEEE Transactions on Computer-Aided Design of Integrated Circuits and Systems, 26*(2), 203–215. doi:10.1109/TCAD.2006.884574

Kuon, I., Tessier, R., & Rose, J. (2007). FPGA Architecture: Survey and Challenges. *Electronic Design Automation, 2*(2), 135–253. doi:10.1561/1000000005

Lassila, O., & Swick, R. R. (1998). *Resource description framework (RDF) model and syntax specification.*

Latfi, F., Descheneaux, C., & Lefebvre, B. (2007). *Habitat intelligent en télé–Santé: ontologie de l'équipement.* Canada: FICCDAT Toronto.

Lee, H., Choi, Y., & Kim, H. (2005). Implementation of tiny hash based on hash algorithm for sensor network. *Proceedings of World Academy of Science, Engineering and Technology, 10*, 135-139.

Lee, Zappaterra, Choi, & Choi. (2014). Securing smart home: Technologies, security challenges, and security requirements. *2014 IEEE Conference on Communications and Network Security*, 67–72.

Lee, S. H., Lee, S., Song, H., & Lee, H. S. (2009). Wireless sensor network design for tactical military applications: Remote largescale environments. In *Proc. IEEE conference in Military Communications* (pp. 1-7). Boston, MA: IEEE. 10.1109/MILCOM.2009.5379900

Li, C., Zhang, B., Yuan, X., Ullah, S., & Vasilakos, A. V. (2018). MC-MAC: A multi-channel based MAC scheme for interference mitigation in WBANs. *Wireless Networks, 24*(3), 719–733. doi:10.100711276-016-1366-0

Lin, J., Yu, W., Zhang, N., Yang, X., Zhang, H., & Zhao, W. (2017). A survey on internet of things: Architecture, enabling technologies, security and privacy, and applications. *IEEE Internet of Things Journal, 4*(5), 1125–1142.

Liu, Xiao, Li, Liang, & Chen. (2012). Cyber Security and Privacy Issues in Smart Grids. *IEEE Communications Surveys and Tutorials, 14*(4), 981–997.

Liu, J., Luo, W., Yang, X., & Wu, L. (2016). Robust model-based fault diagnosis for PEM fuel cell air-feed system. *IEEE Transactions on Industrial Electronics, 63*(5), 3261–3270. doi:10.1109/TIE.2016.2535118

Liu, Y., Ota, K., Zhang, K., Ma, M., Xiong, N., Liu, A., & Long, J. (2018). QTSAC: An energy-efficient MAC protocol for delay minimization in wireless sensor networks. *IEEE Access: Practical Innovations, Open Solutions, 6*, 8273–8291. doi:10.1109/ACCESS.2018.2809501

Li, Y., Li, J., Ren, J., & Wu, J. (2012). *Providing hop-by-hop authentication and source privacy in wireless sensor networks. In IEEE INFOCOM* (pp. 3071–3075). Orlando, FL: IEEE.

Lloret, J., Garcia, M., Bri, D., & Sendra, S. (2009). A wireless sensor network deployment for rural and forest fire detection and verification. *Sensors (Basel), 9*, 8722–8747. PMID:22291533

Loubane, M. (1991). Contribution à l'étude du complexe nématodes-champignons associés aux pourritures racinaires du bananier dans la région du souss-Massa. Engineering diploma Institut Agronomique Vétérinaire Hassan II, Complexe Horticole Agadir.

Lu, R., Lin, X., Zhu, H., Liang, X., & Shen, X. (2012). BECAN: A Bandwidth-Efficient Cooperative Authentication Scheme for Filtering Injected False Data in Wireless Sensor Networks. *IEEE Transactions on Parallel and Distributed Systems*, 32–43.

Machado. (2016). *Machado IOT impacts on supply chain saha machado second place grad, Scribd*. http://apicsterragrande.org/images/articles/Machado__Internet_of_Things_impacts_on_Supply_Chain_Shah_Machado_Second_Place_Grad.pdf

Mahamune & Amdani. (2017). IoT based connected vehiclein smart cities: review and research challenges. *IJERCSE, 4*(4). https://www.technoarete.org/common_abstract/pdf/IJERCSE/v4/i4/Ext_47935.pdf

Majeed, A., & Ali, M. (2018). How Internet-of-Things (IoT) making the university campuses smart? QA higher education (QAHE) perspective. In *2018 IEEE 8th Annual Computing and Communication Workshop and Conference (CCWC)*, (pp. 646–648). IEEE. doi:10.1109/CCWC.2018.8301774

Malik, M., & Sharma, M. (2019). A Novel Approach for Comparative Analysis on Energy Effectiveness of H-MAC and S-MAC Protocols for Wireless Sensor Networks. *Advanced Science, Engineering and Medicine, 11*(1-2), 29–35. doi:10.1166/asem.2019.2326

Mardonova & Choi. (2018). Review of wearable device technology and its application to the mining industry. *Energies, 11*(3), 547.

Market Study Report. (2013, Feb. 21). *Wearable Computing Devices, Like Apple's iWatch, Will Exceed 485 Million Annual Shipments by 2018*. ABI Research.

Markram, H. (1997). Regulation of synaptic efficacy by coincidence of postsynaptic APs and EPSPs. *Science. AAAS, 275*(5297), 213–215. doi:10.1126cience.275.5297.213 PMID:8985014

Martin, A. D., Cano, J. M., Silva, J. F. A., & Vazquez, J. R. (2015). Backstepping control of smart grid-connected distributed photovoltaic power supplies for telecom equipment. *IEEE Transactions on Energy Conversion, 30*(4), 1496–1504. doi:10.1109/TEC.2015.2431613

McGuinness, D. L., & Van Harmelen, F. (2004). OWL web ontology language overview. *W3C recommendation, 10*(10), 2004.

Meddah, N., Ouazzani, T. A., Benkirane, R., & Douira, A. (2011). Etude du pouvoir pathogène de quelques espèces de *Fusarium* sur le bananier sous serre au Maroc. *Bulletin de la Société Royale des Sciences de Liège, 80*, 939–952.

Mehmood, Y., Ahmad, F., & Yaqoob, I. (2017). Internet-of-things-based smart cities: recent advances and challenges. *IEEE Commun Mag, 55*(9), 16–24.

Meirong, X., Yuzhen, W., & Ping, J. (2015, July). Fuzzy graph coloring via semi-tensor product method. In *2015 34th Chinese Control Conference (CCC)* (pp. 973-978). IEEE. 10.1109/ChiCC.2015.7259766

Melgar, J., & Roy, K. W. (1994). Soybean sudden death syndrome cultivar reaction to inoculation in controlled environement and host range and virulence of causal agent. *American Phytopathology Society, 78*, 265–268.

Menadi, A., Abdeddaim, S., Ghamri, A., & Betka, A. (2015, September). Implementation of fuzzy-sliding mode based control of a grid-connected photovoltaic system. *ISA Transactions, 58*, 586–594. doi:10.1016/j.isatra.2015.06.009 PMID:26243440

Miao, R., Dong, Q., Weng, W. Y., Yu, X.Y. (2018). The application model of wearable devices in physical education. *International Conference on Blended Learning*, 311–322.

Michie, D., Spiegelhalter, D., Taylor, C., & Campbell, J. (1994). *Machine Learning, Neural and Statistical Classification*. Ellis Horwood.

Miller, D. E., & Burke, D. W. (1977). Effect of temporary excessive wetting on soil aeration and *Fusarium* root rot of beans. *The Plant Disease Reporter, 61,* 175–179.

Mlayah, A., & Khedher, A. (2018, September). Sliding Mode Control Strategy for Solar Charging of High Energy Lithium Batteries. *International Journal of Renewable Energy Research, 8*(3), 1621–1623.

Mohamad, Teh, Lai, & Chen. (2018). Development of energy storage systems for power network reliability: a review. *Energies, 11*(9).

Mojallizadeh, M. R., & Badamchizadeh, M. A. (2017). Second-order fuzzy sliding-mode control of photovoltaic power generation systems. *Solar Energy, 149,* 332–340. doi:10.1016/j. solener.2017.04.014

Moosavi, S. R., Gia, T. N., Nigussie, E., Rahmani, A. M., Virtanen, S., Tenhunen, H., & Isoaho, J. (2015). End-to-end security scheme for mobility enabled healthcare Internet of Things. *Future Generation Computer Systems, 64,* 108–124. doi:10.1016/j.future.2016.02.020

Moradi, M. H., & Reisi, A. R. (2011). A hybrid maximum power point tracking method for photovoltaic systems. *Solar Energy, 85*(11), 2965–2976. doi:10.1016/j.solener.2011.08.036

Morales, D. P., Garcia, A., Castillo, E., Carvajal, M. A., Banqueri, J., & Palma, A. J. (2011). Flexible ECG Acquisition System Based on Analog and Digital Reconfigurable Devices. *Sensors and Actuators. A, Physical, 165*(2), 261–270. doi:10.1016/j.sna.2010.10.008

Moretti, A., Ferracane, L., Somma, S., Ricci, V., Mulè, G., Susca, A., ... Logrieco, A. F. (2010). Identification, mycotoxin risk and pathogenicity of *Fusarium* species associated with fig endosepsis in Apulia, Italy. *Food Additives and Contaminants, 27*(5), 718–728. doi:10.1080/19440040903573040 PMID:20352549

Morrison, A., Diesmann, M., & Gerstner, W. (2008). Phenomenological models of synaptic plasticity based on spike timing. *Biological Cybernetics, 98*(6), 459–478. doi:10.100700422-008-0233-1 PMID:18491160

Moutacalli, M. T., Bouchard, B., & Bouzouane, A. (2015), Sensors Activation Times Prediction in Smart Home. In *8th ACM International Conference on Pervasive Technologies Related to Assistive Environments.*

Moutacalli, M. T., Bouzouane, A., & Bouchard, B. (2015). The behavioral profiling based on times series forecasting for smart homes assistance. *Journal of Ambient Intelligence and Humanized Computing, 6*(5), 647–659.

Muñoz, S., Ortuno, M. T., Ramírez, J., & Yáñez, J. (2005). Coloring fuzzy graphs. *Omega, Elsevier, 33*(3), 211–221. doi:10.1016/j.omega.2004.04.006

Musolesi, M., Hailes, S., & Mascolo, C. (2005). Adaptive routing for intermittently connected mobile ad hoc networks. *Proceedings of the Sixth IEEE International Symposium on World of wireless mobile and multimedia networks.*

Myint, Gopal, & Aung. (2017). Reconfigurable Smart Water Quality Monitoring System in IoT Environment. *IEEE/ACIS 16th International Conference on Computer and Information Science (ICIS).*

Naiker, S., & Odhav, B. (2004). Mycotic keratitis., profile of Fusarium species and their mycotoxins. Identification, mycotoxin risk and pathogenicity of Fusarium species associated with figendosepsis in Apulia, Italy. *Mycoses, 47,* 50–56. doi:10.1046/j.0933-7407.2003.00936.x PMID:14998400

Nicholls, J.G. (2001). *From Neuron to Brain.* Academic Press.

Ning, P., Liu, A., & Du, W. (2008). Mitigating DOS attacks against broadcast authentication in WSN. *ACM Transactions on Sensor Networks, 4,* 1. doi:10.1145/1325651.1325652

Nucci, M., & Anaissie, E. (2007). *Fusarium* infections in immune-compromised patients. *Clinical Microbiology Reviews, 20*(4), 695–704. doi:10.1128/CMR.00014-07 PMID:17934079

Ojha, T., Misra, S., & Raghuwanshi, N. (2015). Wireless sensor networks for agriculture: The state-of-the-art in practice and future challenges. *Computers and Electronics in Agriculture, 118,* 66-84. Doi:10.1016/j.compag.2015.08.011

Ordóñez, F. J., Iglesias, J. A., De Toledo, P., Ledezma, A., & Sanchis, A. (2013). Online activity recognition using evolving classifiers. *Expert Systems with Applications, 40*(4), 1248–1255.

Oulasvirta, A., Pihlajamaa, A., Perkiö, J., Ray, D., Vähäkangas, T., Hasu, T., ... Myllymäki, P. (2012, September). Long-term effects of ubiquitous surveillance in the home. In *Proceedings of the 2012 ACM Conference on Ubiquitous Computing* (pp. 41-50).

Ozturk, A., Umit, K., Medeni, I., Ucuncu, B., Caylan, M., Akba, F., & Medeni, D.T. (2011). Green ICT (Information and Communication Technologies): a review of academic and practitioner perspectives. *Int. J. EBusiness and EGovernment Stud.*

Pal, Gupta, Tiwari, & Sharma. (2017). IoT Based Air Pollution Monitoring System Using Arduino. *International Research Journal of Engineering and Technology, 4*(10).

Pani, A. K., & Nayak, N. (2019, November). A Short Term Forecasting of PhotoVoltaic Power Generation Using Coupled Based Particle Swarm Optimization Pruned Extreme Learning Machine. *International Journal of Renewable Energy Research, 9,* 1190–1202.

Pardalos, P. M., Mavridou, T., & Xue, J. (1998). The graph coloring problem: a bibliographic survey. In D. Z. Du & P. M. Pardalos (Eds.), *Handbook of combinatorial optimization* (Vol. 2, pp. 331–395). Boston: Kluwer Academic Publishers. doi:10.1007/978-1-4613-0303-9_16

Pascoe, J. (1998, October). Adding generic contextual capabilities to wearable computers. In Digest of papers. second international symposium on wearable computers (cat. no. 98ex215) (pp. 92-99). IEEE.

Pathan, K. A.-S., Lee, H.-W., & Hong, C. S. (2006). Security in wireless sensor networks: issues and challenges. In *The 8th International Conference Advanced Communication Technology* (p. 6). Phoenix Park, South Korea: IEEE.

Patil, Gawande, & Bag. (2017). Smart Agriculture System based on IoT and its Social Impact. *International Journal of Computers and Applications.*

Patterson, D. J., Fox, D., Kautz, H., & Philipose, M. (2005, October). Fine-grained activity recognition by aggregating abstract object usage. In *Ninth IEEE International Symposium on Wearable Computers (ISWC'05)* (pp. 44-51). IEEE.

Pawar, D.J., & Bhaskar, P.C. (2013). FPGA Based FIR Filter Design for Enhancement of ECG Signal by Minimizing Base-line Drift Interference. *International Journal of Current Engineering and Technology*, 1775-1778.

Pegatoquet, A., Trong, N. L., & Michele, M. (2018). A Wake-Up Radio-Based MAC Protocol for Autonomous Wireless Sensor Networks. *IEEE/ACM Transactions on Networking*, 27(1), 56–70. doi:10.1109/TNET.2018.2880797

Pei, X.L., Wang, X., Wang, Y.F., & Li, M.K. (2013). Internet of Things based education: definition, benefits, and challenges. *Applied Mechanics and Materials, 411–414*, 2947–2951. doi:10.4028/www.scientific.net/AMM.411-414.2947

Perrig, A., Szewczyk, R., Tygar, J. D., Wen, V., & Culler, D. E. (2002). SPINS: Security Protocols For Sensor Networks. *Wireless Networks, 8*(5), 521–534. doi:10.1023/A:1016598314198

Pfister, J. P., & Gerstner, W. (2006). Triplets of spikes in a model of spike timing-dependent plasticity. *Journal of Neuroscience. Social Neuroscience, 26*(38), 9673–9682. PMID:16988038

Plantevin, V., Bouzouane, A., & Gaboury, S. (2017). The light node communication framework: A new way to communicate inside smart homes. *Sensors (Basel), 17*(10), 2397.

Porkodi, R., & Bhuvaneswari, V. (2014). The Internet of Things (IoT) applications and communication enabling technology standards: An overview. In *2014 International conference on intelligent computing applications (ICICA)*. IEEE.

Prathibha & Anupama. (2017). IOT based monitoring system in smart agriculture. *2017 International conference*. doi:10.1109/ICRAECT.2017.52

Qiu, Y., Zhou, J., Baek, J., & Lopez, J. (2010). Authentication and key establishment in dynamic wireless sensor network. *Sensors (Basel), 10*(4), 3718–3731. doi:10.3390100403718 PMID:22319321

Radha, S. (2019). ScP Protocol for Wireless Sensor Networks. In *Proceedings of 5th International Conference on Advanced Computing & Communication Systems (ICACCS)*. IEEE.

Rafida, A. R., Faridah, S., Shahrul, A. A., Mazidah, M., & Zamri, I. (2016). Chronoamperometry Measurement For Rapid Cucumber Virus Detection in Plants. *Procedia Chemistry, 20*, 25–28. doi:10.1016/j.proche.2016.07.003

Rahim. (2016). Exploiting heuristic algorithms to efficiently utilize energy management controllers with renewable energy sources. Energy Build.

Rahrah, K., Rekioua, D., Rekioua, T., & Bacha, S. (2015, October). Photovoltaic pumping system in Bejaia climate with battery storage. *International Journal of Hydrogen Energy, 40*(39), 13665–13675. doi:10.1016/j.ijhydene.2015.04.048

Ramesh, Nibi, Kurup, Mohan, Aiswarya, Arsha, & Sarang. (2017). Water Quality Monitoring and Waste Management using IoT. *IEEE Global Humanitarian Technology Conference (GHTC)*.

Ramesh, Nibi, Kurup, Mohan, Aiswarya, Arsha, … Guerrero-Barrantes. (2017). Characterization of biomass pellets from Chlorella vulgaris microalgal production using industrial wastewater. *International Conference in Energy and Sustainability in Small Developing Economies (ES2DE)*.

Ramos, C., Augusto, J. C., & Shapiro, D. (2008). Ambient intelligence—the next step for artificial intelligence. *IEEE Intelligent Systems, 23*(2), 15–18.

Rashidi, P., Cook, D. J., Holder, L. B., & Schmitter-Edgecombe, M. (2010). Discovering activities to recognize and track in a smart environment. *IEEE Transactions on Knowledge and Data Engineering, 23*(4), 527–539.

Ravikumar, M. (2012). Electrocardiogram Signal Processing on FPGA For Emerging Healthcare Applications. *International Journal of Electronics Signals and Systems, 1*(3), 91–96.

Ray, N. K., & Turuk, A. K. (2009). A review of energy-efficient MAC protocols for wireless LANs. In *Proceedings of International Conference on Industrial and Information Systems (ICIIS)*, IEEE. 10.1109/ICIINFS.2009.5429875

Rehman, A., Din, S., Paul, A., & Ahmad, W. (2017). An Algorithm for Alleviating the Effect of Hotspot on Throughput in Wireless Sensor Networks. *Proceedings of the IEEE 42nd Conference on Local Computer Networks Workshops (LCNWorkshops)*, 170–174.

Ricquebourg, V., Menga, D., Durand, D., Marhic, B., Delahoche, L., & Loge, C. (2006, December). The smart home concept: our immediate future. In *2006 1st IEEE international conference on e-learning in industrial electronics* (pp. 23-28). IEEE.

Ricquebourg, V., Delafosse, M., Delahoche, L., Marhic, B., Jolly-Desodt, A., & Menga, D. (2007). *Fault detection by combining redundant sensors: a conflict approach within the tbm framework. Cognitive Systems with Interactive Sensors*. COGIS.

Rinat, G. (2020). Brain machine interface: The accurate interpretation of neurotransmitters' signals targeting the muscles. *International Journal of Applied Research in Bioinformatics, 0102*. doi:10.4018/IJARB.2020

Rinat, G., & Vardan, M. (2019A). Math model of neuron and nervous system research, based on AI constructor creating virtual neural circuits: Theoretical and Methodological Aspects. In V. Mkrttchian, E. Aleshina, & L. Gamidullaeva (Eds.), *Avatar-Based Control, Estimation, Communications, and Development of Neuron Multi-Functional Technology Platforms* (pp. 320–344). Hershey, PA: IGI Global. doi:10.4018/978-1-7998-1581-5.ch015

Rinat, G., & Vardan, M. (2019B). Brain machine interface – for Avatar Control & Estimation in Educational purposes Based on Neural AI plugs: Theoretical and Methodological Aspects. In V. Mkrttchian, E. Aleshina, & L. Gamidullaeva (Eds.), *Avatar-Based Control, Estimation, Communications, and Development of Neuron Multi-Functional Technology Platforms* (pp. 345–360). Hershey, PA: IGI Global. doi:10.4018/978-1-7998-1581-5.ch016

Roberts, F. S. (1979). On the mobile radio frequency assignment problem and the traffic light phasing problem. *Annals of the New York Academy of Sciences*, *319*(1), 466–483. doi:10.1111/j.1749-6632.1979.tb32824.x

Rojo, F. G., Reynoso, M. M., Sofia, M. F., & Torres, A. M. (2006). Biological control by *Trichoderma* species of *Fusarium solani* causing peanut brown root rot under field conditions. *Crop Protection (Guildford, Surrey)*, *26*(4), 549–555. doi:10.1016/j.cropro.2006.05.006

Romberg, M. K., & Davis, R. M. (2007). *Host range and phylogeny of Fusarium solani f. sp. eumartii from potato and tomato in California*. Academic Press.

Rosenfeld, A. (1975). Fuzzy graphs. In *Fuzzy sets and their applications to cognitive and decision processes* (pp. 77–95). Academic Press. doi:10.1016/B978-0-12-775260-0.50008-6

Roy, P. C., Bouchard, B., Bouzouane, A., & Giroux, S. (2013). Ambient Activity Recognition in Smart Environments for Cognitive Assistance. *International Journal of Robotics Applications and Technologies*, *1*(1), 29–56. doi:10.4018/ijrat.2013010103

Ryan, N. S., Pascoe, J., & Morse, D. R. (1998). Enhanced reality fieldwork: the context-aware archaeological assistant. In *Computer applications in archaeology*. Tempus Reparatum.

Sa'nchez-Corcuera, Nun~ez-Marcos, Sesma-Solance, Bilbao-Jayo, Mulero, Zulaika, … Almeida. (2019). Smart cities survey: Technologies, application domains and challenges for the cities of the future. *International Journal of Distributed Sensor Networks*, *15*(6). doi:10.1177/1550147719853984

Sadabadia, H., Ghasemia, M., & Ghaffaria, A. (2007). *A Mathematical Algorithm For ECG Signal denoising Using Window Analysis*. Biomed Pap Med Fac Univ Palacky Olomouc Czech Repub.

Saha, Auddy, Chatterjee, Pal, Pandey, Singh, … Maity. (2017). Pollution Control Using Internet of Things (IoT). *8th Annual Industrial Automation and Electromechanical Engineering Conference (IEMECON)*.

Sakr, S., & Elgammal, A. (2016). Towards a Comprehensive Data Analytics Framework for Smart Healthcare Services. *Big Data Research*, *4*, 44–58. doi:10.1016/j.bdr.2016.05.002

Salman, H. M. (2014). Survey of routing protocols in wireless sensor networks. *International Journal of Sensors and Sensor Networks*, *2*(1), 1–6.

Sarıta, M. (2015). The emergent technological and theoretical paradigms in education: the interrelations of cloud computing (CC), connectivism and Internet of Things (IoT). *Acta Polytechnica Hungarica*, *12*(6), 161–179. doi:10.12700/APH.12.6.2015.6.10

Saxena, N., Roy, A., & Shin, J. (2008). Dynamic duty cycle and adaptive contention window based QoS-MAC protocol for wireless multimedia sensor networks. *Computer Networks, 52*(13), 2532–2542. doi:10.1016/j.comnet.2008.05.009

Schiefer. (2015). Smart Home Definition and Security Threats. *2015 Ninth International Conference on IT Security Incident Management & IT Forensics*, 114–118.

Schilit, B. N., & Theimer, M. M. (1994). Disseminating active map information to mobile hosts. *IEEE Network, 8*(5), 22–32.

Schilit, B., Adams, N., & Want, R. (1994, December). Context-aware computing applications. In *1994 First Workshop on Mobile Computing Systems and Applications* (pp. 85-90). IEEE.

Selim, B., Ekrem, K., & Ali, K. (2015, November). A simpler single-phase single-stage grid connected PV system with maximum power point tracking controller. *Elektronika ir Elektrotechnika, 21*, 44–47.

Selvi. (2018). Difficulties and Data Mining Model For Internet of Things (IoT). *International Journal of Engineering Science Invention*, 44-48. http://www.ijesi.org/papers/NCIOT-2018/Volume-1/8.%2041-45.pdf

Sezer, O. B., Dogdu, E., Ozbayoglu, M., & Onal, A. (2016, December). An extended iot framework with semantics, big data, and analytics. In *2016 IEEE International Conference on Big Data (Big Data)* (pp. 1849-1856). IEEE.

Sharma & Vinod. (2014). SPARK: Personalized Parkinson Disease Interventions through Synergy between a Smartphone and a Smartwatch. In *Design, User Experience, and Usability*. Springer International Publishing.

Sharma, S., Kumar, G., Miishra, D. K., & Mohapatra, D. (2012). Design and Implementation of a Variable Gain Amplifier for Biomedical Signal Acquisition. *International Journal of Advanced Research in Computer Science and Software Engineering, 2*(2), 193–198.

Sherazi, H. H., Raza, L., Alfredo, G., & Gennaro, B. (2018). A comprehensive review of energy harvesting MAC protocols in WSNs: Challenges and tradeoffs. *Ad Hoc Networks, 71*, 117–134. doi:10.1016/j.adhoc.2018.01.004

Shi, X., & Xiao, D. (2013). A reversible watermarking authentication scheme for wireless sensor networks. *Information Sciences, 240*, 173–183. doi:10.1016/j.ins.2013.03.031

Siano. (2016). Demand response and smart grids—A survey. *Renewable & Sustainable Energy Reviews, 30*, 461–478.

Silva, B., Fisher, R.M., Kumar, A., & Hancke, G.P. (2015). Experimental link quality characterization of wireless sensor networks for underground monitoring. *IEEE Transactions on Industrial Informatics, 11*, 1099–1110.

Singh, J. (2001). Fusarium resistance via biotechnology. Final report for the Ontario research enhancement program.

Sinha, R., Vadivelmurugan, I., Basavaiah, M. R., Angappan, S., & Muthappa, S. K. (2019). Impact of drought stress on simultaneously occurring pathogen infection in field-grown chickpea. *Nature Report, 9*(1), 5577. doi:10.103841598-019-41463-z PMID:30944350

Sivakumar, JagabarSathik, Manoj, & Sundararajan. (2016). An assessment on performance of DC–DC converters for renewable energy applications. *Elsevier Renewable and Sustainable Energy Reviews, 58*, 1475-1485.

Son, B., Her, Y.S., & Kim, J.G. (2006). A design and implementation of forest-fires surveillance system based on wireless sensor networks for South Korea Mountains. *Int. J. Comput. Sci. Netw. Secur., 6*, 124–130.

Spinelli, E.M., Martinez, N. H., & Mayosky, M. A. (2001). A Single Supply Biopotential Amplifier. *Medical Engineering & Physics, 23*, 235-238.

Spriggs, E. H., De La Torre, F., & Hebert, M. (2009, June). Temporal segmentation and activity classification from first-person sensing. In *2009 IEEE Computer Society Conference on Computer Vision and Pattern Recognition Workshops* (pp. 17-24). IEEE.

Stankovic. (2014). *Research directions for the Internet of Things*. IEEE.

Stent, G. S. (1973). A physiological mechanism for Hebb's postulate of learning. *Proceedings of the National Academy of Sciences of the United States of America, 70*(4), 997–1001. 10.1073/pnas.70.4.997

Sterbenz, J.P. (2017). Smart city and IoT resilience, survivability, and disruption tolerance: challenges, modelling, and a survey of research opportunities. In *Proceedings of the 9th international workshop on resilient networks design and modeling (RNDM)*. New York: IEEE.

Suberu, M. Y., Mustafa, M. W., & Bashir, N. (2014). Energy storage systems for renewable energy power sector integration and mitigation of intermittency. *Energy Rev., 35*, 499–514.

Sukor, A. S. A., Zakaria, A., Rahim, N. A., Kamarudin, L. M., Setchi, R., & Nishizaki, H. (2019). A hybrid approach of knowledge-driven and data-driven reasoning for activity recognition in smart homes. *Journal of Intelligent & Fuzzy Systems, 36*(5), 4177–4188.

Sundar, Hebbar, & Golla. (2015). Implementing intelligent traffic control system for congestion control, ambulance clearance, and stolen vehicle detection. *IEEE Sensors Journal, 15*(2), 1109–1113.

Suryadevara, N. K., & Mukhopadhyay, S. (2012). Wireless sensor network based home monitoring system for wellness determination of elderly. *IEEE Sensors Journal, 12*(6), 1965–1972. doi:10.1109/JSEN.2011.2182341

Suryadevara, N. K., Mukhopadhyay, S. C., Wang, R., & Rayudu, R. K. (2013). Forecasting the behavior of an elderly using wireless sensors data in a smart home. *Engineering Applications of Artificial Intelligence, 26*(10), 2641–2652.

Tan, W., Wang, Q., Huang, H., Guo, Y., & Zhan, G. (2007). Mine Fire Detection System Based on Wireless Sensor Networks. *Proceedings of the Conference on Information Acquisition (ICIA'07).*

Tawfik, M. M., Selim, H., & Kamal, T. (2010). Human Identification Using QT Signal and QRS Complex of the ECG. *Journal of Electronic and Electrical Engineering, 3*(1), 383–387.

Teodorescu, R., Liserre, M., & Rodriguez, P. (2011, March 15). *Grid Converters for Photovoltaic and Wind Power Systems.* Retrieved from https://www.bookdepository.com/Grid-Converters-for-Photovoltaic-Wind-Power-Systems-Remus-Teodorescu/9780470057513

Tey, Mekhilef, Seyedmahmoudian, Horan, Oo, & Stojcevski. (2018). Improved Differential evolution-based MPPT Algorithm Using SEPIC for PV Systems Under Partial Shading Conditions and Load Variation. *IEEE Transactions on Industrial Informatics, 14*, 22 – 43.

Thierer. (2015). The Internet of Things and Wearable Technology: Addressing Privacy and Security Concerns without Derailing Innovation. *Rich. J. L. & Tech., 21*(6).

Toma & Talpiga. (2018). Secure IOT supply chain management solution using blockchain and smart contrats technology. *International Conference on Security for Information Technology and Communication, SECITC.*

Tonnberg, V. (2016). *Evaluation of resistance against Fusarium root rot in Peas* (Master thesis). Faculty of Landscape Architecture, Horticulture and Crop protection science Swedish University of Agricultural Sciences, Anlarp, Sweden.

Tragardh, E., & Schlegel, T. T. (2006). *High-Frequency ECG.* Academic Press.

Trick, M. (2002). *Computational Series: Graph Coloring and its Generalizations.* Retrieved November 02, 2019, from https://mat.gsia.cmu.edu/COLOR02/

Tsodyks, M., Uziel, A., & Markram, H. (2000). Synchrony generation in recurrent networks with frequency-dependent synapses. The Journal of Neuroscience, 20(1).

Usha & Rukmini. (2016). IOT in connected vehicles: challenges and issues-A review. *International Conference on Signal Processing, Communication, Power and Embedded Systems.* doi:10.1109/SCOPES.2016.7955769

Valtonen, M., Vuorela, T., Kaila, L., & Vanhala, J. (2012). Capacitive indoor positioning and contact sensing for activity recognition in smart homes. *Journal of Ambient Intelligence and Smart Environments, 4*(4), 305–334.

Vaseghi, A., Naser Safaie, N., Bakhshinejad, B., Mohsenifar, A., & Sadeghizadeh, M. (2013). Detection of *Pseudomonas syringae* pathovars by thiol-linked DNA–Gold nanoparticle probes. In G. Rivas & U. Weimar (Eds.), *Sensors and Actuators B: Chemical* (pp. 644–651). Tokyo, Japan: Elsiever. doi:10.1016/j.snb.2013.02.018

Vermensan & Friess. (2013). *Internet-of-Things converging technologies for smart environment and integrated ecosystems.* River Publishers.

Vihervaara, J., & Alapaholuoma, T. (2016). Internet of Things: opportunities for vocational education and training—presentation of the pilot project. In *Proceedings of the 9th International Conference on Computer Supported Education*, (pp. 476–480). SCITEPRESS—Science and Technology Publications. doi:10.5220/0006353204760480

Wai, R., Lin, C., Wu, W., & Huang, H. (2013). Design of backstepping control for high-performance inverter with stand-alone and grid-connected power-supply modes. *IET Power Electronics, 6*(4), 752–762. doi:10.1049/iet-pel.2012.0579

Wang, Y. (2007). Multiplierless CSD Techniques for High Performance FPGA Implementation of Digital Filters (PhD Thesis). University of Oklahoma, Graduate College.

Wang, J., Liu, Z., Zhang, S., & Zhang, X. (2014). Defending collaborative false data injection attacks in wireless sensor networks. *Information Sciences, 254*, 39–53. doi:10.1016/j.ins.2013.08.019

Wang, L., & Li, P. H. C. (2010). Gold nanoparticle-assisted single base-pair mismatch discrimination on a microfluidic microarray device. *Biomicroluidics, 4*(3), 032209. doi:10.1063/1.3463720 PMID:21045930

Wang, P., & Zhigang, L. V. (2012). Design of A Simple 3-Lead ECG Acquisition System Based on MSP430F149. *International Conference on Computer and Automation Engineering, 44*, 86–91.

Wang, W. D., & Liu, X. (2009). ShortPK:a short term public key scheme for broadcast authentication in WSN. *ACM Transactions on Sensor Networks, 4*(1), 29.

Weiser, M. (1991). The Computer for the 21st Century. *Scientific American, 265*(3), 94–105.

Wenning, B.L., Pesch, D., Timm-Giel, A., & Görg, C. (2010). Environmental monitoring aware routing: Making environmental sensor networks more robust. *Telecommun. Syst., 43*, 3–11. https://en.wikipedia.org/wiki/Internet_of_things

Winograd, T. (2001). Architectures for context. *Human-Computer Interaction, 16*(2-4), 401–419.

Wong, K., Zheng, Y., Cao, J., & Wang, S. (2006). A dynamic user authentication scheme for wireless sensor networks. *Proc. IEEE International Conference on Sensor Networks, Ubiquitous, and Trustworthy Computing, 1*, 8. 10.1109/SUTC.2006.1636182

Wu, Lu, Ling, Sun, & Du. (2010). Research on the architecture of Internet of Things. *Proc. 3rd ICACTE, 5*, 484-487.

Xu, L., He, W., & Li, S. (2014). Internet of Things in industries: A survey. *IEEE Transactions on Industrial Informatics, 10*(4), 2233–2243.

Xua, D., Gang, W., Yana, W., & Yanb, X. (2019, January). A novel adaptive command-filtered backstepping sliding mode control for PV grid-connected system with energy storage. *Solar Energy, 178*, 1–17. doi:10.1016/j.solener.2018.12.033

Yadav, S.K., & Mehra, R. (2014). Analysis of Different IIR Filter based on Implementation Cost Performance. *International Journal of Engineering and Advanced Technology, 3*(4), 267-270.

Yang. (2011). Study and application on the architecture and key technologies for IOT. *Proc. ICMT*, 747–751.

Yang, X., Wang, L., Su, J., & Gong, Y. (2018). Hybrid MAC protocol design for mobile wireless sensors networks. *IEEE Sensors Letters*, 2(2), 1–4. doi:10.1109/LSENS.2018.2828339

Yang, Y., Ye, Q., Tung, L., Greenleaf, M., & Li, H. (2018). Integrated size and energy management design of battery storage to enhance grid integration of large-scale PV power plants. *IEEE Transactions on Industrial Electronics*, 65(1), 394–402. doi:10.1109/TIE.2017.2721878

Yao, K. S., Li, S. J., Tzeng, K. C., Cheng, T. C., Chang, C. Y., Chiu, C. Y., ... Lin, Z. P. (2009). Fluorescence Silica Nanoprobe as a Biomarker for Rapid Detection of Plant Pathogens. *Advanced Materials Research*, 79–82, 513–516. . doi:10.4028/www.scientific.net/AMR.79-82.513

Yao, T., Fukunaga, S., & Nakai, T. (2006). Reliable Broadcast authentication in Wireless sensor networks. In *Proc. EUC 2006 Workshops* (pp. 271-280). Seoul, South Korea: Springer. 10.1007/11807964_28

Ye, J., Dobson, S., & McKeever, S. (2012). Situation identification techniques in pervasive computing: A review. *Pervasive and Mobile Computing*, 8(1), 36–66.

Yesilbudak, Colak, & Bayindir. (2018). What are the Current Status and Future Prospects in Solar Irradiance and Solar Power Forecasting? *International Journal of Renewable Energy Research, 8*, 636-646.

Ye, W., Heidemann, J., & Estrin, D. (2002). An energy-efficient MAC protocol for wireless sensor networks. In *Proceedings of Twenty-First Annual Joint Conference of the IEEE Computer and Communications Societies* (vol. 3). IEEE.

Ye, W., Silva, F., & Heidemann, J. (2006). Ultra-Low Duty Cycle MAC with Scheduled Channel Polling, In *Proceedings of the 4th International Conference on Embedded Networked Sensor Systems* (pp. 321-334). ACM. 10.1145/1182807.1182839

Yigitel, M. A., Ozlem, D. I., & Cem, E. (2011). QoS-aware MAC protocols for wireless sensor networks: A survey. *Computer Networks*, 55(8), 1982–2004. doi:10.1016/j.comnet.2011.02.007

Yin, Y., Zeng, Y., Chen, X., & Fan, Y. (2016). The internet of things in healthcare: An overview. *Journal of Industrial Information Integration, 1*, 3–13. doi:10.1016/j.jii.2016.03.004

Yu, J., Shi, P., Dong, W., & Yu, H. (2015). Observer and command-flter-based adaptive fuzzy output feedback control of uncertain nonlinear systems. *IEEE Transactions on Industrial Electronics*, 62(9), 5962–5970. doi:10.1109/TIE.2015.2418317

Yuvaraj. (2016). *Smart supply chain management using IOT*. IEEE, WISPNET. doi:10.1109/WiSPNET.2016.7566196

Zaman, M. T. U., Hossain, D., Arefin, M. T., Rahman, M. A., Islam, S. N., & Haque, A. K. M. F. (2012). Comparative Analysis of De-Noising on ECG Signal. *International Journal of Emerging Technology and Advanced Engineering*, 2(11), 479–486.

Zame, Brehm, Nitica, Richard, & Schweitzer III. (2017). Smart grid and energy storage: policy recommendations. *Renew. Sustain. Energy Rev., 82*(1), 1646–1654.

Zehhar, G., Ouazzani Touhami, A., Badoc, A., & Douira, A. (2006). Effet des *fusarium* des eaux de rizière sur la germination et la croissance des plantules de riz. *Bulletin Societé de Pharmacie de Bordeaux, 145*, 7-18.

Zhamanov, A., Sakhiyeva, Z., Suliyev, R., & Kaldykulova, Z. (2017). IoT smart campus review and implementation of IoT applications into education process of university. In *2017 13th International Conference on Electronics, Computer and Computation (ICECCO)*, (pp. 1–4). IEEE.

Zhang, J., Li, W., Han, N., & Kan, J. (2008). Forest fire detection system based on a ZigBee wireless sensor network. *Frontiers of Forestry in China, 3*, 369–374.

Zhang, J., Song, G., Wang, H., & Meng, T. (2011). Design of a wireless sensor network based monitoring system for home automation. *Proc. International conference on future computer sciences and applications*, 57-60. 10.1109/ICFCSA.2011.20

Zheng, J., & Abbas, J. (2009). *Wireless sensor networks: a networking perspective.* John Wiley & Sons. doi:10.1002/9780470443521

Zhou, G., Huang, C., Yan, T., He, T., Stankovic, J. A., & Abdelzaher, T. F. (2006). Mmsn: Multi-frequency media access control for wireless sensor networks. In *Proceedings of Infocom.* IEEE. doi:10.1109/INFOCOM.2006.250

Zhu, S., Setia, S., & Jajodia, S. (2004). LEAP: efficient security mechanishms in large scale distributed networks. In *10th ACM conference on Computer and communications security* (pp. 62-72). Washington, DC: ACM.

Zikria, Kim, Hahm, Afzal, & Aalsalem. (2019). Internet of Things (IoT) Operating Systems Management: Opportunities, Challenges,and Solution. *Sensors (Basel), 19*, 1793. doi:10.339019081793

Related References

To continue our tradition of advancing information science and technology research, we have compiled a list of recommended IGI Global readings. These references will provide additional information and guidance to further enrich your knowledge and assist you with your own research and future publications.

Aasi, P., Rusu, L., & Vieru, D. (2017). The Role of Culture in IT Governance Five Focus Areas: A Literature Review. *International Journal of IT/Business Alignment and Governance, 8*(2), 42-61. doi:10.4018/IJITBAG.2017070103

Abdrabo, A. A. (2018). Egypt's Knowledge-Based Development: Opportunities, Challenges, and Future Possibilities. In A. Alraouf (Ed.), *Knowledge-Based Urban Development in the Middle East* (pp. 80–101). Hershey, PA: IGI Global. doi:10.4018/978-1-5225-3734-2.ch005

Abu Doush, I., & Alhami, I. (2018). Evaluating the Accessibility of Computer Laboratories, Libraries, and Websites in Jordanian Universities and Colleges. *International Journal of Information Systems and Social Change, 9*(2), 44–60. doi:10.4018/IJISSC.2018040104

Adeboye, A. (2016). Perceived Use and Acceptance of Cloud Enterprise Resource Planning (ERP) Implementation in the Manufacturing Industries. *International Journal of Strategic Information Technology and Applications, 7*(3), 24–40. doi:10.4018/IJSITA.2016070102

Adegbore, A. M., Quadri, M. O., & Oyewo, O. R. (2018). A Theoretical Approach to the Adoption of Electronic Resource Management Systems (ERMS) in Nigerian University Libraries. In A. Tella & T. Kwanya (Eds.), *Handbook of Research on Managing Intellectual Property in Digital Libraries* (pp. 292–311). Hershey, PA: IGI Global. doi:10.4018/978-1-5225-3093-0.ch015

Adhikari, M., & Roy, D. (2016). Green Computing. In G. Deka, G. Siddesh, K. Srinivasa, & L. Patnaik (Eds.), *Emerging Research Surrounding Power Consumption and Performance Issues in Utility Computing* (pp. 84–108). Hershey, PA: IGI Global. doi:10.4018/978-1-4666-8853-7.ch005

Afolabi, O. A. (2018). Myths and Challenges of Building an Effective Digital Library in Developing Nations: An African Perspective. In A. Tella & T. Kwanya (Eds.), *Handbook of Research on Managing Intellectual Property in Digital Libraries* (pp. 51–79). Hershey, PA: IGI Global. doi:10.4018/978-1-5225-3093-0.ch004

Agarwal, R., Singh, A., & Sen, S. (2016). Role of Molecular Docking in Computer-Aided Drug Design and Development. In S. Dastmalchi, M. Hamzeh-Mivehroud, & B. Sokouti (Eds.), *Applied Case Studies and Solutions in Molecular Docking-Based Drug Design* (pp. 1–28). Hershey, PA: IGI Global. doi:10.4018/978-1-5225-0362-0.ch001

Ali, O., & Soar, J. (2016). Technology Innovation Adoption Theories. In L. Al-Hakim, X. Wu, A. Koronios, & Y. Shou (Eds.), *Handbook of Research on Driving Competitive Advantage through Sustainable, Lean, and Disruptive Innovation* (pp. 1–38). Hershey, PA: IGI Global. doi:10.4018/978-1-5225-0135-0.ch001

Alsharo, M. (2017). Attitudes Towards Cloud Computing Adoption in Emerging Economies. *International Journal of Cloud Applications and Computing*, 7(3), 44–58. doi:10.4018/IJCAC.2017070102

Amer, T. S., & Johnson, T. L. (2016). Information Technology Progress Indicators: Temporal Expectancy, User Preference, and the Perception of Process Duration. *International Journal of Technology and Human Interaction*, 12(4), 1–14. doi:10.4018/IJTHI.2016100101

Amer, T. S., & Johnson, T. L. (2017). Information Technology Progress Indicators: Research Employing Psychological Frameworks. In A. Mesquita (Ed.), *Research Paradigms and Contemporary Perspectives on Human-Technology Interaction* (pp. 168–186). Hershey, PA: IGI Global. doi:10.4018/978-1-5225-1868-6.ch008

Anchugam, C. V., & Thangadurai, K. (2016). Introduction to Network Security. In D. G., M. Singh, & M. Jayanthi (Eds.), Network Security Attacks and Countermeasures (pp. 1-48). Hershey, PA: IGI Global. doi:10.4018/978-1-4666-8761-5.ch001

Anchugam, C. V., & Thangadurai, K. (2016). Classification of Network Attacks and Countermeasures of Different Attacks. In D. G., M. Singh, & M. Jayanthi (Eds.), Network Security Attacks and Countermeasures (pp. 115-156). Hershey, PA: IGI Global. doi:10.4018/978-1-4666-8761-5.ch004

Anohah, E. (2016). Pedagogy and Design of Online Learning Environment in Computer Science Education for High Schools. *International Journal of Online Pedagogy and Course Design*, 6(3), 39–51. doi:10.4018/IJOPCD.2016070104

Anohah, E. (2017). Paradigm and Architecture of Computing Augmented Learning Management System for Computer Science Education. *International Journal of Online Pedagogy and Course Design*, 7(2), 60–70. doi:10.4018/IJOPCD.2017040105

Anohah, E., & Suhonen, J. (2017). Trends of Mobile Learning in Computing Education from 2006 to 2014: A Systematic Review of Research Publications. *International Journal of Mobile and Blended Learning*, 9(1), 16–33. doi:10.4018/IJMBL.2017010102

Assis-Hassid, S., Heart, T., Reychav, I., & Pliskin, J. S. (2016). Modelling Factors Affecting Patient-Doctor-Computer Communication in Primary Care. *International Journal of Reliable and Quality E-Healthcare*, 5(1), 1–17. doi:10.4018/IJRQEH.2016010101

Bailey, E. K. (2017). Applying Learning Theories to Computer Technology Supported Instruction. In M. Grassetti & S. Brookby (Eds.), *Advancing Next-Generation Teacher Education through Digital Tools and Applications* (pp. 61–81). Hershey, PA: IGI Global. doi:10.4018/978-1-5225-0965-3.ch004

Balasubramanian, K. (2016). Attacks on Online Banking and Commerce. In K. Balasubramanian, K. Mala, & M. Rajakani (Eds.), *Cryptographic Solutions for Secure Online Banking and Commerce* (pp. 1–19). Hershey, PA: IGI Global. doi:10.4018/978-1-5225-0273-9.ch001

Baldwin, S., Opoku-Agyemang, K., & Roy, D. (2016). Games People Play: A Trilateral Collaboration Researching Computer Gaming across Cultures. In K. Valentine & L. Jensen (Eds.), *Examining the Evolution of Gaming and Its Impact on Social, Cultural, and Political Perspectives* (pp. 364–376). Hershey, PA: IGI Global. doi:10.4018/978-1-5225-0261-6.ch017

Banerjee, S., Sing, T. Y., Chowdhury, A. R., & Anwar, H. (2018). Let's Go Green: Towards a Taxonomy of Green Computing Enablers for Business Sustainability. In M. Khosrow-Pour (Ed.), *Green Computing Strategies for Competitive Advantage and Business Sustainability* (pp. 89–109). Hershey, PA: IGI Global. doi:10.4018/978-1-5225-5017-4.ch005

Basham, R. (2018). Information Science and Technology in Crisis Response and Management. In M. Khosrow-Pour, D.B.A. (Ed.), Encyclopedia of Information Science and Technology, Fourth Edition (pp. 1407-1418). Hershey, PA: IGI Global. doi:10.4018/978-1-5225-2255-3.ch121

Batyashe, T., & Iyamu, T. (2018). Architectural Framework for the Implementation of Information Technology Governance in Organisations. In M. Khosrow-Pour, D.B.A. (Ed.), Encyclopedia of Information Science and Technology, Fourth Edition (pp. 810-819). Hershey, PA: IGI Global. doi:10.4018/978-1-5225-2255-3.ch070

Bekleyen, N., & Çelik, S. (2017). Attitudes of Adult EFL Learners towards Preparing for a Language Test via CALL. In D. Tafazoli & M. Romero (Eds.), *Multiculturalism and Technology-Enhanced Language Learning* (pp. 214–229). Hershey, PA: IGI Global. doi:10.4018/978-1-5225-1882-2.ch013

Bennett, A., Eglash, R., Lachney, M., & Babbitt, W. (2016). Design Agency: Diversifying Computer Science at the Intersections of Creativity and Culture. In M. Raisinghani (Ed.), *Revolutionizing Education through Web-Based Instruction* (pp. 35–56). Hershey, PA: IGI Global. doi:10.4018/978-1-4666-9932-8.ch003

Bergeron, F., Croteau, A., Uwizeyemungu, S., & Raymond, L. (2017). A Framework for Research on Information Technology Governance in SMEs. In S. De Haes & W. Van Grembergen (Eds.), *Strategic IT Governance and Alignment in Business Settings* (pp. 53–81). Hershey, PA: IGI Global. doi:10.4018/978-1-5225-0861-8.ch003

Bhatt, G. D., Wang, Z., & Rodger, J. A. (2017). Information Systems Capabilities and Their Effects on Competitive Advantages: A Study of Chinese Companies. *Information Resources Management Journal*, *30*(3), 41–57. doi:10.4018/IRMJ.2017070103

Bogdanoski, M., Stoilkovski, M., & Risteski, A. (2016). Novel First Responder Digital Forensics Tool as a Support to Law Enforcement. In M. Hadji-Janev & M. Bogdanoski (Eds.), *Handbook of Research on Civil Society and National Security in the Era of Cyber Warfare* (pp. 352–376). Hershey, PA: IGI Global. doi:10.4018/978-1-4666-8793-6.ch016

Boontarig, W., Papasratorn, B., & Chutimaskul, W. (2016). The Unified Model for Acceptance and Use of Health Information on Online Social Networks: Evidence from Thailand. *International Journal of E-Health and Medical Communications*, *7*(1), 31–47. doi:10.4018/IJEHMC.2016010102

Brown, S., & Yuan, X. (2016). Techniques for Retaining Computer Science Students at Historical Black Colleges and Universities. In C. Prince & R. Ford (Eds.), *Setting a New Agenda for Student Engagement and Retention in Historically Black Colleges and Universities* (pp. 251–268). Hershey, PA: IGI Global. doi:10.4018/978-1-5225-0308-8.ch014

Burcoff, A., & Shamir, L. (2017). Computer Analysis of Pablo Picasso's Artistic Style. *International Journal of Art, Culture and Design Technologies*, *6*(1), 1–18. doi:10.4018/IJACDT.2017010101

Related References

Byker, E. J. (2017). I Play I Learn: Introducing Technological Play Theory. In C. Martin & D. Polly (Eds.), *Handbook of Research on Teacher Education and Professional Development* (pp. 297–306). Hershey, PA: IGI Global. doi:10.4018/978-1-5225-1067-3.ch016

Calongne, C. M., Stricker, A. G., Truman, B., & Arenas, F. J. (2017). Cognitive Apprenticeship and Computer Science Education in Cyberspace: Reimagining the Past. In A. Stricker, C. Calongne, B. Truman, & F. Arenas (Eds.), *Integrating an Awareness of Selfhood and Society into Virtual Learning* (pp. 180–197). Hershey, PA: IGI Global. doi:10.4018/978-1-5225-2182-2.ch013

Carlton, E. L., Holsinger, J. W. Jr, & Anunobi, N. (2016). Physician Engagement with Health Information Technology: Implications for Practice and Professionalism. *International Journal of Computers in Clinical Practice, 1*(2), 51–73. doi:10.4018/IJCCP.2016070103

Carneiro, A. D. (2017). Defending Information Networks in Cyberspace: Some Notes on Security Needs. In M. Dawson, D. Kisku, P. Gupta, J. Sing, & W. Li (Eds.), Developing Next-Generation Countermeasures for Homeland Security Threat Prevention (pp. 354-375). Hershey, PA: IGI Global. doi:10.4018/978-1-5225-0703-1.ch016

Cavalcanti, J. C. (2016). The New "ABC" of ICTs (Analytics + Big Data + Cloud Computing): A Complex Trade-Off between IT and CT Costs. In J. Martins & A. Molnar (Eds.), *Handbook of Research on Innovations in Information Retrieval, Analysis, and Management* (pp. 152–186). Hershey, PA: IGI Global. doi:10.4018/978-1-4666-8833-9.ch006

Chase, J. P., & Yan, Z. (2017). Affect in Statistics Cognition. In *Assessing and Measuring Statistics Cognition in Higher Education Online Environments: Emerging Research and Opportunities* (pp. 144–187). Hershey, PA: IGI Global. doi:10.4018/978-1-5225-2420-5.ch005

Chen, C. (2016). Effective Learning Strategies for the 21st Century: Implications for the E-Learning. In M. Anderson & C. Gavan (Eds.), *Developing Effective Educational Experiences through Learning Analytics* (pp. 143–169). Hershey, PA: IGI Global. doi:10.4018/978-1-4666-9983-0.ch006

Chen, E. T. (2016). Examining the Influence of Information Technology on Modern Health Care. In P. Manolitzas, E. Grigoroudis, N. Matsatsinis, & D. Yannacopoulos (Eds.), *Effective Methods for Modern Healthcare Service Quality and Evaluation* (pp. 110–136). Hershey, PA: IGI Global. doi:10.4018/978-1-4666-9961-8.ch006

Cimermanova, I. (2017). Computer-Assisted Learning in Slovakia. In D. Tafazoli & M. Romero (Eds.), *Multiculturalism and Technology-Enhanced Language Learning* (pp. 252–270). Hershey, PA: IGI Global. doi:10.4018/978-1-5225-1882-2.ch015

Cipolla-Ficarra, F. V., & Cipolla-Ficarra, M. (2018). Computer Animation for Ingenious Revival. In F. Cipolla-Ficarra, M. Ficarra, M. Cipolla-Ficarra, A. Quiroga, J. Alma, & J. Carré (Eds.), *Technology-Enhanced Human Interaction in Modern Society* (pp. 159–181). Hershey, PA: IGI Global. doi:10.4018/978-1-5225-3437-2.ch008

Cockrell, S., Damron, T. S., Melton, A. M., & Smith, A. D. (2018). Offshoring IT. In M. Khosrow-Pour, D.B.A. (Ed.), Encyclopedia of Information Science and Technology, Fourth Edition (pp. 5476-5489). Hershey, PA: IGI Global. doi:10.4018/978-1-5225-2255-3.ch476

Coffey, J. W. (2018). Logic and Proof in Computer Science: Categories and Limits of Proof Techniques. In J. Horne (Ed.), *Philosophical Perceptions on Logic and Order* (pp. 218–240). Hershey, PA: IGI Global. doi:10.4018/978-1-5225-2443-4.ch007

Dale, M. (2017). Re-Thinking the Challenges of Enterprise Architecture Implementation. In M. Tavana (Ed.), *Enterprise Information Systems and the Digitalization of Business Functions* (pp. 205–221). Hershey, PA: IGI Global. doi:10.4018/978-1-5225-2382-6.ch009

Das, A., Dasgupta, R., & Bagchi, A. (2016). Overview of Cellular Computing-Basic Principles and Applications. In J. Mandal, S. Mukhopadhyay, & T. Pal (Eds.), *Handbook of Research on Natural Computing for Optimization Problems* (pp. 637–662). Hershey, PA: IGI Global. doi:10.4018/978-1-5225-0058-2.ch026

De Maere, K., De Haes, S., & von Kutzschenbach, M. (2017). CIO Perspectives on Organizational Learning within the Context of IT Governance. *International Journal of IT/Business Alignment and Governance, 8*(1), 32-47. doi:10.4018/IJITBAG.2017010103

Demir, K., Çaka, C., Yaman, N. D., İslamoğlu, H., & Kuzu, A. (2018). Examining the Current Definitions of Computational Thinking. In H. Ozcinar, G. Wong, & H. Ozturk (Eds.), *Teaching Computational Thinking in Primary Education* (pp. 36–64). Hershey, PA: IGI Global. doi:10.4018/978-1-5225-3200-2.ch003

Deng, X., Hung, Y., & Lin, C. D. (2017). Design and Analysis of Computer Experiments. In S. Saha, A. Mandal, A. Narasimhamurthy, S. V, & S. Sangam (Eds.), Handbook of Research on Applied Cybernetics and Systems Science (pp. 264-279). Hershey, PA: IGI Global. doi:10.4018/978-1-5225-2498-4.ch013

Related References

Denner, J., Martinez, J., & Thiry, H. (2017). Strategies for Engaging Hispanic/ Latino Youth in the US in Computer Science. In Y. Rankin & J. Thomas (Eds.), *Moving Students of Color from Consumers to Producers of Technology* (pp. 24–48). Hershey, PA: IGI Global. doi:10.4018/978-1-5225-2005-4.ch002

Devi, A. (2017). Cyber Crime and Cyber Security: A Quick Glance. In R. Kumar, P. Pattnaik, & P. Pandey (Eds.), *Detecting and Mitigating Robotic Cyber Security Risks* (pp. 160–171). Hershey, PA: IGI Global. doi:10.4018/978-1-5225-2154-9.ch011

Dores, A. R., Barbosa, F., Guerreiro, S., Almeida, I., & Carvalho, I. P. (2016). Computer-Based Neuropsychological Rehabilitation: Virtual Reality and Serious Games. In M. Cruz-Cunha, I. Miranda, R. Martinho, & R. Rijo (Eds.), *Encyclopedia of E-Health and Telemedicine* (pp. 473–485). Hershey, PA: IGI Global. doi:10.4018/978-1-4666-9978-6.ch037

Doshi, N., & Schaefer, G. (2016). Computer-Aided Analysis of Nailfold Capillaroscopy Images. In D. Fotiadis (Ed.), *Handbook of Research on Trends in the Diagnosis and Treatment of Chronic Conditions* (pp. 146–158). Hershey, PA: IGI Global. doi:10.4018/978-1-4666-8828-5.ch007

Doyle, D. J., & Fahy, P. J. (2018). Interactivity in Distance Education and Computer-Aided Learning, With Medical Education Examples. In M. Khosrow-Pour, D.B.A. (Ed.), Encyclopedia of Information Science and Technology, Fourth Edition (pp. 5829-5840). Hershey, PA: IGI Global. doi:10.4018/978-1-5225-2255-3.ch507

Elias, N. I., & Walker, T. W. (2017). Factors that Contribute to Continued Use of E-Training among Healthcare Professionals. In F. Topor (Ed.), *Handbook of Research on Individualism and Identity in the Globalized Digital Age* (pp. 403–429). Hershey, PA: IGI Global. doi:10.4018/978-1-5225-0522-8.ch018

Eloy, S., Dias, M. S., Lopes, P. F., & Vilar, E. (2016). Digital Technologies in Architecture and Engineering: Exploring an Engaged Interaction within Curricula. In D. Fonseca & E. Redondo (Eds.), *Handbook of Research on Applied E-Learning in Engineering and Architecture Education* (pp. 368–402). Hershey, PA: IGI Global. doi:10.4018/978-1-4666-8803-2.ch017

Estrela, V. V., Magalhães, H. A., & Saotome, O. (2016). Total Variation Applications in Computer Vision. In N. Kamila (Ed.), *Handbook of Research on Emerging Perspectives in Intelligent Pattern Recognition, Analysis, and Image Processing* (pp. 41–64). Hershey, PA: IGI Global. doi:10.4018/978-1-4666-8654-0.ch002

Filipovic, N., Radovic, M., Nikolic, D. D., Saveljic, I., Milosevic, Z., Exarchos, T. P., ... Parodi, O. (2016). Computer Predictive Model for Plaque Formation and Progression in the Artery. In D. Fotiadis (Ed.), *Handbook of Research on Trends in the Diagnosis and Treatment of Chronic Conditions* (pp. 279–300). Hershey, PA: IGI Global. doi:10.4018/978-1-4666-8828-5.ch013

Fisher, R. L. (2018). Computer-Assisted Indian Matrimonial Services. In M. Khosrow-Pour, D.B.A. (Ed.), Encyclopedia of Information Science and Technology, Fourth Edition (pp. 4136-4145). Hershey, PA: IGI Global. doi:10.4018/978-1-5225-2255-3.ch358

Fleenor, H. G., & Hodhod, R. (2016). Assessment of Learning and Technology: Computer Science Education. In V. Wang (Ed.), *Handbook of Research on Learning Outcomes and Opportunities in the Digital Age* (pp. 51–78). Hershey, PA: IGI Global. doi:10.4018/978-1-4666-9577-1.ch003

García-Valcárcel, A., & Mena, J. (2016). Information Technology as a Way To Support Collaborative Learning: What In-Service Teachers Think, Know and Do. *Journal of Information Technology Research*, *9*(1), 1–17. doi:10.4018/JITR.2016010101

Gardner-McCune, C., & Jimenez, Y. (2017). Historical App Developers: Integrating CS into K-12 through Cross-Disciplinary Projects. In Y. Rankin & J. Thomas (Eds.), *Moving Students of Color from Consumers to Producers of Technology* (pp. 85–112). Hershey, PA: IGI Global. doi:10.4018/978-1-5225-2005-4.ch005

Garvey, G. P. (2016). Exploring Perception, Cognition, and Neural Pathways of Stereo Vision and the Split–Brain Human Computer Interface. In A. Ursyn (Ed.), *Knowledge Visualization and Visual Literacy in Science Education* (pp. 28–76). Hershey, PA: IGI Global. doi:10.4018/978-1-5225-0480-1.ch002

Ghafele, R., & Gibert, B. (2018). Open Growth: The Economic Impact of Open Source Software in the USA. In M. Khosrow-Pour (Ed.), *Optimizing Contemporary Application and Processes in Open Source Software* (pp. 164–197). Hershey, PA: IGI Global. doi:10.4018/978-1-5225-5314-4.ch007

Ghobakhloo, M., & Azar, A. (2018). Information Technology Resources, the Organizational Capability of Lean-Agile Manufacturing, and Business Performance. *Information Resources Management Journal*, *31*(2), 47–74. doi:10.4018/IRMJ.2018040103

Gianni, M., & Gotzamani, K. (2016). Integrated Management Systems and Information Management Systems: Common Threads. In P. Papajorgji, F. Pinet, A. Guimarães, & J. Papathanasiou (Eds.), *Automated Enterprise Systems for Maximizing Business Performance* (pp. 195–214). Hershey, PA: IGI Global. doi:10.4018/978-1-4666-8841-4.ch011

Gikandi, J. W. (2017). Computer-Supported Collaborative Learning and Assessment: A Strategy for Developing Online Learning Communities in Continuing Education. In J. Keengwe & G. Onchwari (Eds.), *Handbook of Research on Learner-Centered Pedagogy in Teacher Education and Professional Development* (pp. 309–333). Hershey, PA: IGI Global. doi:10.4018/978-1-5225-0892-2.ch017

Gokhale, A. A., & Machina, K. F. (2017). Development of a Scale to Measure Attitudes toward Information Technology. In L. Tomei (Ed.), *Exploring the New Era of Technology-Infused Education* (pp. 49–64). Hershey, PA: IGI Global. doi:10.4018/978-1-5225-1709-2.ch004

Grace, A., O'Donoghue, J., Mahony, C., Heffernan, T., Molony, D., & Carroll, T. (2016). Computerized Decision Support Systems for Multimorbidity Care: An Urgent Call for Research and Development. In M. Cruz-Cunha, I. Miranda, R. Martinho, & R. Rijo (Eds.), *Encyclopedia of E-Health and Telemedicine* (pp. 486–494). Hershey, PA: IGI Global. doi:10.4018/978-1-4666-9978-6.ch038

Gupta, A., & Singh, O. (2016). Computer Aided Modeling and Finite Element Analysis of Human Elbow. *International Journal of Biomedical and Clinical Engineering*, *5*(1), 31–38. doi:10.4018/IJBCE.2016010104

H., S. K. (2016). Classification of Cybercrimes and Punishments under the Information Technology Act, 2000. In S. Geetha, & A. Phamila (Eds.), *Combating Security Breaches and Criminal Activity in the Digital Sphere* (pp. 57-66). Hershey, PA: IGI Global. doi:10.4018/978-1-5225-0193-0.ch004

Hafeez-Baig, A., Gururajan, R., & Wickramasinghe, N. (2017). Readiness as a Novel Construct of Readiness Acceptance Model (RAM) for the Wireless Handheld Technology. In N. Wickramasinghe (Ed.), *Handbook of Research on Healthcare Administration and Management* (pp. 578–595). Hershey, PA: IGI Global. doi:10.4018/978-1-5225-0920-2.ch035

Hanafizadeh, P., Ghandchi, S., & Asgarimehr, M. (2017). Impact of Information Technology on Lifestyle: A Literature Review and Classification. *International Journal of Virtual Communities and Social Networking*, *9*(2), 1–23. doi:10.4018/IJVCSN.2017040101

Harlow, D. B., Dwyer, H., Hansen, A. K., Hill, C., Iveland, A., Leak, A. E., & Franklin, D. M. (2016). Computer Programming in Elementary and Middle School: Connections across Content. In M. Urban & D. Falvo (Eds.), *Improving K-12 STEM Education Outcomes through Technological Integration* (pp. 337–361). Hershey, PA: IGI Global. doi:10.4018/978-1-4666-9616-7.ch015

Haseski, H. İ., Ilic, U., & Tuğtekin, U. (2018). Computational Thinking in Educational Digital Games: An Assessment Tool Proposal. In H. Ozcinar, G. Wong, & H. Ozturk (Eds.), *Teaching Computational Thinking in Primary Education* (pp. 256–287). Hershey, PA: IGI Global. doi:10.4018/978-1-5225-3200-2.ch013

Hee, W. J., Jalleh, G., Lai, H., & Lin, C. (2017). E-Commerce and IT Projects: Evaluation and Management Issues in Australian and Taiwanese Hospitals. *International Journal of Public Health Management and Ethics*, 2(1), 69–90. doi:10.4018/IJPHME.2017010104

Hernandez, A. A. (2017). Green Information Technology Usage: Awareness and Practices of Philippine IT Professionals. *International Journal of Enterprise Information Systems*, 13(4), 90–103. doi:10.4018/IJEIS.2017100106

Hernandez, A. A., & Ona, S. E. (2016). Green IT Adoption: Lessons from the Philippines Business Process Outsourcing Industry. *International Journal of Social Ecology and Sustainable Development*, 7(1), 1–34. doi:10.4018/IJSESD.2016010101

Hernandez, M. A., Marin, E. C., Garcia-Rodriguez, J., Azorin-Lopez, J., & Cazorla, M. (2017). Automatic Learning Improves Human-Robot Interaction in Productive Environments: A Review. *International Journal of Computer Vision and Image Processing*, 7(3), 65–75. doi:10.4018/IJCVIP.2017070106

Horne-Popp, L. M., Tessone, E. B., & Welker, J. (2018). If You Build It, They Will Come: Creating a Library Statistics Dashboard for Decision-Making. In L. Costello & M. Powers (Eds.), *Developing In-House Digital Tools in Library Spaces* (pp. 177–203). Hershey, PA: IGI Global. doi:10.4018/978-1-5225-2676-6.ch009

Hossan, C. G., & Ryan, J. C. (2016). Factors Affecting e-Government Technology Adoption Behaviour in a Voluntary Environment. *International Journal of Electronic Government Research*, 12(1), 24–49. doi:10.4018/IJEGR.2016010102

Hu, H., Hu, P. J., & Al-Gahtani, S. S. (2017). User Acceptance of Computer Technology at Work in Arabian Culture: A Model Comparison Approach. In M. Khosrow-Pour (Ed.), *Handbook of Research on Technology Adoption, Social Policy, and Global Integration* (pp. 205–228). Hershey, PA: IGI Global. doi:10.4018/978-1-5225-2668-1.ch011

Huie, C. P. (2016). Perceptions of Business Intelligence Professionals about Factors Related to Business Intelligence input in Decision Making. *International Journal of Business Analytics*, 3(3), 1–24. doi:10.4018/IJBAN.2016070101

Hung, S., Huang, W., Yen, D. C., Chang, S., & Lu, C. (2016). Effect of Information Service Competence and Contextual Factors on the Effectiveness of Strategic Information Systems Planning in Hospitals. *Journal of Global Information Management*, 24(1), 14–36. doi:10.4018/JGIM.2016010102

Ifinedo, P. (2017). Using an Extended Theory of Planned Behavior to Study Nurses' Adoption of Healthcare Information Systems in Nova Scotia. *International Journal of Technology Diffusion*, 8(1), 1–17. doi:10.4018/IJTD.2017010101

Ilie, V., & Sneha, S. (2018). A Three Country Study for Understanding Physicians' Engagement With Electronic Information Resources Pre and Post System Implementation. *Journal of Global Information Management*, 26(2), 48–73. doi:10.4018/JGIM.2018040103

Inoue-Smith, Y. (2017). Perceived Ease in Using Technology Predicts Teacher Candidates' Preferences for Online Resources. *International Journal of Online Pedagogy and Course Design*, 7(3), 17–28. doi:10.4018/IJOPCD.2017070102

Islam, A. A. (2016). Development and Validation of the Technology Adoption and Gratification (TAG) Model in Higher Education: A Cross-Cultural Study Between Malaysia and China. *International Journal of Technology and Human Interaction*, 12(3), 78–105. doi:10.4018/IJTHI.2016070106

Islam, A. Y. (2017). Technology Satisfaction in an Academic Context: Moderating Effect of Gender. In A. Mesquita (Ed.), *Research Paradigms and Contemporary Perspectives on Human-Technology Interaction* (pp. 187–211). Hershey, PA: IGI Global. doi:10.4018/978-1-5225-1868-6.ch009

Jamil, G. L., & Jamil, C. C. (2017). Information and Knowledge Management Perspective Contributions for Fashion Studies: Observing Logistics and Supply Chain Management Processes. In G. Jamil, A. Soares, & C. Pessoa (Eds.), *Handbook of Research on Information Management for Effective Logistics and Supply Chains* (pp. 199–221). Hershey, PA: IGI Global. doi:10.4018/978-1-5225-0973-8.ch011

Jamil, G. L., Jamil, L. C., Vieira, A. A., & Xavier, A. J. (2016). Challenges in Modelling Healthcare Services: A Study Case of Information Architecture Perspectives. In G. Jamil, J. Poças Rascão, F. Ribeiro, & A. Malheiro da Silva (Eds.), *Handbook of Research on Information Architecture and Management in Modern Organizations* (pp. 1–23). Hershey, PA: IGI Global. doi:10.4018/978-1-4666-8637-3.ch001

Janakova, M. (2018). Big Data and Simulations for the Solution of Controversies in Small Businesses. In M. Khosrow-Pour, D.B.A. (Ed.), Encyclopedia of Information Science and Technology, Fourth Edition (pp. 6907-6915). Hershey, PA: IGI Global. doi:10.4018/978-1-5225-2255-3.ch598

Jha, D. G. (2016). Preparing for Information Technology Driven Changes. In S. Tiwari & L. Nafees (Eds.), *Innovative Management Education Pedagogies for Preparing Next-Generation Leaders* (pp. 258–274). Hershey, PA: IGI Global. doi:10.4018/978-1-4666-9691-4.ch015

Jhawar, A., & Garg, S. K. (2018). Logistics Improvement by Investment in Information Technology Using System Dynamics. In A. Azar & S. Vaidyanathan (Eds.), *Advances in System Dynamics and Control* (pp. 528–567). Hershey, PA: IGI Global. doi:10.4018/978-1-5225-4077-9.ch017

Kalelioğlu, F., Gülbahar, Y., & Doğan, D. (2018). Teaching How to Think Like a Programmer: Emerging Insights. In H. Ozcinar, G. Wong, & H. Ozturk (Eds.), *Teaching Computational Thinking in Primary Education* (pp. 18–35). Hershey, PA: IGI Global. doi:10.4018/978-1-5225-3200-2.ch002

Kamberi, S. (2017). A Girls-Only Online Virtual World Environment and its Implications for Game-Based Learning. In A. Stricker, C. Calongne, B. Truman, & F. Arenas (Eds.), *Integrating an Awareness of Selfhood and Society into Virtual Learning* (pp. 74–95). Hershey, PA: IGI Global. doi:10.4018/978-1-5225-2182-2.ch006

Kamel, S., & Rizk, N. (2017). ICT Strategy Development: From Design to Implementation – Case of Egypt. In C. Howard & K. Hargiss (Eds.), *Strategic Information Systems and Technologies in Modern Organizations* (pp. 239–257). Hershey, PA: IGI Global. doi:10.4018/978-1-5225-1680-4.ch010

Kamel, S. H. (2018). The Potential Role of the Software Industry in Supporting Economic Development. In M. Khosrow-Pour, D.B.A. (Ed.), Encyclopedia of Information Science and Technology, Fourth Edition (pp. 7259-7269). Hershey, PA: IGI Global. doi:10.4018/978-1-5225-2255-3.ch631

Karon, R. (2016). Utilisation of Health Information Systems for Service Delivery in the Namibian Environment. In T. Iyamu & A. Tatnall (Eds.), *Maximizing Healthcare Delivery and Management through Technology Integration* (pp. 169–183). Hershey, PA: IGI Global. doi:10.4018/978-1-4666-9446-0.ch011

Related References

Kawata, S. (2018). Computer-Assisted Parallel Program Generation. In M. Khosrow-Pour, D.B.A. (Ed.), Encyclopedia of Information Science and Technology, Fourth Edition (pp. 4583-4593). Hershey, PA: IGI Global. doi:10.4018/978-1-5225-2255-3.ch398

Khanam, S., Siddiqui, J., & Talib, F. (2016). A DEMATEL Approach for Prioritizing the TQM Enablers and IT Resources in the Indian ICT Industry. *International Journal of Applied Management Sciences and Engineering, 3*(1), 11–29. doi:10.4018/IJAMSE.2016010102

Khari, M., Shrivastava, G., Gupta, S., & Gupta, R. (2017). Role of Cyber Security in Today's Scenario. In R. Kumar, P. Pattnaik, & P. Pandey (Eds.), *Detecting and Mitigating Robotic Cyber Security Risks* (pp. 177–191). Hershey, PA: IGI Global. doi:10.4018/978-1-5225-2154-9.ch013

Khouja, M., Rodriguez, I. B., Ben Halima, Y., & Moalla, S. (2018). IT Governance in Higher Education Institutions: A Systematic Literature Review. *International Journal of Human Capital and Information Technology Professionals, 9*(2), 52–67. doi:10.4018/IJHCITP.2018040104

Kim, S., Chang, M., Choi, N., Park, J., & Kim, H. (2016). The Direct and Indirect Effects of Computer Uses on Student Success in Math. *International Journal of Cyber Behavior, Psychology and Learning, 6*(3), 48–64. doi:10.4018/IJCBPL.2016070104

Kiourt, C., Pavlidis, G., Koutsoudis, A., & Kalles, D. (2017). Realistic Simulation of Cultural Heritage. *International Journal of Computational Methods in Heritage Science, 1*(1), 10–40. doi:10.4018/IJCMHS.2017010102

Korikov, A., & Krivtsov, O. (2016). System of People-Computer: On the Way of Creation of Human-Oriented Interface. In V. Mkrttchian, A. Bershadsky, A. Bozhday, M. Kataev, & S. Kataev (Eds.), *Handbook of Research on Estimation and Control Techniques in E-Learning Systems* (pp. 458–470). Hershey, PA: IGI Global. doi:10.4018/978-1-4666-9489-7.ch032

Köse, U. (2017). An Augmented-Reality-Based Intelligent Mobile Application for Open Computer Education. In G. Kurubacak & H. Altinpulluk (Eds.), *Mobile Technologies and Augmented Reality in Open Education* (pp. 154–174). Hershey, PA: IGI Global. doi:10.4018/978-1-5225-2110-5.ch008

Lahmiri, S. (2018). Information Technology Outsourcing Risk Factors and Provider Selection. In M. Gupta, R. Sharman, J. Walp, & P. Mulgund (Eds.), *Information Technology Risk Management and Compliance in Modern Organizations* (pp. 214–228). Hershey, PA: IGI Global. doi:10.4018/978-1-5225-2604-9.ch008

Landriscina, F. (2017). Computer-Supported Imagination: The Interplay Between Computer and Mental Simulation in Understanding Scientific Concepts. In I. Levin & D. Tsybulsky (Eds.), *Digital Tools and Solutions for Inquiry-Based STEM Learning* (pp. 33–60). Hershey, PA: IGI Global. doi:10.4018/978-1-5225-2525-7.ch002

Lau, S. K., Winley, G. K., Leung, N. K., Tsang, N., & Lau, S. Y. (2016). An Exploratory Study of Expectation in IT Skills in a Developing Nation: Vietnam. *Journal of Global Information Management, 24*(1), 1–13. doi:10.4018/JGIM.2016010101

Lavranos, C., Kostagiolas, P., & Papadatos, J. (2016). Information Retrieval Technologies and the "Realities" of Music Information Seeking. In I. Deliyannis, P. Kostagiolas, & C. Banou (Eds.), *Experimental Multimedia Systems for Interactivity and Strategic Innovation* (pp. 102–121). Hershey, PA: IGI Global. doi:10.4018/978-1-4666-8659-5.ch005

Lee, W. W. (2018). Ethical Computing Continues From Problem to Solution. In M. Khosrow-Pour, D.B.A. (Ed.), Encyclopedia of Information Science and Technology, Fourth Edition (pp. 4884-4897). Hershey, PA: IGI Global. doi:10.4018/978-1-5225-2255-3.ch423

Lehto, M. (2016). Cyber Security Education and Research in the Finland's Universities and Universities of Applied Sciences. *International Journal of Cyber Warfare & Terrorism, 6*(2), 15–31. doi:10.4018/IJCWT.2016040102

Lin, C., Jalleh, G., & Huang, Y. (2016). Evaluating and Managing Electronic Commerce and Outsourcing Projects in Hospitals. In A. Dwivedi (Ed.), *Reshaping Medical Practice and Care with Health Information Systems* (pp. 132–172). Hershey, PA: IGI Global. doi:10.4018/978-1-4666-9870-3.ch005

Lin, S., Chen, S., & Chuang, S. (2017). Perceived Innovation and Quick Response Codes in an Online-to-Offline E-Commerce Service Model. *International Journal of E-Adoption, 9*(2), 1–16. doi:10.4018/IJEA.2017070101

Liu, M., Wang, Y., Xu, W., & Liu, L. (2017). Automated Scoring of Chinese Engineering Students' English Essays. *International Journal of Distance Education Technologies, 15*(1), 52–68. doi:10.4018/IJDET.2017010104

Luciano, E. M., Wiedenhöft, G. C., Macadar, M. A., & Pinheiro dos Santos, F. (2016). Information Technology Governance Adoption: Understanding its Expectations Through the Lens of Organizational Citizenship. *International Journal of IT/Business Alignment and Governance, 7*(2), 22-32. doi:10.4018/IJITBAG.2016070102

Mabe, L. K., & Oladele, O. I. (2017). Application of Information Communication Technologies for Agricultural Development through Extension Services: A Review. In T. Tossy (Ed.), *Information Technology Integration for Socio-Economic Development* (pp. 52–101). Hershey, PA: IGI Global. doi:10.4018/978-1-5225-0539-6.ch003

Manogaran, G., Thota, C., & Lopez, D. (2018). Human-Computer Interaction With Big Data Analytics. In D. Lopez & M. Durai (Eds.), *HCI Challenges and Privacy Preservation in Big Data Security* (pp. 1–22). Hershey, PA: IGI Global. doi:10.4018/978-1-5225-2863-0.ch001

Margolis, J., Goode, J., & Flapan, J. (2017). A Critical Crossroads for Computer Science for All: "Identifying Talent" or "Building Talent," and What Difference Does It Make? In Y. Rankin & J. Thomas (Eds.), *Moving Students of Color from Consumers to Producers of Technology* (pp. 1–23). Hershey, PA: IGI Global. doi:10.4018/978-1-5225-2005-4.ch001

Mbale, J. (2018). Computer Centres Resource Cloud Elasticity-Scalability (CRECES): Copperbelt University Case Study. In S. Aljawarneh & M. Malhotra (Eds.), *Critical Research on Scalability and Security Issues in Virtual Cloud Environments* (pp. 48–70). Hershey, PA: IGI Global. doi:10.4018/978-1-5225-3029-9.ch003

McKee, J. (2018). The Right Information: The Key to Effective Business Planning. In *Business Architectures for Risk Assessment and Strategic Planning: Emerging Research and Opportunities* (pp. 38–52). Hershey, PA: IGI Global. doi:10.4018/978-1-5225-3392-4.ch003

Mensah, I. K., & Mi, J. (2018). Determinants of Intention to Use Local E-Government Services in Ghana: The Perspective of Local Government Workers. *International Journal of Technology Diffusion*, 9(2), 41–60. doi:10.4018/IJTD.2018040103

Mohamed, J. H. (2018). Scientograph-Based Visualization of Computer Forensics Research Literature. In J. Jeyasekar & P. Saravanan (Eds.), *Innovations in Measuring and Evaluating Scientific Information* (pp. 148–162). Hershey, PA: IGI Global. doi:10.4018/978-1-5225-3457-0.ch010

Moore, R. L., & Johnson, N. (2017). Earning a Seat at the Table: How IT Departments Can Partner in Organizational Change and Innovation. *International Journal of Knowledge-Based Organizations*, 7(2), 1–12. doi:10.4018/IJKBO.2017040101

Mtebe, J. S., & Kissaka, M. M. (2016). Enhancing the Quality of Computer Science Education with MOOCs in Sub-Saharan Africa. In J. Keengwe & G. Onchwari (Eds.), *Handbook of Research on Active Learning and the Flipped Classroom Model in the Digital Age* (pp. 366–377). Hershey, PA: IGI Global. doi:10.4018/978-1-4666-9680-8.ch019

Mukul, M. K., & Bhattaharyya, S. (2017). Brain-Machine Interface: Human-Computer Interaction. In E. Noughabi, B. Raahemi, A. Albadvi, & B. Far (Eds.), *Handbook of Research on Data Science for Effective Healthcare Practice and Administration* (pp. 417–443). Hershey, PA: IGI Global. doi:10.4018/978-1-5225-2515-8.ch018

Na, L. (2017). Library and Information Science Education and Graduate Programs in Academic Libraries. In L. Ruan, Q. Zhu, & Y. Ye (Eds.), *Academic Library Development and Administration in China* (pp. 218–229). Hershey, PA: IGI Global. doi:10.4018/978-1-5225-0550-1.ch013

Nabavi, A., Taghavi-Fard, M. T., Hanafizadeh, P., & Taghva, M. R. (2016). Information Technology Continuance Intention: A Systematic Literature Review. *International Journal of E-Business Research*, *12*(1), 58–95. doi:10.4018/IJEBR.2016010104

Nath, R., & Murthy, V. N. (2018). What Accounts for the Differences in Internet Diffusion Rates Around the World? In M. Khosrow-Pour, D.B.A. (Ed.), Encyclopedia of Information Science and Technology, Fourth Edition (pp. 8095-8104). Hershey, PA: IGI Global. doi:10.4018/978-1-5225-2255-3.ch705

Nedelko, Z., & Potocan, V. (2018). The Role of Emerging Information Technologies for Supporting Supply Chain Management. In M. Khosrow-Pour, D.B.A. (Ed.), Encyclopedia of Information Science and Technology, Fourth Edition (pp. 5559-5569). Hershey, PA: IGI Global. doi:10.4018/978-1-5225-2255-3.ch483

Ngafeeson, M. N. (2018). User Resistance to Health Information Technology. In M. Khosrow-Pour, D.B.A. (Ed.), Encyclopedia of Information Science and Technology, Fourth Edition (pp. 3816-3825). Hershey, PA: IGI Global. doi:10.4018/978-1-5225-2255-3.ch331

Nozari, H., Najafi, S. E., Jafari-Eskandari, M., & Aliahmadi, A. (2016). Providing a Model for Virtual Project Management with an Emphasis on IT Projects. In C. Graham (Ed.), *Strategic Management and Leadership for Systems Development in Virtual Spaces* (pp. 43–63). Hershey, PA: IGI Global. doi:10.4018/978-1-4666-9688-4.ch003

Nurdin, N., Stockdale, R., & Scheepers, H. (2016). Influence of Organizational Factors in the Sustainability of E-Government: A Case Study of Local E-Government in Indonesia. In I. Sodhi (Ed.), *Trends, Prospects, and Challenges in Asian E-Governance* (pp. 281–323). Hershey, PA: IGI Global. doi:10.4018/978-1-4666-9536-8.ch014

Odagiri, K. (2017). Introduction of Individual Technology to Constitute the Current Internet. In *Strategic Policy-Based Network Management in Contemporary Organizations* (pp. 20–96). Hershey, PA: IGI Global. doi:10.4018/978-1-68318-003-6.ch003

Okike, E. U. (2018). Computer Science and Prison Education. In I. Biao (Ed.), *Strategic Learning Ideologies in Prison Education Programs* (pp. 246–264). Hershey, PA: IGI Global. doi:10.4018/978-1-5225-2909-5.ch012

Olelewe, C. J., & Nwafor, I. P. (2017). Level of Computer Appreciation Skills Acquired for Sustainable Development by Secondary School Students in Nsukka LGA of Enugu State, Nigeria. In C. Ayo & V. Mbarika (Eds.), *Sustainable ICT Adoption and Integration for Socio-Economic Development* (pp. 214–233). Hershey, PA: IGI Global. doi:10.4018/978-1-5225-2565-3.ch010

Oliveira, M., Maçada, A. C., Curado, C., & Nodari, F. (2017). Infrastructure Profiles and Knowledge Sharing. *International Journal of Technology and Human Interaction*, *13*(3), 1–12. doi:10.4018/IJTHI.2017070101

Otarkhani, A., Shokouhyar, S., & Pour, S. S. (2017). Analyzing the Impact of Governance of Enterprise IT on Hospital Performance: Tehran's (Iran) Hospitals – A Case Study. *International Journal of Healthcare Information Systems and Informatics*, *12*(3), 1–20. doi:10.4018/IJHISI.2017070101

Otunla, A. O., & Amuda, C. O. (2018). Nigerian Undergraduate Students' Computer Competencies and Use of Information Technology Tools and Resources for Study Skills and Habits' Enhancement. In M. Khosrow-Pour, D.B.A. (Ed.), Encyclopedia of Information Science and Technology, Fourth Edition (pp. 2303-2313). Hershey, PA: IGI Global. doi:10.4018/978-1-5225-2255-3.ch200

Özçınar, H. (2018). A Brief Discussion on Incentives and Barriers to Computational Thinking Education. In H. Ozcinar, G. Wong, & H. Ozturk (Eds.), *Teaching Computational Thinking in Primary Education* (pp. 1–17). Hershey, PA: IGI Global. doi:10.4018/978-1-5225-3200-2.ch001

Pandey, J. M., Garg, S., Mishra, P., & Mishra, B. P. (2017). Computer Based Psychological Interventions: Subject to the Efficacy of Psychological Services. *International Journal of Computers in Clinical Practice*, *2*(1), 25–33. doi:10.4018/IJCCP.2017010102

Parry, V. K., & Lind, M. L. (2016). Alignment of Business Strategy and Information Technology Considering Information Technology Governance, Project Portfolio Control, and Risk Management. *International Journal of Information Technology Project Management*, *7*(4), 21–37. doi:10.4018/IJITPM.2016100102

Patro, C. (2017). Impulsion of Information Technology on Human Resource Practices. In P. Ordóñez de Pablos (Ed.), *Managerial Strategies and Solutions for Business Success in Asia* (pp. 231–254). Hershey, PA: IGI Global. doi:10.4018/978-1-5225-1886-0.ch013

Patro, C. S., & Raghunath, K. M. (2017). Information Technology Paraphernalia for Supply Chain Management Decisions. In M. Tavana (Ed.), *Enterprise Information Systems and the Digitalization of Business Functions* (pp. 294–320). Hershey, PA: IGI Global. doi:10.4018/978-1-5225-2382-6.ch014

Paul, P. K. (2016). Cloud Computing: An Agent of Promoting Interdisciplinary Sciences, Especially Information Science and I-Schools – Emerging Techno-Educational Scenario. In L. Chao (Ed.), *Handbook of Research on Cloud-Based STEM Education for Improved Learning Outcomes* (pp. 247–258). Hershey, PA: IGI Global. doi:10.4018/978-1-4666-9924-3.ch016

Paul, P. K. (2018). The Context of IST for Solid Information Retrieval and Infrastructure Building: Study of Developing Country. *International Journal of Information Retrieval Research*, 8(1), 86–100. doi:10.4018/IJIRR.2018010106

Paul, P. K., & Chatterjee, D. (2018). iSchools Promoting "Information Science and Technology" (IST) Domain Towards Community, Business, and Society With Contemporary Worldwide Trend and Emerging Potentialities in India. In M. Khosrow-Pour, D.B.A. (Ed.), Encyclopedia of Information Science and Technology, Fourth Edition (pp. 4723-4735). Hershey, PA: IGI Global. doi:10.4018/978-1-5225-2255-3.ch410

Pessoa, C. R., & Marques, M. E. (2017). Information Technology and Communication Management in Supply Chain Management. In G. Jamil, A. Soares, & C. Pessoa (Eds.), *Handbook of Research on Information Management for Effective Logistics and Supply Chains* (pp. 23–33). Hershey, PA: IGI Global. doi:10.4018/978-1-5225-0973-8.ch002

Pineda, R. G. (2016). Where the Interaction Is Not: Reflections on the Philosophy of Human-Computer Interaction. *International Journal of Art, Culture and Design Technologies*, 5(1), 1–12. doi:10.4018/IJACDT.2016010101

Pineda, R. G. (2018). Remediating Interaction: Towards a Philosophy of Human-Computer Relationship. In M. Khosrow-Pour (Ed.), *Enhancing Art, Culture, and Design With Technological Integration* (pp. 75–98). Hershey, PA: IGI Global. doi:10.4018/978-1-5225-5023-5.ch004

Poikela, P., & Vuojärvi, H. (2016). Learning ICT-Mediated Communication through Computer-Based Simulations. In M. Cruz-Cunha, I. Miranda, R. Martinho, & R. Rijo (Eds.), *Encyclopedia of E-Health and Telemedicine* (pp. 674–687). Hershey, PA: IGI Global. doi:10.4018/978-1-4666-9978-6.ch052

Related References

Qian, Y. (2017). Computer Simulation in Higher Education: Affordances, Opportunities, and Outcomes. In P. Vu, S. Fredrickson, & C. Moore (Eds.), *Handbook of Research on Innovative Pedagogies and Technologies for Online Learning in Higher Education* (pp. 236–262). Hershey, PA: IGI Global. doi:10.4018/978-1-5225-1851-8.ch011

Radant, O., Colomo-Palacios, R., & Stantchev, V. (2016). Factors for the Management of Scarce Human Resources and Highly Skilled Employees in IT-Departments: A Systematic Review. *Journal of Information Technology Research*, 9(1), 65–82. doi:10.4018/JITR.2016010105

Rahman, N. (2016). Toward Achieving Environmental Sustainability in the Computer Industry. *International Journal of Green Computing*, 7(1), 37–54. doi:10.4018/IJGC.2016010103

Rahman, N. (2017). Lessons from a Successful Data Warehousing Project Management. *International Journal of Information Technology Project Management*, 8(4), 30–45. doi:10.4018/IJITPM.2017100103

Rahman, N. (2018). Environmental Sustainability in the Computer Industry for Competitive Advantage. In M. Khosrow-Pour (Ed.), *Green Computing Strategies for Competitive Advantage and Business Sustainability* (pp. 110–130). Hershey, PA: IGI Global. doi:10.4018/978-1-5225-5017-4.ch006

Rajh, A., & Pavetic, T. (2017). Computer Generated Description as the Required Digital Competence in Archival Profession. *International Journal of Digital Literacy and Digital Competence*, 8(1), 36–49. doi:10.4018/IJDLDC.2017010103

Raman, A., & Goyal, D. P. (2017). Extending IMPLEMENT Framework for Enterprise Information Systems Implementation to Information System Innovation. In M. Tavana (Ed.), *Enterprise Information Systems and the Digitalization of Business Functions* (pp. 137–177). Hershey, PA: IGI Global. doi:10.4018/978-1-5225-2382-6.ch007

Rao, Y. S., Rauta, A. K., Saini, H., & Panda, T. C. (2017). Mathematical Model for Cyber Attack in Computer Network. *International Journal of Business Data Communications and Networking*, 13(1), 58–65. doi:10.4018/IJBDCN.2017010105

Rapaport, W. J. (2018). Syntactic Semantics and the Proper Treatment of Computationalism. In M. Danesi (Ed.), *Empirical Research on Semiotics and Visual Rhetoric* (pp. 128–176). Hershey, PA: IGI Global. doi:10.4018/978-1-5225-5622-0.ch007

Raut, R., Priyadarshinee, P., & Jha, M. (2017). Understanding the Mediation Effect of Cloud Computing Adoption in Indian Organization: Integrating TAM-TOE- Risk Model. *International Journal of Service Science, Management, Engineering, and Technology, 8*(3), 40–59. doi:10.4018/IJSSMET.2017070103

Regan, E. A., & Wang, J. (2016). Realizing the Value of EHR Systems Critical Success Factors. *International Journal of Healthcare Information Systems and Informatics, 11*(3), 1–18. doi:10.4018/IJHISI.2016070101

Rezaie, S., Mirabedini, S. J., & Abtahi, A. (2018). Designing a Model for Implementation of Business Intelligence in the Banking Industry. *International Journal of Enterprise Information Systems, 14*(1), 77–103. doi:10.4018/IJEIS.2018010105

Rezende, D. A. (2016). Digital City Projects: Information and Public Services Offered by Chicago (USA) and Curitiba (Brazil). *International Journal of Knowledge Society Research, 7*(3), 16–30. doi:10.4018/IJKSR.2016070102

Rezende, D. A. (2018). Strategic Digital City Projects: Innovative Information and Public Services Offered by Chicago (USA) and Curitiba (Brazil). In M. Lytras, L. Daniela, & A. Visvizi (Eds.), *Enhancing Knowledge Discovery and Innovation in the Digital Era* (pp. 204–223). Hershey, PA: IGI Global. doi:10.4018/978-1-5225-4191-2.ch012

Riabov, V. V. (2016). Teaching Online Computer-Science Courses in LMS and Cloud Environment. *International Journal of Quality Assurance in Engineering and Technology Education, 5*(4), 12–41. doi:10.4018/IJQAETE.2016100102

Ricordel, V., Wang, J., Da Silva, M. P., & Le Callet, P. (2016). 2D and 3D Visual Attention for Computer Vision: Concepts, Measurement, and Modeling. In R. Pal (Ed.), *Innovative Research in Attention Modeling and Computer Vision Applications* (pp. 1–44). Hershey, PA: IGI Global. doi:10.4018/978-1-4666-8723-3.ch001

Rodriguez, A., Rico-Diaz, A. J., Rabuñal, J. R., & Gestal, M. (2017). Fish Tracking with Computer Vision Techniques: An Application to Vertical Slot Fishways. In M. S., & V. V. (Eds.), Multi-Core Computer Vision and Image Processing for Intelligent Applications (pp. 74-104). Hershey, PA: IGI Global. doi:10.4018/978-1-5225-0889-2.ch003

Romero, J. A. (2018). Sustainable Advantages of Business Value of Information Technology. In M. Khosrow-Pour, D.B.A. (Ed.), Encyclopedia of Information Science and Technology, Fourth Edition (pp. 923-929). Hershey, PA: IGI Global. doi:10.4018/978-1-5225-2255-3.ch079

Related References

Romero, J. A. (2018). The Always-On Business Model and Competitive Advantage. In N. Bajgoric (Ed.), *Always-On Enterprise Information Systems for Modern Organizations* (pp. 23–40). Hershey, PA: IGI Global. doi:10.4018/978-1-5225-3704-5.ch002

Rosen, Y. (2018). Computer Agent Technologies in Collaborative Learning and Assessment. In M. Khosrow-Pour, D.B.A. (Ed.), Encyclopedia of Information Science and Technology, Fourth Edition (pp. 2402-2410). Hershey, PA: IGI Global. doi:10.4018/978-1-5225-2255-3.ch209

Rosen, Y., & Mosharraf, M. (2016). Computer Agent Technologies in Collaborative Assessments. In Y. Rosen, S. Ferrara, & M. Mosharraf (Eds.), *Handbook of Research on Technology Tools for Real-World Skill Development* (pp. 319–343). Hershey, PA: IGI Global. doi:10.4018/978-1-4666-9441-5.ch012

Roy, D. (2018). Success Factors of Adoption of Mobile Applications in Rural India: Effect of Service Characteristics on Conceptual Model. In M. Khosrow-Pour (Ed.), *Green Computing Strategies for Competitive Advantage and Business Sustainability* (pp. 211–238). Hershey, PA: IGI Global. doi:10.4018/978-1-5225-5017-4.ch010

Ruffin, T. R. (2016). Health Information Technology and Change. In V. Wang (Ed.), *Handbook of Research on Advancing Health Education through Technology* (pp. 259–285). Hershey, PA: IGI Global. doi:10.4018/978-1-4666-9494-1.ch012

Ruffin, T. R. (2016). Health Information Technology and Quality Management. *International Journal of Information Communication Technologies and Human Development*, 8(4), 56–72. doi:10.4018/IJICTHD.2016100105

Ruffin, T. R., & Hawkins, D. P. (2018). Trends in Health Care Information Technology and Informatics. In M. Khosrow-Pour, D.B.A. (Ed.), Encyclopedia of Information Science and Technology, Fourth Edition (pp. 3805-3815). Hershey, PA: IGI Global. doi:10.4018/978-1-5225-2255-3.ch330

Safari, M. R., & Jiang, Q. (2018). The Theory and Practice of IT Governance Maturity and Strategies Alignment: Evidence From Banking Industry. *Journal of Global Information Management*, 26(2), 127–146. doi:10.4018/JGIM.2018040106

Sahin, H. B., & Anagun, S. S. (2018). Educational Computer Games in Math Teaching: A Learning Culture. In E. Toprak & E. Kumtepe (Eds.), *Supporting Multiculturalism in Open and Distance Learning Spaces* (pp. 249–280). Hershey, PA: IGI Global. doi:10.4018/978-1-5225-3076-3.ch013

Sanna, A., & Valpreda, F. (2017). An Assessment of the Impact of a Collaborative Didactic Approach and Students' Background in Teaching Computer Animation. *International Journal of Information and Communication Technology Education*, *13*(4), 1–16. doi:10.4018/IJICTE.2017100101

Savita, K., Dominic, P., & Ramayah, T. (2016). The Drivers, Practices and Outcomes of Green Supply Chain Management: Insights from ISO14001 Manufacturing Firms in Malaysia. *International Journal of Information Systems and Supply Chain Management*, *9*(2), 35–60. doi:10.4018/IJISSCM.2016040103

Scott, A., Martin, A., & McAlear, F. (2017). Enhancing Participation in Computer Science among Girls of Color: An Examination of a Preparatory AP Computer Science Intervention. In Y. Rankin & J. Thomas (Eds.), *Moving Students of Color from Consumers to Producers of Technology* (pp. 62–84). Hershey, PA: IGI Global. doi:10.4018/978-1-5225-2005-4.ch004

Shahsavandi, E., Mayah, G., & Rahbari, H. (2016). Impact of E-Government on Transparency and Corruption in Iran. In I. Sodhi (Ed.), *Trends, Prospects, and Challenges in Asian E-Governance* (pp. 75–94). Hershey, PA: IGI Global. doi:10.4018/978-1-4666-9536-8.ch004

Siddoo, V., & Wongsai, N. (2017). Factors Influencing the Adoption of ISO/IEC 29110 in Thai Government Projects: A Case Study. *International Journal of Information Technologies and Systems Approach*, *10*(1), 22–44. doi:10.4018/IJITSA.2017010102

Sidorkina, I., & Rybakov, A. (2016). Computer-Aided Design as Carrier of Set Development Changes System in E-Course Engineering. In V. Mkrttchian, A. Bershadsky, A. Bozhday, M. Kataev, & S. Kataev (Eds.), *Handbook of Research on Estimation and Control Techniques in E-Learning Systems* (pp. 500–515). Hershey, PA: IGI Global. doi:10.4018/978-1-4666-9489-7.ch035

Sidorkina, I., & Rybakov, A. (2016). Creating Model of E-Course: As an Object of Computer-Aided Design. In V. Mkrttchian, A. Bershadsky, A. Bozhday, M. Kataev, & S. Kataev (Eds.), *Handbook of Research on Estimation and Control Techniques in E-Learning Systems* (pp. 286–297). Hershey, PA: IGI Global. doi:10.4018/978-1-4666-9489-7.ch019

Simões, A. (2017). Using Game Frameworks to Teach Computer Programming. In R. Alexandre Peixoto de Queirós & M. Pinto (Eds.), *Gamification-Based E-Learning Strategies for Computer Programming Education* (pp. 221–236). Hershey, PA: IGI Global. doi:10.4018/978-1-5225-1034-5.ch010

Sllame, A. M. (2017). Integrating LAB Work With Classes in Computer Network Courses. In H. Alphin Jr, R. Chan, & J. Lavine (Eds.), *The Future of Accessibility in International Higher Education* (pp. 253–275). Hershey, PA: IGI Global. doi:10.4018/978-1-5225-2560-8.ch015

Smirnov, A., Ponomarev, A., Shilov, N., Kashevnik, A., & Teslya, N. (2018). Ontology-Based Human-Computer Cloud for Decision Support: Architecture and Applications in Tourism. *International Journal of Embedded and Real-Time Communication Systems*, 9(1), 1–19. doi:10.4018/IJERTCS.2018010101

Smith-Ditizio, A. A., & Smith, A. D. (2018). Computer Fraud Challenges and Its Legal Implications. In M. Khosrow-Pour, D.B.A. (Ed.), Encyclopedia of Information Science and Technology, Fourth Edition (pp. 4837-4848). Hershey, PA: IGI Global. doi:10.4018/978-1-5225-2255-3.ch419

Sohani, S. S. (2016). Job Shadowing in Information Technology Projects: A Source of Competitive Advantage. *International Journal of Information Technology Project Management*, 7(1), 47–57. doi:10.4018/IJITPM.2016010104

Sosnin, P. (2018). Figuratively Semantic Support of Human-Computer Interactions. In *Experience-Based Human-Computer Interactions: Emerging Research and Opportunities* (pp. 244–272). Hershey, PA: IGI Global. doi:10.4018/978-1-5225-2987-3.ch008

Spinelli, R., & Benevolo, C. (2016). From Healthcare Services to E-Health Applications: A Delivery System-Based Taxonomy. In A. Dwivedi (Ed.), *Reshaping Medical Practice and Care with Health Information Systems* (pp. 205–245). Hershey, PA: IGI Global. doi:10.4018/978-1-4666-9870-3.ch007

Srinivasan, S. (2016). Overview of Clinical Trial and Pharmacovigilance Process and Areas of Application of Computer System. In P. Chakraborty & A. Nagal (Eds.), *Software Innovations in Clinical Drug Development and Safety* (pp. 1–13). Hershey, PA: IGI Global. doi:10.4018/978-1-4666-8726-4.ch001

Srisawasdi, N. (2016). Motivating Inquiry-Based Learning Through a Combination of Physical and Virtual Computer-Based Laboratory Experiments in High School Science. In M. Urban & D. Falvo (Eds.), *Improving K-12 STEM Education Outcomes through Technological Integration* (pp. 108–134). Hershey, PA: IGI Global. doi:10.4018/978-1-4666-9616-7.ch006

Stavridi, S. V., & Hamada, D. R. (2016). Children and Youth Librarians: Competencies Required in Technology-Based Environment. In J. Yap, M. Perez, M. Ayson, & G. Entico (Eds.), *Special Library Administration, Standardization and Technological Integration* (pp. 25–50). Hershey, PA: IGI Global. doi:10.4018/978-1-4666-9542-9.ch002

Sung, W., Ahn, J., Kai, S. M., Choi, A., & Black, J. B. (2016). Incorporating Touch-Based Tablets into Classroom Activities: Fostering Children's Computational Thinking through iPad Integrated Instruction. In D. Mentor (Ed.), *Handbook of Research on Mobile Learning in Contemporary Classrooms* (pp. 378–406). Hershey, PA: IGI Global. doi:10.4018/978-1-5225-0251-7.ch019

Syväjärvi, A., Leinonen, J., Kivivirta, V., & Kesti, M. (2017). The Latitude of Information Management in Local Government: Views of Local Government Managers. *International Journal of Electronic Government Research, 13*(1), 69–85. doi:10.4018/IJEGR.2017010105

Tanque, M., & Foxwell, H. J. (2018). Big Data and Cloud Computing: A Review of Supply Chain Capabilities and Challenges. In A. Prasad (Ed.), *Exploring the Convergence of Big Data and the Internet of Things* (pp. 1–28). Hershey, PA: IGI Global. doi:10.4018/978-1-5225-2947-7.ch001

Teixeira, A., Gomes, A., & Orvalho, J. G. (2017). Auditory Feedback in a Computer Game for Blind People. In T. Issa, P. Kommers, T. Issa, P. Isaías, & T. Issa (Eds.), *Smart Technology Applications in Business Environments* (pp. 134–158). Hershey, PA: IGI Global. doi:10.4018/978-1-5225-2492-2.ch007

Thompson, N., McGill, T., & Murray, D. (2018). Affect-Sensitive Computer Systems. In M. Khosrow-Pour, D.B.A. (Ed.), Encyclopedia of Information Science and Technology, Fourth Edition (pp. 4124-4135). Hershey, PA: IGI Global. doi:10.4018/978-1-5225-2255-3.ch357

Trad, A., & Kalpić, D. (2016). The E-Business Transformation Framework for E-Commerce Control and Monitoring Pattern. In I. Lee (Ed.), *Encyclopedia of E-Commerce Development, Implementation, and Management* (pp. 754–777). Hershey, PA: IGI Global. doi:10.4018/978-1-4666-9787-4.ch053

Triberti, S., Brivio, E., & Galimberti, C. (2018). On Social Presence: Theories, Methodologies, and Guidelines for the Innovative Contexts of Computer-Mediated Learning. In M. Marmon (Ed.), *Enhancing Social Presence in Online Learning Environments* (pp. 20–41). Hershey, PA: IGI Global. doi:10.4018/978-1-5225-3229-3.ch002

Related References

Tripathy, B. K. T. R., S., & Mohanty, R. K. (2018). Memetic Algorithms and Their Applications in Computer Science. In S. Dash, B. Tripathy, & A. Rahman (Eds.), Handbook of Research on Modeling, Analysis, and Application of Nature-Inspired Metaheuristic Algorithms (pp. 73-93). Hershey, PA: IGI Global. doi:10.4018/978-1-5225-2857-9.ch004

Turulja, L., & Bajgoric, N. (2017). Human Resource Management IT and Global Economy Perspective: Global Human Resource Information Systems. In M. Khosrow-Pour (Ed.), *Handbook of Research on Technology Adoption, Social Policy, and Global Integration* (pp. 377–394). Hershey, PA: IGI Global. doi:10.4018/978-1-5225-2668-1.ch018

Unwin, D. W., Sanzogni, L., & Sandhu, K. (2017). Developing and Measuring the Business Case for Health Information Technology. In K. Moahi, K. Bwalya, & P. Sebina (Eds.), *Health Information Systems and the Advancement of Medical Practice in Developing Countries* (pp. 262–290). Hershey, PA: IGI Global. doi:10.4018/978-1-5225-2262-1.ch015

Vadhanam, B. R. S., M., Sugumaran, V., V., V., & Ramalingam, V. V. (2017). Computer Vision Based Classification on Commercial Videos. In M. S., & V. V. (Eds.), Multi-Core Computer Vision and Image Processing for Intelligent Applications (pp. 105-135). Hershey, PA: IGI Global. doi:10.4018/978-1-5225-0889-2.ch004

Valverde, R., Torres, B., & Motaghi, H. (2018). A Quantum NeuroIS Data Analytics Architecture for the Usability Evaluation of Learning Management Systems. In S. Bhattacharyya (Ed.), *Quantum-Inspired Intelligent Systems for Multimedia Data Analysis* (pp. 277–299). Hershey, PA: IGI Global. doi:10.4018/978-1-5225-5219-2.ch009

Vassilis, E. (2018). Learning and Teaching Methodology: "1:1 Educational Computing. In K. Koutsopoulos, K. Doukas, & Y. Kotsanis (Eds.), *Handbook of Research on Educational Design and Cloud Computing in Modern Classroom Settings* (pp. 122–155). Hershey, PA: IGI Global. doi:10.4018/978-1-5225-3053-4.ch007

Wadhwani, A. K., Wadhwani, S., & Singh, T. (2016). Computer Aided Diagnosis System for Breast Cancer Detection. In Y. Morsi, A. Shukla, & C. Rathore (Eds.), *Optimizing Assistive Technologies for Aging Populations* (pp. 378–395). Hershey, PA: IGI Global. doi:10.4018/978-1-4666-9530-6.ch015

Wang, L., Wu, Y., & Hu, C. (2016). English Teachers' Practice and Perspectives on Using Educational Computer Games in EIL Context. *International Journal of Technology and Human Interaction*, *12*(3), 33–46. doi:10.4018/IJTHI.2016070103

Watfa, M. K., Majeed, H., & Salahuddin, T. (2016). Computer Based E-Healthcare Clinical Systems: A Comprehensive Survey. *International Journal of Privacy and Health Information Management*, 4(1), 50–69. doi:10.4018/IJPHIM.2016010104

Weeger, A., & Haase, U. (2016). Taking up Three Challenges to Business-IT Alignment Research by the Use of Activity Theory. *International Journal of IT/Business Alignment and Governance*, 7(2), 1-21. doi:10.4018/IJITBAG.2016070101

Wexler, B. E. (2017). Computer-Presented and Physical Brain-Training Exercises for School Children: Improving Executive Functions and Learning. In B. Dubbels (Ed.), *Transforming Gaming and Computer Simulation Technologies across Industries* (pp. 206–224). Hershey, PA: IGI Global. doi:10.4018/978-1-5225-1817-4.ch012

Williams, D. M., Gani, M. O., Addo, I. D., Majumder, A. J., Tamma, C. P., Wang, M., ... Chu, C. (2016). Challenges in Developing Applications for Aging Populations. In Y. Morsi, A. Shukla, & C. Rathore (Eds.), *Optimizing Assistive Technologies for Aging Populations* (pp. 1–21). Hershey, PA: IGI Global. doi:10.4018/978-1-4666-9530-6.ch001

Wimble, M., Singh, H., & Phillips, B. (2018). Understanding Cross-Level Interactions of Firm-Level Information Technology and Industry Environment: A Multilevel Model of Business Value. *Information Resources Management Journal*, 31(1), 1–20. doi:10.4018/IRMJ.2018010101

Wimmer, H., Powell, L., Kilgus, L., & Force, C. (2017). Improving Course Assessment via Web-based Homework. *International Journal of Online Pedagogy and Course Design*, 7(2), 1–19. doi:10.4018/IJOPCD.2017040101

Wong, Y. L., & Siu, K. W. (2018). Assessing Computer-Aided Design Skills. In M. Khosrow-Pour, D.B.A. (Ed.), Encyclopedia of Information Science and Technology, Fourth Edition (pp. 7382-7391). Hershey, PA: IGI Global. doi:10.4018/978-1-5225-2255-3.ch642

Wongsurawat, W., & Shrestha, V. (2018). Information Technology, Globalization, and Local Conditions: Implications for Entrepreneurs in Southeast Asia. In P. Ordóñez de Pablos (Ed.), *Management Strategies and Technology Fluidity in the Asian Business Sector* (pp. 163–176). Hershey, PA: IGI Global. doi:10.4018/978-1-5225-4056-4.ch010

Yang, Y., Zhu, X., Jin, C., & Li, J. J. (2018). Reforming Classroom Education Through a QQ Group: A Pilot Experiment at a Primary School in Shanghai. In H. Spires (Ed.), *Digital Transformation and Innovation in Chinese Education* (pp. 211–231). Hershey, PA: IGI Global. doi:10.4018/978-1-5225-2924-8.ch012

Yilmaz, R., Sezgin, A., Kurnaz, S., & Arslan, Y. Z. (2018). Object-Oriented Programming in Computer Science. In M. Khosrow-Pour, D.B.A. (Ed.), Encyclopedia of Information Science and Technology, Fourth Edition (pp. 7470-7480). Hershey, PA: IGI Global. doi:10.4018/978-1-5225-2255-3.ch650

Yu, L. (2018). From Teaching Software Engineering Locally and Globally to Devising an Internationalized Computer Science Curriculum. In S. Dikli, B. Etheridge, & R. Rawls (Eds.), *Curriculum Internationalization and the Future of Education* (pp. 293–320). Hershey, PA: IGI Global. doi:10.4018/978-1-5225-2791-6.ch016

Yuhua, F. (2018). Computer Information Library Clusters. In M. Khosrow-Pour, D.B.A. (Ed.), Encyclopedia of Information Science and Technology, Fourth Edition (pp. 4399-4403). Hershey, PA: IGI Global. doi:10.4018/978-1-5225-2255-3.ch382

Zare, M. A., Taghavi Fard, M. T., & Hanafizadeh, P. (2016). The Assessment of Outsourcing IT Services using DEA Technique: A Study of Application Outsourcing in Research Centers. *International Journal of Operations Research and Information Systems*, 7(1), 45–57. doi:10.4018/IJORIS.2016010104

Zhao, J., Wang, Q., Guo, J., Gao, L., & Yang, F. (2016). An Overview on Passive Image Forensics Technology for Automatic Computer Forgery. *International Journal of Digital Crime and Forensics*, 8(4), 14–25. doi:10.4018/IJDCF.2016100102

Zimeras, S. (2016). Computer Virus Models and Analysis in M-Health IT Systems: Computer Virus Models. In A. Moumtzoglou (Ed.), *M-Health Innovations for Patient-Centered Care* (pp. 284–297). Hershey, PA: IGI Global. doi:10.4018/978-1-4666-9861-1.ch014

Zlatanovska, K. (2016). Hacking and Hacktivism as an Information Communication System Threat. In M. Hadji-Janev & M. Bogdanoski (Eds.), *Handbook of Research on Civil Society and National Security in the Era of Cyber Warfare* (pp. 68–101). Hershey, PA: IGI Global. doi:10.4018/978-1-4666-8793-6.ch004

About the Contributors

Salahddine Krit received the Hability Physics-Infromatics from the Faculty of Sciences, University Ibn Zohr Agadir morocco in 2015, the B.S. and Ph.D degrees in Software Engineering from Sidi Mohammed Ben Abdellah university, Fez, Morroco in 2004 and 2009, respectively. During 2002-2008, he worked as an engineer Team leader in audio and power management Integrated Circuits (ICs) Research, Design, simulation and layout of analog and digital blocks dedicated for mobile phone and satellite communication systems using Cadence, Eldo, Orcad, VHDL-AMS technology. He is currently a professor of Informatics with Polydisciplinary Faculty of Ouarzazate, Ibn Zohr university, Agadir, Morroco. His research interests include Wireless Sensor Networks (Software and Hardware), computer engineering and wireless communications, Genetic Algorithms, Gender and ICT.

Valentina Emilia Balas is currently Full Professor in the Department of Automatics and Applied Software at the Faculty of Engineering, University "Aurel Vlaicu" Arad (Romania). She holds a Ph.D. in Applied Electronics and Telecommunications from Polytechnic University of Timisoara. She is author of more than 300 research papers in refereed journals and International Conferences. Her research interests are in Intelligent Systems, Fuzzy Control, Soft Computing, Smart Sensors, Information Fusion, Modeling and Simulation.She is the Editor-in Chief to International Journal of Advanced Intelligence Paradigms (IJAIP) and IJCSysE, member in Editorial Board member of several national and international journals and is evaluator expert for national and international projects. She participated in many international conferences as General Chair, Organizer, Session Chair and member in International Program Committee. She is a member of EUSFLAT, ACM and a Senior Member IEEE, member in TC – Fuzzy Systems (IEEE CIS), member in TC - Emergent Technologies (IEEE CIS), member in TC – Soft Computing (IEEE SMCS).

Mohamed Elhoseny is currently an Assistant Professor at the Faculty of Computers and Information, Mansoura University. Dr. Elhoseny has been appointed as an ACM Distinguished Speaker from 2019 to 2022. Collectively, Dr. Elhoseny authored/co-authored over 95 ISI Journal articles in high-ranked and prestigious journals such as IEEE Transactions on Industrial Informatics (IEEE), IEEE Transactions on Reliability (IEEE), Future Generation Computer Systems (Elsevier), and Neural Computing and Applications (Springer). Besides, Dr. Elhoseny authored/edited several international books. His research interests include Smart Cities, Network Security, Artificial Intelligence, Internet of Things, and Intelligent Systems. Dr. Elhoseny serves as the Editor-in-Chief of International Journal of Smart Sensor Technologies and Applications (IGI Global). Moreover, he is an Associate Editor of many journals such as IEEE Journal of Biomedical and Health Informatics (IEEE), IEEE Access (IEEE), Scientific Reports (Nature), IEEE Future Directions (IEEE), Remote Sensing (MDPI), and International Journal of E-services and Mobile Applications(IGI Global), Applied Intelligence (Springer). Moreover, he served as the co-chair, the publication chair, the program chair, and a track chair for several international conferences published by IEEE and Springer. Dr. Elhoseny is the Editor-in-Chief of the Studies in Distributed Intelligence Springer Book Series, the Editor-in-Chief of The Sensors Communication for Urban Intelligence CRC Press-Taylor& Francis Book Series, and the Editor-in-Chief of The Distributed Sensing and Intelligent Systems CRC Press-Taylor& Francis Book Series. He was granted several awards by diverse funding bodies such as the Young Researcher Award in Artificial Intelligence from the Federation of Arab Scientific Research Councils in 2019, Obada International Prize for young distinguished scientists 2020, the Egypt National Prize for Young Researchers in 2018, the best Ph.D. thesis in Mansoura University in 2015, the SRGE best young researcher award in 2017, and the membership of The Egyptian Young Academy of Science (EYAS) in 2019. Besides, he is a TPC Member or Reviewer in 50+ International Conferences and Workshops. Furthermore, he has been reviewing papers for 80+ International Journals including IEEE Communications Magazine, IEEE Transactions on Intelligent Transportation Systems, IEEE Sensors Letters, IEEE Communication Letters, Elsevier Computer Communications, Computer Networks, Sustainable Cities and Society, Wireless Personal Communications, and Expert Systems with Applications. Dr. Elhoseny has been invited as a guest in many media programs to comment on technologies and related issues.

* * *

Mohamed Achouri is a Distinguished Professor of Mycology and Phytopthology for Research in Diagnostic & Control approaches in plant protection, Department of plant protection, IAV Hassan II CHA. Expertise in Plant Microbiology, Phytopathology, Microbial Ecology, Ecophysiology.

Jihane Alami Chentoufi is a Professor at the IBN Tofail University, Computer Science Department.

Mohamed Amine Basmassi is currently Ph.D. student in the informatics, systems and optimization laboratory in Ibn-Tofail University, Kenitra, Morocco, supervised by Prof. Alami Chentoufi Jihane. In 2015, he received the M.S degree in Computer Engineering from the Abdelmalek Essaadi University, Tetouan, Morocco. He had the B.S. degree in fundamental mathematics and computer science from Cadi Ayyad University, Marrakech-Safi, Morocco in 2012. His research interests fall in meta-heuristic algorithms and combinatorial optimization.

Lamia Benameur is a Professor and Ph.D. supervisor in Abdelmalik Essaadi University, Morocco, Faculty of Sciences Tetuan, since 2011. She received her M.S and PhD degrees in computer science in 2003 and 2010, respectively. Her research interests include Metaheuristics and multi-objective optimization problem.

Omar Bouattane was born in Figuig City, Morocco in 1962. He received his PhD from the University Hassan II of Casablanca, MOROCCO in 2001 in Parallel Computing and Image processing. He currently serves as a full Professor in the Department of Electrical Engineering at "Ecole Normale Superieure de l'enseignement technique" ENSET of Mohammedia. He has more than 150 scientific publications in various domains of Computational Intelligence, high performance computing, image processing and renewable energy. He has registered 6 national and PCT international Patents regarding to the networking technology and signal synthesis. He was awarded as the owner of the best PCT patent in Morocco on 2011. Since 2012, He was the head of the laboratory of Signals Distributed Systems and Artificial Intelligence. He involved his laboratory in several partnership activities and developed many funded projects in Morocco and in his university. Overall, Prof. Bouattane work has received more than 300 citations. Prof. Bouattane has been the principal investigator and leader in 4 academic projects, funded either publicly or privately, in the USA, Canada and France. He was the supervisor from Morocco of a partnership program entitled "Linkage for entrepreneurship achievement program" funded by the USAID and HED of USA from December 2012 to December 2014. He supervised beside his US partners a scholarship program named "study adroad, Moroccain culture" December 2016 to December 2018. All his academic and research activities are in the researchgate portal at https://www.researchgate.net/profile/Omar_Bouattane/contributions.

Hassan Boubaker is Distinguished Professor of Plant Biology and Phytopathology, Department of Plant Biology, Ibn Zohr University Agadir. Expertise in Plant and Fungi interactions, Diseases patho-systems Management and Biocontrol, Phytopathology, Molecular Ecology,

Sidina Boudaakat received the B.S. degree from Mohamed V University of Rabat, Morocco in 2013, and MS degree in Abdelmalek Essaadi University, Tetouan, Morocco, in 2015. Since 2016 he has been working toward the Ph.D. degrees on Computational Intelligence in Urban Traffic To solved problems related to the management of road traffic and road safety at Laboratory SSDIA in Normal School of Technical Education (NSTE), Mohammadia, Morocco.

Noureddine Chtaina is a Distinguished Professor of Plant Protection and IPM Approaches, Early Monitoring, and Phytopthology for Research in Diagnostic & Control approaches in plant protection, Department of Production and Protection in Plant Biology, IAV Hassan II Rabat. Expertise in Pesticides safe use and Pests/diseases Management and Biocontrol, Phytopathology, Molecular Ecology.

Rinat Galiautdinov is a Principal Software Developer and Architect having the expertise in Information Technology and Computer Science. Mr. Galiautdinov is also an expert in Banking/Financial industry as well as in Neurobiological sphere. Mr. Rinat Galiautdinov works on the number of highly important researches as an independent researcher.

Asha Karegowda is currently working as Associate Professor, Dept of MCA, Siddaganga Institute of Technology form last 20 years. Has authored few books on C and Data structures using C, currently writing books on Python, Data mining and Data structures using C++. Area of interest include Data mining, WSN, Remote sensing, Bio inspired computing. Has published papers both in international conferences and journals. Guiding 3 research scholars in the area of image processing and remote sensing. Handling subjects for both PG and UG students: Python, Data structures C and C++, Data mining, Data Analytics.

Wafaa Mokhtari is a Doctor in Agriculture and food Sciences. She is expertizing in Plant health and protection, she loves sciences. Her research was on Molecular characterization of Trichoderma spp and their Biocontrol Potential against economical important soil borne pathogens. Some of her Interests also include eco-system interaction, biocontrol concept and last but not least new state of the art technologies to investigate in sustainable bio-resources. Growing up as a city kid in Agadir, Morocco, Wafaa was a late-bloomer when it comes to discovery of the green wild nature and its striking bio-resources. She got her Bachelor degree of sciences in biology at Ibn Zohr University, Agadir then flight to the Greek Island Chania to pursue her post-graduate studies in food quality & chemistry in the Mediterranean Institute of Agronomy. In 2008 she started her Master degree of sciences in biotechnology at Al Akhawayn University. During these whole 7 years, it was becoming clear her love and curiosity for the wild nature. After her Master degree graduation, she was welcome to integrate different National & International Institutions & Organizations. Working in Dossiers and Projects; Action plan she get more devoted to sustainable nature and economic balance. She is interested in doing more research, tackling more challenges in R&D and fetches more on how concepts in the wild nature may balance the "ecolo-econo-system" where she lives "the Globe!".

Ambika Nagaraj has completed her MCA in the year 2001 and M.Phil in the year 2008. She completed her Ph.D. from Bharathiar University, Coimbatore in the year 2015. She has 12 years of experience. She is working as faculty in the Department of computer applications, SSMRV College, Bangalore. She has published a significant amount of papers in International papers and Journals.

Manoj Nayak is Professor & former Head of Department in Mechanical Engineering Department, Faculty of Engineering, Manav Rachna International institute of Research & Studies. His research area is Manufacturing & Automation. He has 25 years of Industrial and Academic experience and many research publications in national and international journal of repute.

Seema Nayak is presently working as Professor and Head of department (ECE) in IIMT COE, Greater Noida. She has 21 years academic experience. She has published many papers in International Conferences and Journals. Her research area is Signal Processing and Digital System Design. She is member of IEEE and IEI.

Pankaj Pathak received B.Sc. Physics and M.Sc. Physics specialization in electronics, all from Rohilkhand University, India in 1992 and 1994 respectively. He received his Ph.D. degree in electronic Science from University of Delhi in 2000. He also completed a postgraduate level course (M. Tech. - Computer Science) from Institution of Electronics & Telecommunication Engineers, Delhi. He is currently working with G. L. Bajaj Institute of Technology & Management along with his industrial consultations to Nanotrics Innovations Pvt. Ltd., India. He is actively involved in improving industry-academia interactions and encouraging innovations in academic research. He joined department of electronic science, University of Delhi as Research Scholar/Lecturer from 1996 to 2000. He has 14 years of electronics & IT research and product development experience with leading organizations. He was listed in the Who's Who in the World in 2011. He is life corporate member of Institution of Electronics & Telecommunication Engineers and member of Semiconductor Society of India. His research interests include semiconductor/microwave communication devices, wireless & multimedia communication systems, embedded & VLSI system design. For Publication details, please visit: http://orcid.org/0000-0003-2642-1500.

Ahmed Rebbani received the B.S. degree in Electronics in 1988, the M.S. degree in Applied Electronics in 1992 from the ENSET Institute, Mohammedia, Morocco. He received the DEA diploma in information processing in 1997 from the the faculty of sciences Ben Msik, Casablanca, Morocco. has his Ph.D. degree in 2014 In electrical energy storage from the faculty of sciences and techniques Mohammedia, Morocco. He is now a teacher and researcher at the University Hassan II, ENSET Institute. His research is focused on Internet of things and renewable energy.

Abdellah Remah is a Professor of Virology and Phytopthology for Research in Diagnostic in plant protection, Department of plant protection, IAV Hassan II CHA. Expertise in Plant Microbiology, Phytopathology.

Tapaswini Samant is working as Associate professor in school of Electronics Engineering, KIIT Deemed to be University. She completed Ph.D. in communication. Also worked in the area of wireless sensor networks.

Index

A

activity recognition 148-149, 151, 155, 157
Artificial Intelligence 25-26, 28, 38, 51, 85, 103, 152
Authentication Protocol 133

C

Chromatic Number 55-58, 60-61, 64

D

DC/AC Inverter 168-169, 171, 182
DC/DC Boost 168-170, 182, 193, 195
Denoising 1, 15-16
Digital Filter 3, 5, 12, 18-20

E

ECG 1-3, 5-6, 10, 13, 15-21, 90
Energy efficiency 12, 76, 200, 202, 205-206, 210, 215
Evolutionary Algorithm 56, 64

F

FIR Filter 4-5, 12, 17, 20-21
Forge Attack 133-134, 139-140, 143
FPGA 1, 6-14, 19-21
Frame-Slotted MAC Protocols 215
Future Trends 18, 65
Fuzzy Graph 54-58, 61-62, 64

G

Genetic Algorithm 55-58, 60-61, 64
Graph Coloring 54-57, 60-61
Greedy Algorithm 57, 64
Greedy Sequential Algorithm 55-57, 60-61

H

Hybrid Genetic Approach 54-55, 60

I

IIR Filter 4-5, 12, 19, 21
Internet of Things 25, 51, 65-66, 69, 75, 85, 89, 95, 97, 156, 200
IoT Applications 65, 68, 75-76, 86
IPM 114-115

L

Lyapunov 168, 170-171, 180, 195

M

MAC Protocols 200, 202-208, 210-212, 214-215
MPPT 168-171, 176-177, 179, 193, 195
Multichannel MAC Protocols 214-215

N

Nervous Circuit 25
Neuron 25, 29-35, 39-40, 43-46, 48-51

O

ontology 148-150, 152-154, 157-158, 160-162
Optimization 55-56, 64, 200

P

Plant Pathogens 115-116, 118
Plant Protection 114-115
PV 116, 168-171, 173, 175-179, 181, 185-187, 190-191, 193, 195

R

Replay Attack 140

S

Sliding Mode 168, 170-171, 176-179, 192-193, 195
smart home 77-79, 84, 98-99, 104, 131, 148-151, 157, 163
Synapse 31, 34-38, 40, 44-45, 50
Synchronous MAC 212

T

Trichoderma 115, 121-126

W

Wireless Sensor Network 69, 73, 97, 201

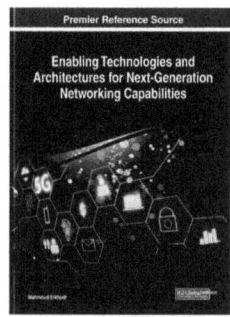

Ensure Quality Research is Introduced to the Academic Community

Become an IGI Global Reviewer for Authored Book Projects

Premier Reference Source

Emerging GIS Applications for Emergency and Disaster Management

Premier Reference Source

Managerial Strategies and Green Solutions for Project Sustainability

Premier Reference Source

Comparative Approaches to Using R and Python for Statistical Data Analysis

Premier Reference Source

Solutions for High-Touch Communications in a High-Tech World

The overall success of an authored book project is dependent on quality and timely reviews.

In this competitive age of scholarly publishing, constructive and timely feedback significantly expedites the turnaround time of manuscripts from submission to acceptance, allowing the publication and discovery of forward-thinking research at a much more expeditious rate. Several IGI Global authored book projects are currently seeking highly-qualified experts in the field to fill vacancies on their respective editorial review boards:

Applications and Inquiries may be sent to:
development@igi-global.com

Applicants must have a doctorate (or an equivalent degree) as well as publishing and reviewing experience. Reviewers are asked to complete the open-ended evaluation questions with as much detail as possible in a timely, collegial, and constructive manner. All reviewers' tenures run for one-year terms on the editorial review boards and are expected to complete at least three reviews per term. Upon successful completion of this term, reviewers can be considered for an additional term.

If you have a colleague that may be interested in this opportunity, we encourage you to share this information with them.